LONDON MATHEMATICAL SOCIETY MONOGRAPHS
NEW SERIES

Editors: P. M. Cohn and B. E. Johnson

LONDON MATHEMATICAL SOCIETY MONOGRAPHS
NEW SERIES

Previous volumes of the LMS Monographs were published by Academic Press to whom all enquiries should be addressed. Volumes in the New Series will be published by Oxford University Press throughout the world.

NEW SERIES

1. *Diophantine inequalities*   R. C. Baker

Diophantine Inequalities

R. C. Baker

Reader in Pure Mathematics, Royal Holloway and Bedford New College, University of London

CLARENDON PRESS · OXFORD
1986

Oxford University Press, Walton Street, Oxford OX2 6DP
Oxford New York Toronto
Delhi Bombay Calcutta Madras Karachi
Kuala Lumpur Singapore Hong Kong Tokyo
Nairobi Dar es Salaam Cape Town
Melbourne Auckland
and associated companies in
Beirut Berlin Ibadan Nicosia

Oxford is a trade mark of Oxford University Press

Published in the United States
by Oxford University Press, New York

British Library Cataloguing in Publication Data
Baker, R. C.
Diophantine inequalities.
1. Diophantine analysis
I. Title
512.74   QA242
ISBN 0-19-853545-7

Library of Congress Cataloging in Publication Data
Baker, R. C. (Roger Clive), 1947–
Diophantine inequalities.
(London Mathematical Society monographs; new ser.
no. 1)
Bibliography: p.
Includes index.
1. Diophantine analysis.   2. Inequalities
(Mathematics)   I. Title.   II. Series.
QA242.B35 1986   512'.73   85-31066
ISBN 0-19-853545-7

Typeset and printed by The Universities Press (Belfast) Ltd., Northern Ireland

Preface

In his Springer lecture notes on Diophantine approximation, Wolfgang Schmidt (1980*b*) lists a number of topics that are excluded: Weyl sums, nonlinear approximation, and Diophantine inequalities involving forms in many variables. The topics mentioned hang together; they are certainly connected with the Hardy–Littlewood method, but have their own special flavour.

Schmidt himself wrote a beautiful set of lectures on fractional parts of polynomials, published by the American Mathematical Society in 1977. Many developments have taken place since then; the most rich and exciting of these are due to Schmidt, in his work on forms in many variables. In the present book I give an overview of the subject, which takes into account these developments. Most of the principal results discussed are analogous for higher-degree polynomials of Dirichlet's theorem on integer multiples $n\theta$ $(n = 1, \dots, N)$ of an arbitrary point in Euclidean space.

Since Bob Vaughan (1981) has recently written an excellent book on the Hardy–Littlewood method, I have referred to it quite often. In general, I have aimed to make the book easily accessible to postgraduate students. The style is leisurely, and proofs are given in detail.

London   R. C. B.
June 1985

Contents

Notation

Let $\mathbb{R}$, $\mathbb{C}$ denote the real and complex numbers respectively and $\mathbb{Z}$ the integers. As usual, $\mathbb{R}^h$ denotes Euclidean h-dimensional space and $\mathbb{Z}^h$ the set of points in $\mathbb{R}^h$ with integer coordinates.

We denote by k a natural number, and $K = 2^{k-1}$. In Chapters 1–8, $k \geq 2$. Lemmas, theorems, etc., in which ε appears, are true for every positive ε. The symbol

$$F \ll G$$

may also be written $G \gg F$ or $F = O(G)$. It means that for the function F and the positive function G we have

$$|F| \leq CG.$$

Here C (and C_0, C_1, C_2, ...) depend at most on k and ε in Chapters 3–5, and on other parameters elsewhere. Sometimes dependence is made explicit. Otherwise, we set up the conventions in the relevant chapter. We write $\delta = \varepsilon/C_0$, and η for a multiple of δ of the form $C(k)\delta$, not necessarily the same at each occurrence. By suitable choice of C_0 we ensure that η is small compared with ε, say $\eta < \varepsilon/(C_0^{1/2})$, right through each chapter.

For real α, we write $e(\alpha) = e^{2\pi i\alpha}$, $e_q(\alpha) = e(\alpha/q)$ for $q = 1, 2, \ldots$, and

$$\|\alpha\| = \min_{n \in \mathbb{Z}} |\alpha - n|.$$

We also split up α as

$$\alpha = [\alpha] + \{\alpha\},$$

where $[\alpha] \in \mathbb{Z}$ and $0 \leq \{\alpha\} < 1$.

We denote by N a positive integer. The expression $f(x)$ (at least in Chapters 1–8) stands for a polynomial with real coefficients without constant term,

$$f(x) = \alpha_k x^k + \cdots + \alpha_1 x.$$

We write

$$S_h(f) = \sum_{n=1}^{N} e(hf(n))$$

(dependence on N is suppressed) and $S(f) = S_1(f)$.

In $\mathbb{R}^h$ we write **ab** for the ordinary inner product and $|\mathbf{x}| = (\mathbf{xx})^{1/2}$; but in Chapter 14 and thereafter, $|\mathbf{x}|$ denotes $\max(|x_1|, \ldots, |x_h|)$. Lebesgue measure is written $m(\cdots)$. On the real line, $[a, b) = \{x: a \leqq x < b\}$ is used, and similarly for other bounded intervals. (We also use $(a_1, \ldots, a_n)$ for the greatest common divisor of integers $a_1, \ldots, a_n$; this should not cause confusion). The Cartesian product $[0, 1)^h$ is written U_h.

We write $L^1(a, b)$ for the set of functions Lebesgue integrable on (a, b) and define $L^1(\mathbb{R}^h)$ similarly. We use the notation

$$\hat{F}(\mathbf{y}) = \int_{\mathbb{R}^h} F(\mathbf{x})e(-\mathbf{xy})\,d\mathbf{x} \qquad (\mathbf{y} \in \mathbb{R}^h)$$

for the Fourier transform of F in $L^1(\mathbb{R}^h)$, but

$$\hat{g}(\mathbf{n}) = \int_{U_h} g(\mathbf{x})e(-\mathbf{xn})\,d\mathbf{x} \qquad (\mathbf{n} \in \mathbb{Z}^h)$$

for g in $L^1(U_h)$ of period one in each variable. Again, this should not lead to ambiguity.

The cardinality of a finite set $\mathcal{A}$ is written $|\mathcal{A}|$. The notation $A + B$ denotes $\{a + b: a \in A, b \in B\}$.

The field with q elements is denoted by $\mathbb{F}_q$. We denote by $R[X_1, \ldots, X_s] = R[\mathbf{X}]$ the ring of polynomials, with coefficients in a given commutative ring R, in variables $X_1, \ldots, X_s$, and by $R[[\mathbf{X}]]$ the corresponding ring of formal power series in $\mathbf{X} = (X_1, \ldots, X_s)$.

For a subset $\mathcal{M}$ of $\mathbb{Z}^h$, we write

$$z_P(\mathcal{M}) = \sum_{|x_1| \leqq P} \cdots \sum_{|x_h| \leqq P; \mathbf{x} \in \mathcal{M}} 1.$$

1
Introduction

1.1 Fractional parts of a polynomial

The notion of a uniformly distributed sequence will probably be familiar to the reader. Let $x_1, \ldots, x_N, \ldots$ be a sequence in $[0, 1)$ and suppose that the number of $j \leqq N$ for which

$$x_j \in [0, \alpha)$$

is denoted by $Z(N, \alpha)$. If

$$\lim_{N \to \infty} N^{-1} Z(N, \alpha) = \alpha$$

for all α in $[0, 1)$, we say that (x_n) is a *uniformly distributed* sequence. It was shown by Weyl (1916) that for any polynomial of the form

$$f(x) = \alpha_k x^k + \cdots + \alpha_1 x \tag{1.1}$$

the fractional parts $(\{f(n)\})$ form a uniformly distributed sequence, provided that at least one of $\alpha_k, \ldots, \alpha_1$ is irrational. In this book, however, we shall be concerned only with finding small values of $f(n)$ modulo one. The first theorem of this type is due to Vinogradov (1927); he showed that for any real θ,

$$\min_{1 \leqq n \leqq N} \|n^k \theta\| < N^{\varepsilon - k/(Kk+1)}$$

provided that $N > C_1 (k, \varepsilon)$. The existence of such a bound had been conjectured by Hardy and Littlewood (1914).

Later this was improved by Heilbronn (1948), who replaced the exponent $-(2/5) + \varepsilon$ by $-(1/2) + \varepsilon$ in the case $k = 2$. Danicic (1957) gave the following extension of Heilbronn's theorem:

$$\min_{1 \leqq n \leqq N} \|n^k \theta\| < N^{-(1/K)+\varepsilon} \tag{1.2}$$

for $N > C_2(k, \varepsilon)$. No better result is known for $2 \leqq k \leqq 8$, although we can do much better for $k \geqq 9$.

Davenport (1967) proposed the problem of obtaining the analogue of (1.2) for a sequence $f(n)$, with f as in (1.1), in place of $n^k \theta$. He gave a weaker inequality in this direction; Schmidt (1977a) solved the case $k = 2$

of Davenport's problem. The general case is due to the author (1981*a*, 1982*b*):

$$\min_{1\leq n\leq N} \|f(n)\| < N^{-(1/K)+\varepsilon} \tag{1.3}$$

for $N > C_3(k, \varepsilon)$.

One might ask for inequalities of the type (say)

$$\min_{1\leq n\leq N} \|f(n) - \lambda\| < N^{-\sigma} \tag{1.4}$$

where $0 < \sigma < 1$. In fact, it is easily seen that intervals (modulo one) centred on zero play a special role. Suppose that $\alpha_1, \ldots, \alpha_k$ have a simultaneous approximation by rationals with the same small denominator q; let's say $q < N^\sigma/2$, and

$$\left|\alpha_j - \frac{a_j}{q}\right| < \frac{1}{2kqN^j} \qquad (j = 1, \ldots, k)$$

where $a_1, \ldots, a_k$ are integers. In this case

$$\left|\sum_{j=1}^{k} n^j \alpha_j - \frac{m}{q}\right| < \frac{1}{2q}$$

for $n = 1, \ldots, N$ and integer $m = m(n)$. Hence

$$\min_{1\leq n\leq N} \left\|f(n) - \frac{1}{2q}\right\| > \frac{1}{2q} > N^{-\sigma}.$$

However, we do know that $\min_{1\leq n\leq N} \|f(n)\|$ is small in this case, because

$$\|f(q)\| \leq \sum_{j=1}^{k} q^{j-1} \|\alpha_j q\| < \sum_{j=1}^{k} N^{j-1} \frac{1}{2kN^j} = N^{-1}.$$

So the lack of an inhomogeneous inequality (1.4) need not preclude a good homogeneous inequality.

Of course, we could ask whether there are infinitely many n for which

$$\|f(n) - \lambda\| < n^{-\sigma}.$$

This type of question will not be considered in this book: see Hooley (1979) for an interesting result, and Hall and Tenenbaum (1982), Tenenbaum (to appear) for a strengthening of Hooley's method. Related questions such as the distribution of αp modulo one (where p runs through the prime numbers) have a considerable literature. See the papers of Harman (1981, 1983*a*, 1983*b*, 1984), where references to earlier work of Vinogradov and Vaughan may be found; and Heath-Brown (1984).

The principle of proof of (1.3) can easily be indicated. Let us write

$M = [N^{(1/K)-\varepsilon}] + 1$, where $N > C_3(k, \varepsilon)$. Assume that we have a non-negative auxiliary function

$$\psi(x) = \sum_{m=-\infty}^{\infty} a_m e(mx) \qquad (1.5)$$

of period one, with an absolutely convergent Fourier series, constructed in such a way that

$$\psi(x) = 0 \qquad \text{for } \|x\| \geq M^{-1}.$$

Let us suppose that none of the distances $\|f(n)\|$ $(n = 1, \ldots, N)$ are smaller than M^{-1}; then

$$\psi(f(n)) = 0 \qquad (n = 1, \ldots, N)$$

and indeed on summing over n we obtain

$$\sum_{m=-\infty}^{\infty} a_m \sum_{n=1}^{N} e(mf(n)) = 0. \qquad (1.6)$$

In particular, then,

$$\sum_{m \neq 0} |a_m| \left| \sum_{n=1}^{N} e(mf(n)) \right| \geq a_0 N. \qquad (1.7)$$

With careful choice of ψ, we can guarantee that

$$\sum_{|m| > MN^\delta} |a_m| < \frac{a_0}{2}, \qquad (1.8)$$

where $\delta = \varepsilon/C_0(k)$. It is easy to deduce from (1.7), (1.8) that

$$\sum_{0 < |m| \leq MN^\delta} |a_m| \left| \sum_{n=1}^{N} e(mf(n)) \right| \geq a_0 N - \sum_{|m| > MN^\delta} |a_m| N > a_0 N/2.$$

Indeed, since $\psi \geq 0$ we know that

$$|a_m| = \left| \int_0^1 \psi(x) e(mx) \, dx \right| \leq a_0.$$

Thus we have

$$\sum_{0 < |m| \leq MN^\delta} \left| \sum_{n=1}^{N} e(mf(n)) \right| \geq \frac{N}{2}. \qquad (1.9)$$

The 'Weyl sums'

$$S_m(f) = \sum_{n=1}^{N} e(mf(n)) \qquad (1.10)$$

are not well understood, at least for $k \geq 3$, but we do know that from inequalities of the type (1.9), with reasonably small M, it is possible to draw conclusions about good rational approximations to the coefficients of f. With the M specified above, one can deduce that there exists a natural number q such that

$$q < M^k N^\varepsilon \leqq N \tag{1.11}$$

$$\|\alpha_j q\| < M^{k-1} N^{-j+\varepsilon} \qquad (j = 1, \ldots, k). \tag{1.12}$$

The key to this deduction is the use of Weyl's inequality (see Chapter 3). Now, of course,

$$\|f(q)\| \leqq q^{k-1} \|\alpha_k q\| + \cdots + q \|\alpha_2 q\| + \|\alpha_1 q\|$$

$$\leqq \sum_{j=1}^{k} N^{j-1} M^{k-1} N^{-j+\varepsilon} < M^{-1},$$

since $kM^k \leqq N^{1-\varepsilon}$. This contradicts our initial hypothesis, since $q \leqq N$. Therefore there must be solutions of (1.3) after all.

Apart from the construction of ψ, which is considered in Chapter 2, the missing step is the deduction of (1.11), (1.12) from (1.9). This is quite difficult. We will give the details in Chapter 3 for the simplest cases (i) $\alpha_{k-1} = \cdots = \alpha_1 = 0$, and (ii) $k = 2$, and postpone the discussion of the general case to Chapter 5.

1.2 Vinogradov's methods

For $k \geqq 14$ the method of Vinogradov (1954), (1971) for comparing a trigonometric sum (1.10) with a mean value

$$\left\{ \int_0^1 \cdots \int_0^1 \left| \sum_{n=1}^{N} e(\alpha_{k-1} n^{k-1} + \cdots + \alpha_1 n) \right|^{2b} d\alpha_1 \cdots d\alpha_{k-1} \right\}^{1/2b}$$

leads to a stronger inequality than (1.3). We may replace $-(1/K) + \varepsilon$ by $-1/J$, where $J = J(k)$ is asymptotically equal to $8k^2 \log k$ for large k. We can in fact deduce (1.11), (1.12) from (1.9) with $M = N^{1/J}$. Some of the ideas for the proof of (1.3) were drawn from this earlier work of Vinogradov. For this reason, we discuss Vinogradov's method in Chapter 4, before proving (1.3) for $k \geqq 3$.

Another of Vinogradov's methods is based on the estimation of special trigonometric sums of the type

$$\sum_{u,v} e(\alpha u v^k) \tag{1.13}$$

where u and v vary over suitably chosen sets. In the special case $f(n) = \alpha n^k$, this gives an even stronger result than the 'mean value method' and takes over from the 'Weyl inequality method' for $k \geqq 9$. The result obtained is of the form

$$\min_{1 \le n \le N} \|\alpha n^k\| < N^{-1/U} \tag{1.14}$$

for $N > C_4(k, \varepsilon)$, where $U = U(k)$ is asymptotically equal to $4k \log k$. (Incidentally, I conjecture that the left-hand side of (1.14) is $O(N^{-1+\varepsilon})$, so this is not a very satisfactory result!). The method appears to break down if we try to treat a general polynomial. (It yields results with an extra hypothesis on one of the coefficients. See Vinogradov (1954), Chapter 5.) The use of sums (1.13) also seems to be unhelpful in questions of simultaneous approximation, discussed below. The inequality (1.14) is proved in Chapter 6.

1.3 Simultaneous approximations

The simplest question of this type was considered by Danicic (1957) in his thesis. He proved that for any real θ, ϕ the inequality

$$\max (\|n^2\theta\|, \|n^2\phi\|) < N^{-(1/9)+\varepsilon}$$

has a solution $n \le N$, whenever $N > C_5(k, \varepsilon)$. The quantity $\frac{1}{9}$ was subsequently improved to $\frac{1}{8}$ by Danicic himself (1959), to $\frac{1}{7}$ by Liu (1970b) and to $\frac{1}{6}$ by Schmidt (1977b).

It seems natural to conjecture the following extension of (1.3).

Conjecture A. *Let $N > C_6(h, k, \varepsilon)$. Let $\psi_1, \ldots, \psi_h$ be numbers satisfying $0 < \psi_i < 1$ $(i = 1, \ldots, h)$ and*

$$\psi_1 \cdots \psi_h \ge N^{-(1/K)+\varepsilon}. \tag{1.15}$$

Then, if $f_1, \ldots, f_h$ are polynomials of degree at most k without constant term, there is a natural number $n \le N$ satisfying

$$\|f_i(n)\| < \psi_i \quad (i = 1, \ldots, h).$$

For instance, we shall prove this conjecture in Chapter 8 for $h = 2$, $k = 3$, and, more generally, whenever

$$(h - 1)k + 1 \le K. \tag{1.16}$$

The stumbling block for larger h is our inability to sharpen a 'determinant argument' of Schmidt. We will go into this question thoroughly in Chapter 7. For the moment, we make a few general observations.

Although we are particularly interested in an outcome of the type

$$(f_1(n), \ldots, f_h(n)) \in B + \Lambda \quad (n \le N),$$

where B is a convex body and Λ is the integer lattice, Schmidt (1977b)

realized that it is helpful to consider a general lattice Λ in Euclidean space $\mathbb{R}^h$. This permits a suitable result to be obtained by induction on h. Earlier authors—for example Liu (1975), Cook (1973, 1975) and Baker and Gajraj (1976a) carried out the induction on the number of polynomials, for integer lattices only, and obtained far worse results.

Schmidt considered only quadratic polynomials, for $h \geq 2$. For any $\theta_1, \ldots, \theta_h$ and $\phi_1, \ldots, \phi_h$, he showed that there is an $n \leq N$ such that

$$\max_{1 \leq i \leq h} \|n^2 \theta_i + n \phi_i\| < N^{-(1/2h^2)+\varepsilon}.$$

Here we suppose that $N > C_7(h, \varepsilon)$. He also gave the sharper exponent $-1/(h^2 + h) + \varepsilon$ in the case $\phi_1 = \cdots = \phi_h = 0$. This (at present unbeaten) result is proved for general quadratic polynomials in Chapter 7. In the following chapter we prove the inequality of Baker and Harman (1984b),

$$\max_{1 \leq i \leq h} \|f_i(n)\| < N^{-1/(h^2 + h\tau) + \varepsilon} \qquad (n \leq N), \tag{1.17}$$

where $f_1, \ldots, f_h$ are as in Conjecture A, and $h > 6K$, $N > C_8(h, k, \varepsilon)$. Here $\tau = (k + 1)/2$. This is really quite close to Schmidt's exponent for $n^2\theta$. Again, the key to further progress would be an improvement of Schmidt's 'determinant argument'.

In Chapter 8 we also give a proof of the above conjecture for the case $f_i(n) = n^k \theta_i$ $(i = 1, \ldots, h)$, for

$$h \leq K/2.$$

This covers far more cases than (1.16); for instance, if $k = 7$, h may run up to 64 rather than 10. The improvement is based on an ingenious lemma of Harman (1982). We shall obtain parallel theorems for large k, by Vinogradov's mean value method.

1.4 Functions of many variables

If we consider polynomials $f(x_1, \ldots, x_s)$ in many variables, rather than functions of a single variable as in (1.3), we may be able to make

$$|f(x_1, \ldots, x_s)|$$

small at a nonzero integer point, rather than merely small modulo one. For the sake of simplicity, we consider only *forms* in the present book, that is, polynomials $f(x_1, \ldots, x_s)$ homogeneous of degree k in $x_1, \ldots, x_s$.

We are obliged to restrict the forms f considered, if we do not work modulo one. A good example of this type of result is the theorem of Birch–Davenport–Ridout: every real indefinite quadratic form $Q(x_1, \ldots, x_s)$ with $\det Q \neq 0$, $s \geq 21$, takes arbitrarily small values at nonzero integer points. That is,

$$|Q(n_1, \ldots, n_s)| < \varepsilon \qquad (1.18)$$

for some $\mathbf{n} = (n_1, \ldots, n_s)$ in $\mathbb{Z}^s$, $\mathbf{n} \neq \mathbf{0}$. One can replace (1.18) by an inequality of the form

$$|Q(n_1, \ldots, n_s)| < N^{-V(s)+\varepsilon}$$

for $N > C_9(Q, \varepsilon)$, where

$$0 < \max(|n_1|, \ldots, |n_s|) \leqq N. \qquad (1.19)$$

See Gajraj (1976a) and Cook (1983) for calculations of $V(s)$. However, the principal interest here is in reducing the number of variables: perhaps 5 variables, rather than 21, would suffice. The series of papers leading up to (1.18), with the exception of Ridout (1958), is collected in Davenport (1977), and it would take too much space to prove (1.18) here.

Danicic (1958) considered the problem of values of quadratic forms modulo one. Let $Q(x_1, \ldots, x_s)$ be a quadratic form with real coefficients. For $N > C_{10}(s, \varepsilon)$ there are integers $n_1, \ldots, n_s$ satisfying (1.19) for which

$$\|Q(n_1, \ldots, n_s)\| < N^{-(s/(s+1))+\varepsilon}. \qquad (1.20)$$

Danicic's method was adapted from Davenport's work on (1.18) (Davenport (1956), (1958)). Schinzel, Schlickewei, and Schmidt (1980) devised a new approach to inequalities of type (1.20). One begins by showing that a congruence

$$F(y_1, \ldots, y_s) \equiv 0 \pmod{m}. \qquad (1.21)$$

where $m > C_{11}(s)$ and F is a quadratic form with integer coefficients, has small solutions. That is, there exists $(y_1, \ldots, y_s)$ in $\mathbb{Z}^s$ satisfying (1.21) with

$$0 < \max(|y_1|, \ldots, |y_s|) < m^{(1/2)+\gamma(s)}. \qquad (1.22)$$

Here $\gamma(s)$ tends to zero as $s \to \infty$. This can be used to deduce an inequality.

$$\|Q(n_1, \ldots, n_s)\| < N^{-2+\lambda(s)+\varepsilon}$$

for $N > C_{12}(s, \varepsilon)$ and $\mathbf{n}$ satisfying (1.19); $\lambda(s)$ also tends to zero as $s \to \infty$. In Chapter 9 we present the best results of this type known at present; (1.20) remains unbeaten for $s \leqq 3$.

Baker and Harman (1981) discussed simultaneous inequalities for quadratic forms. Let $Q_1, \ldots, Q_h$ be real quadratic forms in s variables where $s > C_{13}(h, \varepsilon)$. Then for $N > C_{14}(h, \varepsilon)$ there is an integer point $\mathbf{n}$ with (1.19) for which

$$\|Q_j(\mathbf{n})\| < N^{-(2/h)+\varepsilon} \qquad (j = 1, \ldots, h) \qquad (1.23)$$

The limiting exponent $-2/h$ is 'best possible' as we explain below. This work is considered in Chapter 10.

Naturally, one would like to extend the discussion to forms of degree greater than two. The simplest form of degree k is an 'additive' form

$$F(\mathbf{x}) = \lambda_1 x_1^k + \cdots + \lambda_s x_s^k. \tag{1.24}$$

The analogue of (1.23) turns out to be: there is an integer point $\mathbf{n}$ with (1.19), and all $n_i \geqq 0$, for which

$$\|F_j(\mathbf{n})\| < N^{-(k/h)+\varepsilon} \qquad (j = 1, \ldots, h). \tag{1.25}$$

Here $F_1, \ldots, F_h$ are real additive forms as in (1.24), with $s > C_{15}(h, \varepsilon)$, and $N > C_{16}(h, k, \varepsilon)$. Once again, one needs to start with small solutions of congruences. In fact an ingredient of the proof is a solution of a diagonal *equation*

$$c_1 x_1^k + \cdots + c_s x_s^k = 0 \tag{1.26}$$

($c_i \in \mathbb{Z}$, not all of the same sign), where the x_i are bounded by a power of $\max |c_i|$. This is work of Pitman (1971*b*) and Schmidt (1979*a*) although it has its origins in work on (1.18): see the editors' notes in Davenport (1977). We consider equations (1.26) in Chapter 11 and apply the results to simultaneous congruences

$$F_j(\mathbf{y}) \equiv 0 \pmod{m} \qquad (j = 1, \ldots, h) \tag{1.27}$$

in Chapter 12. Here $F_1, \ldots, F_h$ are forms as in (1.24), but now with integer coefficients, and the condition (1.22) is replaced by a similar one with limiting exponent $1/k$. The case $k = 2$ was studied in Baker (1980*b*): in fact general quadratic forms are considered there. An extra idea needed for $k \geqq 3$ was supplied by Glyn Harman in 1979 (unpublished). Actually, only the case $h = 1$ of (1.27) is required for (1.25), which is proved in Chapter 10.

For general forms of degree k,

$$F(\mathbf{x}) = \sum_{\substack{i_1 + \cdots + i_s = k \\ i_1 \geqq 0, \ldots, i_s \geqq 0}} \alpha(i_1, \ldots, i_s) x_1^{i_1} \cdots x_s^{i_s},$$

there is a dichotomy, depending on whether k is even or odd, and we discuss this in the following two sections.

1.5 Forms of odd degree

Let $F_1, \ldots, F_h$ be real forms in $\mathbf{x} = (x_1, \ldots, x_s)$ of odd degrees $\leqq k$. In the case where $F_1, \ldots, F_h$ have rational coefficients, it was shown by Birch (1957) that the equations

$$F_j(n_1, \ldots, n_s) = 0 \qquad (j = 1, \ldots, h)$$

have a simultaneous solution in integers $n_1, \ldots, n_s$ not all zero, provided that $s > C_{17}(k, h)$. See also Vaughan (1981).

It was conjectured by Birch and Davenport that for forms $F_1, \ldots, F_h$ as above, with *real* coefficients, the simultaneous inequalities

$$|F_j(n_1, \ldots, n_s)| < 1 \qquad (j = 1, \ldots, h)$$

are soluble nontrivially in integers, provided that $s > C_{18}(k, h)$. The case $k = 3$, $h = 1$ was given by Pitman (1968) with $C_{18} = (1314)^{256}$. The general case, which is very difficult, was proved by Schmidt (1979a, b). In fact he proved more:

Let $F_1(\mathbf{x}), \ldots, F_h(\mathbf{x})$ be forms of odd degrees $\leq k$ in $\mathbf{x} = (x_1, \ldots, x_s)$. Let $E > 0$ and suppose that $s > C_{19}(k, h, E)$, and $N > C_{20}(k, h, E)$. Then the simultaneous inequalities

$$|F_j(n_1, \ldots, n_s)| < H(F_j)N^{-E} \qquad (j = 1, \ldots, h) \tag{1.28}$$

are soluble in integers $n_1, \ldots, n_s$ satisfying (1.19).

Here $H(F_j)$, which denotes the maximum absolute value of the coefficients of F_j, could obviously be dispensed with if we considered $\|F_j\|$ instead of $|F_j|$.

The inequalities (1.28) will be considered in Chapter 14, after setting up a stronger form of our earlier results about (1.26) in Chapter 13. What is needed is, in fact, a solution of (1.26) with

$$0 < \max |x_j| < (\max (1, |c_1|, \ldots, |c_s|))^{\varepsilon}$$

The fact that x_i^k takes both signs, for odd k, is crucial!

We mention here the important recent work of Schmidt (1982a, b, c, d) on systems of cubic equations, for which we have no space in this book.

1.6 Forms of even degree

We only consider inequalities modulo one and congruences in the present book; see Davenport (1977) and Schmidt (1982e, 1985) for systems of equations of even degree. One might well make the following

Conjecture B. For $h \geq 1$, there exist $C_i(k, h, \varepsilon)$ $(i = 21, 22)$ as follows. Let $F_1, \ldots, F_h$ be forms with real coefficients in $s > C_{21}$ variables with even degrees between 1 and k. Then given $N > C_{22}$ there is an integer point $\mathbf{n}$ with (1.19) for which

$$\|F_i(\mathbf{n})\| < N^{-(2/h)+\varepsilon} \qquad (i = 1, \ldots, h).$$

The nearest we can get to such a conjecture at present is Schmidt's remarkable theorem on congruences (Schmidt (1984b; to appear)):

For $h \geqq 1$ there exist $C_i = C_i(k, h, \varepsilon)$ $(i = 23, 24)$ as follows. Let $F_1, \ldots, F_h$ be forms with integer coefficients in $s > C_{23}$ variables with even degrees between 1 and k. Then given $m \geqq C_{24}$ there is an integer point $(y_1, \ldots, y_s)$ with

$$\left. \begin{array}{l} 0 < \max (|y_1|, \ldots, |y_s|) \leqq m^{(1/2)+\varepsilon}, \\ F_i(y_1, \ldots, y_s) \equiv 0 \pmod{m} \qquad (i = 1, \ldots, h). \end{array} \right\} \tag{1.29}$$

This will be proved in Chapter 18 after studying appropriate exponential sums in Chapters 15–17.

The case $h = 1$ of Conjecture B implies the case $h = 1$ of (1.29), as we now proceed to show. Let $F(\mathbf{x})$ be a form of even degree k with integer coefficients in $s > C_{21}$ variables, let $m > C_{22}^2$, and put $G(\mathbf{x}) = m^{-1}F(\mathbf{x})$. Set $N = [m^{1/2+\varepsilon}]$. There is an integer point $\mathbf{y}$ with $0 < \max (|y_1|, \ldots, |y_s|) \leqq m^{(1/2)+\varepsilon}$ and with $\|G(y)\| < N^{-2+\varepsilon} < m^{-1}$, hence with $F(\mathbf{y}) \equiv 0 \pmod{n}$. But the general case of (1.29) does not seem to follow from the general case of Conjecture B.

If true, Conjecture B is essentially best possible. This may be seen as follows. Using the Borel–Cantelli lemma one finds that, for given $h \geqq 1$, $l \geqq 1$, almost all h-tuples $(\alpha_1, \ldots, \alpha_h)$, in the sense of Lebesgue measure, have

$$\max (\|\alpha_1 x^l\|, \ldots, \|\alpha_h x^l\|) > C_{25}(\boldsymbol{\alpha})x^{-1/h}(\log x)^{-2/h} \tag{1.30}$$

for $x = 1, 2, \ldots$; see the proof of Theorem 3A of Schmidt (1980b), Chapter III. Setting $l = k/2$ and

$$F_i(\mathbf{x}) = \alpha_i(x_1^2 + \cdots + x_s^2)^l,$$

we have that (1.19) implies

$$\max_i \|F_i(\mathbf{n})\| > C_{26}(\boldsymbol{\alpha}, s)N^{-2/h}(\log N)^{-2/h}.$$

As for (1.25), setting $l = 1$ and $x = x_1^k + \cdots + x_s^k$ we get forms $F_i(\mathbf{x}) = \alpha_i(x_1^k + \cdots + x_s^k)$ for which

$$\max_i \|F_i(\mathbf{n})\| > C_{27}(\boldsymbol{\alpha}, s)N^{-k/h}(\log N)^{-2/h}$$

for $n_i \geqq 0$, $0 < \max_i n_i \leqq N$. One also sees directly from (1.30) that the exponent in (1.17) cannot be improved beyond $-1/h$, even in the special case $f_i(x) = \alpha_i x^k$.

Similar remarks apply to congruences. We can use forms $(x_1^2 + \cdots + x_s^2)^{k/2}$ and $x_1^k + \cdots + x_s^k$ to show that our limiting exponents $\frac{1}{2}$ (in the case of (1.29)) and $1/k$ (in the case of (1.27)) are best possible, even when $h = 1$.

The level of difficulty of the material rises steadily in the last few chapters. In particular, a fair acquaintance with algebraic geometry is required for the results described in this section. One should point out, however, that it is quite routine nowadays to use much deeper material from algebraic geometry in studying Diophantine equations by analytic methods; see Heath-Brown (1983) for a beautiful example of this. Baker (1983) discusses small solutions of congruences in rather few variables by a similar method; unfortunately one can only tackle 'non-singular' and diagonal forms in this way.

In all the results of Sections 1.5 and 1.6, very large numbers of variables are required. It is difficult, although possible in principle, to write down bounds for constants such as $C_{18}(5, 1)$. It would be nice to have better bounds—and simpler proofs!

2
Fourier analysis

2.1 Vinogradov's auxiliary function

The function ψ that we needed in (1.5) can be obtained by 'smoothing' the indicator function of an interval a number of times.

Lemma 2.1. *Let* $0 < \Delta \leq \frac{1}{4}$. *Let* r *be a natural number. There is a function* $\psi(x)$ *of period one such that*

$$\psi(x) \geq 0 \text{ for all } x; \tag{2.1}$$

$$\psi(x) = 0 \text{ for } \|x\| \geq 2\Delta; \tag{2.2}$$

$$\psi(x) = 2\Delta + \sum_{\substack{m=-\infty \\ m \neq 0}}^{\infty} a_m e(mx) \text{ for all } x, \tag{2.3}$$

where

$$|a_m| \leq C_1(r) |m|^{-r-1} \Delta^{-r} \qquad (m \neq 0). \tag{2.4}$$

Proof. Let $\psi_0(x)$ be the function of period one defined by

$$\psi_0(x) = \begin{cases} 1 & (-\Delta \leq x < \Delta) \\ 0 & (\Delta \leq x < 1 - \Delta). \end{cases}$$

Then the Fourier coefficients of ψ_0 are

$$\hat{\psi}_0(m) = \int_{-1/2}^{1/2} \psi_0(x) e(-mx) \, dx = \frac{\sin 2\pi m \Delta}{\pi m} \qquad (m \neq 0);$$

of course $\hat{\psi}(0) = 2\Delta$.

Define γ by $r\gamma = \Delta$, and define the functions $\psi_1, \ldots, \psi_r$, all of period one, by the recurrence relation

$$\psi_l(x) = \frac{1}{2\gamma} \int_{-\gamma}^{\gamma} \psi_{l-1}(x + z) \, dz \qquad (l = 1, \ldots, r).$$

We shall prove by induction that

(i) $\psi_l(x) = 0$ in the interval $\Delta + l\gamma < x < 1 - \Delta - l\gamma$,
(ii) $\psi_l(x) \geq 0$ for all x,

(iii)   For $l \geq 1$, we have

$$\psi_l(x) = \sum_{m=-\infty}^{\infty} \hat{\psi}_l(m) e(mx)$$

for all real x, with $\hat{\psi}_l(0) = 2\Delta$ and

$$|\hat{\psi}_l(m)| \leq C_2(l, r) |m|^{-l-1} \Delta^{-l} \qquad (m \neq 0). \qquad (2.5)$$

Now (i) and (ii) hold for $\psi_0(x)$. Suppose they hold for $\psi_{l-1}(x)$ where $l \geq 1$. Since $\psi_l(x)$ is the average of ψ_{l-1} in the interval $(x - \gamma, x + \gamma)$ it follows at once that (i) and (ii) hold for $\psi_l(x)$.

Now for (iii). We have

$$\hat{\psi}_l(m) = \int_{-1/2}^{1/2} \frac{1}{2\gamma} \int_{-\gamma}^{\gamma} \psi_{l-1}(x + z) e(-mx) \, dz \, dx$$

$$= \frac{1}{2\gamma} \int_{-\gamma}^{\gamma} dz \int_{-1/2}^{1/2} \psi_{l-1}(x + z) e(-mx) \, dx$$

$$= \hat{\psi}_{l-1}(m) \frac{1}{2\gamma} \int_{-\gamma}^{\gamma} e(mz) \, dz.$$

This immediately gives $\hat{\psi}_l(0) = 2\Delta$ and enables us to prove (2.5) by induction on l, since

$$|\hat{\psi}_l(m)| = |\hat{\psi}_l(m)| \left| \frac{\sin 2\pi m\gamma}{2\pi m\gamma} \right|$$

$$\leq \frac{r}{2\pi m\Delta} |\hat{\psi}_{l-1}(m)|.$$

In particular, $\sum_m |\hat{\psi}_l(m)| < \infty$ for $l \geq 1$. Since $\psi_1, \psi_2, \ldots$ are continuous it follows that $\psi_l(x)$ is the sum of its Fourier series for $l \geq 1$; see for example Zygmund (1968).

This proves (iii) for $\psi_l(x)$.

To complete the proof we note that $\psi = \psi_r$ does all that is required.

The auxiliary function ψ was first used by Vinogradov; see Vinogradov (1954) Chapter I, for example. We can use it to prove the following lemma.

Lemma 2.2. *Let $x_1, \ldots, x_N$ be real numbers. Let $M > C_3(\varepsilon)$, and suppose that $\|x_n\| \geq M^{-1}$ $(n = 1, \ldots, N)$. Then*

$$\sum_{1 \leq m \leq M^{1+\varepsilon}} \left| \sum_{n=1}^{N} e(mx_n) \right| > \frac{N}{4}. \qquad (2.6)$$

Proof. Let $\Delta = 1/(2M)$, $r = [2/\varepsilon] + 1$. We have established in Lemma

2.1 that there is a non-negative function $\psi(x)$ of period one such that

$$\psi(x) = 0 \qquad (\|x\| \geqq 2\Delta), \tag{2.7}$$

$$\psi(x) = \sum_{m=-\infty}^{\infty} a_m e(mx) \qquad \text{(all real } x),$$

where

$$a_0 = 2\Delta, \qquad |a_m| \leqq C_4(\varepsilon) |m|^{-r-1} \Delta^{-r}.$$

Consequently

$$\sum_{|m|>M^{1+\varepsilon}} |a_m| \leqq C_4(\varepsilon) \sum_{|m|>M^{1+\varepsilon}} |m|^{-r-1} \Delta^{-r}$$

$$\leqq C_5(\varepsilon)(M^{1+\varepsilon})^{-r}\Delta^{-r}$$

$$< C_6(\varepsilon)M^{-r\varepsilon} < C_6(\varepsilon)M^{-2} < a_0/2,$$

since $r\varepsilon > 2$ and $M > C_3(\varepsilon)$.

From (2.7) and our hypothesis on the x_n,

$$\psi(x_n) = 0 \qquad (n = 1, \ldots, N).$$

Summing over n gives

$$\sum_{m=-\infty}^{\infty} a_m \sum_{n=1}^{N} e(mx_n) = 0.$$

We can now complete the proof just as in Section 1.1. We have

$$\sum_{m \neq 0} |a_m| \left| \sum_{n=1}^{N} e(mx_n) \right| \geqq a_0 N,$$

$$\sum_{|m|>M^{1+\varepsilon}} |a_m| \left| \sum_{n=1}^{N} e(mx_n) \right| \leqq N \sum_{|m|>M^{1+\varepsilon}} |a_m| < \frac{a_0 N}{2}.$$

Since all $|a_m| \leqq a_0$, we find that

$$\sum_{0<|m|\leqq M^{1+\varepsilon}} \left| \sum_{n=1}^{N} e(mx_n) \right| > \frac{N}{2}.$$

The proof may be completed by observing that replacing m by $-m$ leaves $|\sum_{n=1}^{N} e(mx_n)|$ unchanged.

2.2 Construction of trigonometric polynomials

We are now going to prove a 'best possible' version of Lemma 2.2, with a trigonometric polynomial $T_1(x)$ more or less in the role of $\psi(x)$. Unfortunately $T_1(x)$, which arises from work of Beurling and Selberg, takes both signs. This is a nuisance when one comes to write down an extension of Lemma 2.2 to h dimensions (see Chapters 3, 7). We shall use a multidimensional version of ψ in those chapters.

Let $z \in \mathbb{C}$. We write sgn $z = 1$ for Re $z \geq 0$, sgn $z = -1$ for Re $z < 0$.

Lemma 2.3.   *Let*

$$F(z) = \left(\frac{\sin \pi z}{\pi}\right)^2 \left(\sum_{n=0}^{\infty} \frac{1}{(z-n)^2} - \sum_{n=1}^{\infty} \frac{1}{(z+n)^2} + \frac{2}{z}\right).$$

Then

(i) $F(z)$ *is entire,*

(ii) $F(x) \geq \text{sgn } x$ *for real* x,

(iii) $F(z) = \text{sgn } z + 0\left(\frac{1}{|z|} e^{2\pi |\text{Im } z|}\right),$

(iv) $\int_{-\infty}^{\infty} (F(x) - \text{sgn } x)\, dx = 1.$

Here, and throughout the remainder of the chapter, the implied constants are numerical.

Proof.   Assertion (i) is obvious. Next, we note the identity

$$\left(\frac{\sin \pi z}{\pi}\right)^2 \sum_{n=-\infty}^{\infty} \frac{1}{(z-n)^2} = 1 \qquad (2.8)$$

(this is an easy application of Theorem 3.2.1 of Jameson (1970), for instance). Now for $x > 0$,

$$\sum_{n=1}^{\infty} \frac{1}{(x+n)^2} \leq \sum_{n=1}^{\infty} \int_{x+n-1}^{x+n} \frac{du}{u^2} = \frac{1}{x} = \sum_{n=0}^{\infty} \int_{x+n}^{x+n+1} \frac{du}{u^2} \leq \sum_{n=0}^{\infty} \frac{1}{(x+n)^2}.$$

$$(2.9)$$

From (2.8),

$$F(z) - \text{sgn } z = \left(\frac{\sin \pi z}{\pi}\right)^2 \left(\frac{2}{z} - 2 \sum_{n=1}^{\infty} \frac{1}{(z+n)^2}\right) \qquad (2.10)$$

for Re $z > 0$, and

$$F(z) - \text{sgn } z = \left(\frac{\sin \pi z}{\pi}\right)^2 \left(\frac{2}{z} + 2 \sum_{n=0}^{\infty} \frac{1}{(z-n)^2}\right) \qquad (2.11)$$

for Re $z < 0$. We deduce (ii) easily from (2.9)–(2.11). To get (iii), we observe first that

$$\sin^2 \pi z = 0(e^{2\pi |\text{Im } z|}). \qquad (2.12)$$

Moreover, for $\alpha > 0$, $\beta > 0$,

$$\sum_{n=0}^{\infty} \frac{1}{(\alpha+n)^2 + \beta^2} \leq \frac{1}{\alpha^2 + \beta^2} + \min\left(\int_0^{\infty} \frac{dx}{(\alpha+x)^2}, \int_0^{\infty} \frac{dx}{x^2 + \beta^2}\right)$$

$$= \frac{1}{\alpha^2 + \beta^2} + \min\left(\frac{1}{\alpha}, \frac{\pi}{2\beta}\right).$$

Thus

$$\sum_{n=1}^{\infty} \frac{1}{|z+n|^2} = 0\left(\frac{1}{|z|}\right) \qquad \text{for Re } z \geq 0, \tag{2.13}$$

while

$$\sum_{n=0}^{\infty} \frac{1}{|z-n|^2} = 0\left(\frac{1}{|z|}\right) \qquad \text{for Re } z < 0. \tag{2.14}$$

Now (iii) follows on combining (2.10)–(2.13) and (2.14).

As for the final assertion, we note that the integrand is non-negative. Moreover,

$$\int_{-A}^{A} (F(x) - \text{sgn } x)\, dx = \int_0^A (F(x) + F(-x))\, dx$$

$$= \int_0^A \left(\frac{\sin \pi x}{\pi}\right)^2 \frac{2}{x^2}\, dx$$

after a short computation. The last integral tends to 1 as $A \to \infty$ (see, for example Jameson (1970), p. 127).

For any interval I in $\mathbb{R}$, we write $\chi_I(x)$ for the indicator function of I.

Lemma 2.4. *Given any interval $I = [a, b]$, there are continuous functions $G_1(x)$, $G_2(x)$ in $L^1(\mathbb{R})$ such that*

$$G_1(x) \leq \chi_I(x) \leq G_2(x),$$
$$\hat{G}_i(t) = 0 \qquad \text{for } |t| \geq 1 \quad (i = 1, 2),$$

and

$$\int_{-\infty}^{\infty} (\chi_I(x) - G_1(x))\, dx = 1, \qquad \int_{-\infty}^{\infty} (G_2(x) - \chi_I(x))\, dx = 1.$$

Proof. With F as in Lemma 2.3, take

$$G_2(x) = \tfrac{1}{2}(F(x - a) + F(b - x)).$$

Then

$$G_2(x) \geq \tfrac{1}{2}(\text{sgn } (x - a) + \text{sgn } (b - x)) = \chi_I(x),$$

and

$$\int_{-\infty}^{\infty} (G_2(x) - \chi_I(x))\, dx = \tfrac{1}{2} \int_{-\infty}^{\infty} (F(x-a) - \operatorname{sgn}(x-a))\, dx$$

$$+ \tfrac{1}{2} \int_{-\infty}^{\infty} (F(b-x) - \operatorname{sgn}(b-x))\, dx = 1.$$

Hence $G_2 \in L^1(\mathbb{R})$. The function G_2 is continuous, as the restriction of an entire function. Note that we have avoided the problems arising from F not being Lebesgue integrable!

If $t > 1$, then

$$\hat{G}_2(t) = \int_{-\infty}^{\infty} G_2(x) e(-tx)\, dx = 0. \tag{2.15}$$

To see this, we shall show for a fixed $t > 1$ that

$$J(A, B) = \int_{-A}^{B} G_2(x) e(-tx)\, dx = O\left(\frac{1}{A} + \frac{1}{B}\right) \tag{2.16}$$

as $A \to \infty$ and $B \to \infty$. First of all, $J(A, B)$ may be expressed as the sum of two integrals along vertical line segments $[-A, -A - iU]$, $[B - iU, B]$, where $U > 0$, and a third integral along a horizontal segment $[-A - iU, B - iU]$. The latter integral has modulus at most

$$\int_{-A'}^{B'} |F(x - iU)|\, e^{-2\pi t U}\, dx = O\left(\int_{-A'}^{B'} e^{2\pi U} \cdot e^{-2\pi t U}\, dx\right)$$

from Lemma 2.2(iii); here $A' = A + \max(|a|, |b|)$ and similarly for B'. Letting $U \to \infty$ this tends to 0 and $J(A, B)$ is the sum of two integrals over vertical half lines. Now on the left one of these lines, let $z = -A + iy$, then

$$F(z - a) = -1 + O\left(\frac{1}{A} e^{-2\pi y}\right) \qquad \text{for } A > |a|,$$

$$F(b - z) = 1 + O\left(\frac{1}{A} e^{-2\pi y}\right) \qquad \text{for } A > |b|,$$

so that

$$G_2(z) = O\left(\frac{1}{A} e^{-2\pi y}\right),$$

and the integral over the left vertical line is

$$O\left(A^{-1} \int_{-\infty}^{0} e^{-2\pi y} e^{2\pi t y}\, dy\right) = O(A^{-1}).$$

It is now easy to complete the proof of (2.16) and (2.15).

For $t < -1$, we get (2.15) from the equation $\hat{G}_2(-t) = \overline{\hat{G}_2(t)}$: also, by continuity of $\hat{G}_2$, (2.15) holds for $t = \pm 1$.

For the function $G_1(x)$ we take

$$G_1(x) = -\tfrac{1}{2}(F(a - x) + F(x - b))$$

and establish the properties we need in a similar way.

Lemma 2.5. *For any $\gamma > 0$ and any interval $I = [a, b]$ there are continuous functions $H_1(x)$, $H_2(x)$ in $L^1(\mathbb{R})$ such that*

$$H_1(x) \leq \chi_I(x) \leq H_2(x), \quad \hat{H}_i(t) = 0 \quad \text{for } |t| \geq \gamma$$

($i = 1, 2$) and

$$\int_{-\infty}^{\infty} (\chi_I(x) - H_1(x)) \, dx = \int_{-\infty}^{\infty} (H_2(x) - \chi_I(x)) \, dx = \gamma^{-1}.$$

Proof. Let $G_1(x) \leq \chi_{[\gamma a, \gamma b]}(x) \leq G_2(x)$ as in Lemma 2.4 and put $H_i(x) = G_i(\gamma x)$. The desired properties follow at once.

Lemma 2.6. *Let $f \in L^1(\mathbb{R})$. Then the series*

$$F(x) = \sum_{n=-\infty}^{\infty} f(n + x)$$

is absolutely convergent, and has period one, for almost all x. Moreover,

$$\hat{F}(k) = \hat{f}(k)$$

for all integers k.

Proof. The first sentence follows easily from

$$\int_0^1 \sum_{n=-\infty}^{\infty} |f(n + x)| \, dx = \int_{-\infty}^{\infty} |f(t)| \, dt < \infty.$$

As for the final assertion, we may write

$$\int_0^1 F(x)e(-kx) \, dx = \sum_{n=-\infty}^{\infty} \int_0^1 f(n + x)e(-kx) \, dx$$

$$= \int_{-\infty}^{\infty} f(t)e(-kt) \, dt.$$

Lemma 2.7. *Let L be a natural number. For any interval $I = [a, b]$ or (a, b), etc., with length $b - a < 1$, write $\phi_I(x) = \sum_{n=-\infty}^{\infty} \chi_I(x + n)$. There are trigonometric polynomials*

$$T_i(x) = \sum_{m=-L}^{L} \hat{T}_i(m)e(mx) \quad (i = 1, 2)$$

such that

$$T_1(x) \leqq \phi_I(x) \leqq T_2(x) \qquad \text{for all } x, \tag{2.17}$$

$$\hat{T}_1(0) = b - a - \frac{1}{L+1}, \qquad \hat{T}_2(0) = b - a + \frac{1}{L+1}. \tag{2.18}$$

Proof. Take $\gamma = L + 1$ in Lemma 2.5, and put $V_i(x) = \sum_{n=-\infty}^{\infty} H_i(x + n)$. By Lemma 2.6, $V_i \in L^1(0, 1)$, $\hat{V}_i(h) = 0$ for $|h| \geqq L + 1$. Consequently

$$V_i(x) = \sum_{m=-L}^{L} \hat{V}_i(m)e(mx)$$

almost everywhere (see Zygmund (1968), Chapter 1, Theorem 6.2). We write

$$T_i(x) = \sum_{m=-L}^{L} \hat{V}_i(m)e(mx).$$

From $\chi_I(x) \geqq H_1(x)$ we find that

$$\phi_I(x) = \sum_{n=-\infty}^{\infty} \chi_I(x + n) \geqq V_1(x)$$

for almost all x. It is easily deduced that $\phi_I(x) \geqq T_1(x)$ for all x by a continuity argument; similarly $\phi_I(x) \leqq T_2(x)$ for all x. Moreover,

$$\hat{T}_1(0) = \int_0^1 V_1(x)\, \mathrm{d}x = \int_{-\infty}^{\infty} H_1(x)\, \mathrm{d}x = \int_{-\infty}^{\infty} \chi_I(x)\, \mathrm{d}x - \gamma^{-1};$$

similarly for $\hat{T}_2(0)$. This proves Lemma 2.7.

2.3   Lower bounds for trigonometric sums

The following well-known theorem, although not directly useful in this book, has some interesting related applications; see for example Baker (1981c), Schmidt (1964), Baker and Kolesnik (1985).

Theorem 2.1.   *(Erdös–Turán (1948a, b)). Let $x_1, \ldots, x_N$ be real, let $I = [a, b]$, or (a, b), etc., be an arbitrary interval of length $b - a < 1$ and let L be a positive integer. Then*

$$\sum_{\substack{n=1 \\ x_n \in I \,(\mathrm{mod}\, 1)}}^{N} 1 - N(b - a) \leqq \frac{N}{L+1} + 3 \sum_{m=1}^{L} \frac{1}{m} \left| \sum_{n=1}^{N} e(mx_n) \right|.$$

Proof. In the notation of the previous lemma, we have

$$\sum_{n=1}^{N} \phi_I(x_n) \geq \sum_{n=1}^{N} T_1(x_n)$$

$$= N(b-a) - \frac{N}{L+1} + \sum_{\substack{m=-L \\ m \neq 0}}^{L} \hat{T}_1(m) \sum_{n=1}^{N} e(mx_n)$$

$$\geq N(b-a) - \frac{N}{L+1} - \sum_{0<|m|\leq L} |\hat{T}_1(m)| \left| \sum_{n=1}^{N} e(mx_n) \right| \quad (2.19)$$

To estimate $\hat{T}_1(m)$ we simply use the inequality

$$|\hat{g}_1(m) - \hat{g}_2(m)| = \left| \int_0^1 (g_1(x) - g_2(x))e(-mx) \right| dx \leq \int_0^1 |g_1(x) - g_2(x)| \, dx.$$

Now

$$\hat{\phi}_I(m) = e(-\tfrac{1}{2}m(a+b)) \frac{\sin \pi(b-a)m}{\pi m},$$

so that

$$|\hat{T}_1(m)| \leq \int_0^1 (\phi_I(x) - T_1(x)) \, dx + |\hat{\phi}_I(m)|$$

$$\leq \frac{1}{L+1} + \left| \frac{\sin \pi(b-a)m}{\pi m} \right| \leq \frac{3}{2m}. \quad (2.20)$$

It is an easy matter to deduce that

$$\sum_{n=1}^{N} \phi_I(x_n) - N(b-a) \geq \frac{-N}{L+1} - 3 \sum_{m=1}^{L} \frac{1}{m} \left| \sum_{n=1}^{N} e(mx_n) \right|,$$

and the corresponding upper bound is proved in a similar way.

A variant of this argument for short intervals was pointed out to the author by H. L. Montgomery. It gives the promised strengthening of Lemma 2.2.

Theorem 2.2. *Let* $\|x_j\| \geq M^{-1}$ $(j = 1, \ldots, N)$ *for a real sequence* $x_1, \ldots, x_N$. *Then*

$$\sum_{1 \leq m \leq M} \left| \sum_{n=1}^{N} e(mx_n) \right| > \frac{N}{6}.$$

Proof. Let $L = [M]$ and apply (2.19) with $I = (-1/M, 1/M)$, so that $b - a = 2/M$.

Since $|\sin \alpha| \leqq |\alpha|$, we find that

$$0 = \sum_{\substack{n=1 \\ x_n \in I \ (\text{mod } 1)}}^{N} 1 \geqq N(b-a) - \frac{N}{L+1} - 2 \sum_{m=1}^{L} \left(\frac{1}{L+1} + (b-a) \right) \left| \sum_{n=1}^{N} e(mx_n) \right|.$$

Now

$$b - a - \frac{1}{L+1} > \frac{1}{M}, \qquad \frac{1}{L+1} + (b-a) < \frac{3}{M}.$$

The lemma follows.

This result is surprisingly close to best possible. Let N be a multiple of a natural number L and let $M = 2L$. Let the x_n be the points $j/L - 1/2L$ ($j = 1, \ldots, L$) repeated N/L times each. Then

$$\sum_{n=1}^{N} e(mx_n) = e\left(-\frac{m}{2L} \right) \frac{N}{L} \sum_{j=1}^{N} e\left(\frac{mj}{L} \right) = 0$$

for $m = 1, \ldots, L - 1$, so that

$$\sum_{1 \leqq m < M/2} \left| \sum_{n=1}^{N} e(mx_n) \right| = 0$$

although $\|x_n\| \geqq M^{-1}$ for $n = 1, \ldots, N$. Note also that

$$\sum_{m=1}^{M} \left| \sum_{n=1}^{N} e(mx_n) \right| = 2N.$$

For a delicate application of the ideas in Sections 2.2 and 2.3 to the large sieve, see Montgomery (1978).

3
Heilbronn's theorem and Schmidt's extension

3.1 The basic technique

Let α and β be real numbers with the property that

$$\|\alpha n^2 + \beta n\| \geq M^{-1} \qquad (n = 1, \ldots, N)$$

where M is an integer, $2 \leq M \leq N$. We know from Theorem 2.2 that

$$\sum_{m=1}^{M} |S_m| > \frac{N}{6}, \qquad (3.1)$$

where

$$S_m = \sum_{n=1}^{N} e(m(\alpha n^2 + \beta n)).$$

To obtain an upper bound for S_m, we square the modulus:

$$|S_m|^2 = \sum_{n_1=1}^{N} \sum_{n_2=1}^{N} e(m(n_2 - n_1)(\alpha(n_1 + n_2) + \beta)).$$

Putting $u = n_1 + n_2$, $v = n_2 - n_1$, we obtain

$$|S_m|^2 = \sum_{u,v} e(mv(\alpha u + \beta))$$

where the sum is over integers u, v with $u \equiv v \pmod{2}$ and

$$0 < u + v \leq 2N, \qquad 0 < u - v \leq 2N.$$

The N summands with $v = 0$ give a contribution N. Since the substitution $v \to -v$ changes each summand into its complex conjugate, we have

$$|S_m|^2 = N + 2 \operatorname{Re} \sum_{u,v} e(mv(\alpha u + \beta)) \qquad (3.2)$$

where the sum is over u, v with $u \equiv v \pmod{2}$ and

$$0 < v < N, \qquad v < u \leq 2N - v. \qquad (3.3)$$

Before going further, we observe that for any real γ, θ,

$$\left| \sum_{x=U+1}^{U+V} e(\gamma x + \theta) \right| = \frac{|e(\gamma(V+1)) - 1|}{|e(\gamma) - 1|}$$

$$\leq \frac{1}{|\sin \pi \gamma|} \leq \frac{1}{2\|\gamma\|};$$

indeed, we have

$$\left| \sum_{x=U+1}^{U+V} e(\gamma x + \theta) \right| \leq \min\left(V, \frac{1}{2\|\gamma\|}\right). \tag{3.4}$$

We apply this to the double sum in (3.2). For fixed v, the terms $mv(\alpha u + \beta)$ with integers $u \equiv v \pmod 2$ form an arithmetic progression with common difference $2mv\alpha$, and length $N - v$, by (3.3). The sum of $e(mv(\alpha u + \beta))$ over the terms of this arithmetic progression is

$$\ll \min\left(N, \|2mv\alpha\|^{-1}\right).$$

(We recall that in Chapters 3–5 implied constants depend at most on k, ε.) Summing over v in $0 < v < N$, we obtain Weyl's estimate

$$|S_m|^2 \ll N + \sum_{v=1}^{N} \min\left(N, \|2mv\alpha\|^{-1}\right)$$

$$\ll \sum_{v=1}^{N} \min\left(N, \|2mv\alpha\|^{-1}\right). \tag{3.5}$$

Combining this with (3.1), we find that

$$\sum_{m=1}^{M} \sum_{v=1}^{N} \min\left(N, \|2mv\alpha\|^{-1}\right) \gg \sum_{m=1}^{M} |S_m|^2 \gg N^2 M^{-1}, \tag{3.6}$$

by Cauchy's inequality. Putting $z = 2mv$, we observe that for $z \leq 2MN \leq 2N^2$ the number of divisors of z is $\ll z^\delta \ll N^{2\delta}$ (see Hardy and Wright (1979), Theorem 315). Thus

$$\sum_{z=1}^{2MN} \min\left(N, \|z\alpha\|^{-1}\right) \gg N^{2-2\delta} M^{-1}. \tag{3.7}$$

We now reverse the roles of u, v above; this idea is due to Schmidt (1977a). The inequalities (3.3) may be rewritten as

$$0 < u < 2N, \qquad 0 < v \leq \min(u - 1, 2N - u).$$

For fixed u, the terms $mv(\alpha u + \beta)$ with $v \equiv u \pmod 2$ form an arithmetic progression with common difference $2m(\alpha u + \beta)$. Carrying out the summation over v first in (3.2),

$$|S_m|^2 \ll N + \sum_{u=1}^{2N} \min(N, \|2m(\alpha u + \beta)\|^{-1})$$

$$\ll \sum_{u=1}^{2N} \min(N, \|2m(\alpha u + \beta)\|^{-1}). \tag{3.8}$$

We now deduce from the second inequality in (3.6) that

$$\sum_{m=1}^{2M} \sum_{u=1}^{2N} \min(N, \|m(\alpha u + \beta)\|^{-1}) \gg N^2 M^{-1}. \tag{3.9}$$

3.2 Lemmas on reciprocal sums

We assemble in this section some upper bounds for sums such as

$$\sum_{i=1}^{R} \min(Y, \|\gamma_i\|^{-1})$$

which we now know to be relevant to finding small values of $\alpha n^2 + \beta n$ (mod 1).

Lemma 3.1. *Let $\rho > 0$, and let $\gamma_1, \ldots, \gamma_R$ be reals with*

$$\|\gamma_i - \gamma_j\| \geq \rho \quad \text{for} \quad i \neq j,$$

and with $\|\gamma_1\| = \min(\|\gamma_1\|, \ldots, \|\gamma_R\|)$. Then

$$\sum_{i=2}^{R} \|\gamma_i\|^{-1} \ll \rho^{-1} \log R.$$

Proof. We may suppose that $\|\gamma_1\| \leq \cdots \leq \|\gamma_R\|$. Then it is easy to see that $\|\gamma_i\| \geq (i-1)\rho/2$ for $i = 2, 3, \ldots, R$. The lemma follows.

Lemma 3.2. *Suppose α, β are real and $\left|\alpha - \dfrac{a}{q}\right| < q^{-2}$, where $(a, q) = 1$. Then*

$$\sum_{z=1}^{R} \min(N, \|\alpha z + \beta\|^{-1}) \ll (N + q \log q)\left(\frac{R}{q} + 1\right).$$

Proof. The result is obvious if $q = 1$. If $q > 1$, we consider the contribution in the above sum from the integers z in a block $\mathcal{B}$ of at most $\frac{1}{2}q$ consecutive integers. If $z_1, z_2 \in \mathcal{B}$ with $z_1 \neq z_2$, we have

$$\|(\alpha z_2 + \beta) - (\alpha z_1 + \beta)\| \geq \left\|\frac{a(z_2 - z_1)}{q}\right\| - \frac{|z_2 - z_1|}{q^2}$$

$$\geq \frac{1}{q} - \frac{1}{2q} = \frac{1}{2q}.$$

From Lemma 3.1, then,

$$\sum_{z \in \mathscr{B}} \min (N, \|\alpha z + \beta\|^{-1}) \ll N + q \log q,$$

and the result follows, since the interval $1 \leqq z \leqq R$ may be divided into

$$\ll (R/q) + 1$$

such blocks $\mathscr{B}$.

Lemma 3.3. *Suppose that $N > C_1(\delta)$ and*

$$\sum_{z=1}^{R} \min (N, \|z\alpha + \beta\|^{-1}) \gg B,$$

where $B \gg N^{1+\delta} + R^{1+\delta}$. Then there are coprime integers q and a with

$$1 \leqq q \ll NRB^{-1}, \qquad |\alpha q - a| < N^{\delta}B^{-1}. \tag{3.10}$$

Proof. By Dirichlet's theorem (Schmidt (1980*b*), Chapter I) there are coprime integers q, a with

$$1 \leqq q \leqq BN^{-\delta} \quad \text{and} \quad \|\alpha q\| = |\alpha q - a| < B^{-1}N^{\delta}.$$

The hypothesis of the present lemma, together with Lemma 3.2, yields

$$B \ll (N + q \log q)(Rq^{-1} + 1) = N + R \log q + q \log q + NRq^{-1}.$$

Here N and $R \log q \leqq R \log B$ are smaller than $B^{1-(\delta/2)}$, and so is $q \log q$. It follows that

$$B \ll NRq^{-1} \quad \text{and} \quad q \ll NRB^{-1}.$$

Let us turn back to the problem of $\alpha n^2 + \beta n$ for a moment.

Lemma 3.4. *Let $N > C_1(\delta)$. Suppose that α and β are real and that*

$$\|\alpha n^2 + \beta n\| \geqq M^{-1} \qquad (n = 1, \ldots, N)$$

where M is an integer, $2 \leqq M \ll N^{1/2-\varepsilon}$. Then there are coprime integers q and a with

$$1 \leqq q \leqq M^2 N^{3\delta}, \qquad |\alpha q - a| < MN^{3\delta-2}.$$

Proof. From (3.7) we know that

$$\sum_{z=1}^{2MN} \min (N, \|z\alpha\|^{-1}) \gg N^{2-2\delta}M^{-1} = B,$$

say. Lemma 3.3 may be applied with $R = 2MN$, since

$$B \gg (2MN)^{1+\delta}.$$

Thus the natural number q in (3.10) satisfies

$$q \ll NRB^{-1} \ll N^2M(N^{2-2\delta}M^{-1})^{-1} \ll M^2N^{2\delta}.$$

while

$$\|\alpha q\| < N^\delta B^{-1} = N^{3\delta-2}M.$$

Since $N > C_1(\delta)$, the lemma follows.

An immediate consequence of Lemma 2.4 is

Theorem 3.1 (*Heilbronn*). *Let $N > C_2(\varepsilon)$. Then for any real α there is a natural number $n \leqq N$ satisfying*

$$\|\alpha n^2\| < N^{-(1/2)+\varepsilon}.$$

Proof. Suppose if possible that no such n exists. Applying Lemma 3.4 with $\beta = 0$, $M = [N^{1/2-\varepsilon}] + 1$, we obtain a natural number $q \leqq M^2N^\varepsilon \leqq N$ such that

$$\|\alpha q^2\| \leqq q |\alpha q - a| < M^3N^{\varepsilon-2} < M^{-1}.$$

This contradicts our hypothesis, and the theorem is proved.

More subtle arguments are needed to obtain the same quality of result for $\alpha n^2 + \beta n$. We continue with reciprocal sums, following Schmidt (1977*a*).

Lemma 3.5. *Let ρ, σ be real, and R a natural number with $R|\rho| \leqq 1$. Write*

$$\theta = \min_{1 \leqq j \leqq R} \|\rho j + \sigma\|, \qquad \Delta = \max_{1 \leqq j \leqq R} \|\rho j + \sigma\|.$$

Then

$$\sum_{j=1}^{R} \|\rho j + \sigma\|^{-1} = \theta^{-1} + O(\Delta^{-1}R \log R). \tag{3.11}$$

Proof. Write the numbers $\rho j + \sigma$ with $j = 1, \ldots, R$ as $\gamma_1, \ldots, \gamma_R$, arranged such that $\|\gamma_1\| = \theta = \min (\|\gamma_1\|, \ldots, \|\gamma_R\|)$. We have $\|\gamma_i - \gamma_j\| \geqq |\rho|$ for $i \neq j$, because $R|\rho| \leqq 1$. Thus if $\rho \geqq \Delta/(2R)$, the desired conclusion follows from Lemma 3.1. On the other hand, if $|\rho| < \Delta/(2R)$, then $\|\rho j + \sigma\| > \frac{1}{2}\Delta$ $(j = 1, \ldots, R)$ and the sum in (3.11) is bounded by $2\Delta^{-1}R$.

Lemma 3.6. *Suppose that r, t are coprime,*

$$1 \leqq t \leqq N \quad and \quad \|\zeta t\| = |\zeta t - r| < (2N)^{-1}.$$

Then

$$\sum_{j=1}^{t} \min\left(N, \|\zeta j + \mu\|^{-1}\right) \ll \min\left(N, t\,\|\mu t\|^{-1}\right) + t \log t. \qquad (3.12)$$

Proof. Writing $\zeta = (r/t) + \lambda$, we have $|j\lambda| < (2N)^{-1}$ for $j = 1, \ldots, t$. Hence, for these integers j,

$$\min\left(N, \|\zeta j + \mu\|^{-1}\right) \leqq 2 \min\left(N, \|(rj/t) + \mu\|^{-1}\right)$$

as we see on considering the cases $\|\zeta j + \mu\| < 1/N$ and $\|\zeta j + \mu\| \geqq 1/N$. Thus the sum in (3.12) is

$$\leqq 2 \sum_{j=1}^{t} \min\left(N, \|(rj)/t + \mu\|^{-1}\right)$$

$$= 2 \sum_{j=1}^{t} \min\left(N, \|(j/t) + \mu\|^{-1}\right).$$

We now apply Lemma 3.5 with $R = t$, $\rho = 1/t$, $\sigma = \mu$, and obtain a bound

$$\ll \min\left(N, \theta^{-1}\right) + \Delta^{-1} t \log t.$$

In our particular situation, θ is the distance from μ to the nearest integer multiple of $1/t$, or $\theta = t^{-1}\|\mu t\|$. If $t \geqq 2$, then $\Delta \geqq \frac{1}{4}$, and we are done. Of course we are also done when $t = 1$.

Lemma 3.7. *Suppose that r, t are coprime. Let N and H be natural numbers,*

$$1 \leqq t < N \leqq H, \qquad \|\zeta t\| = |\zeta t - r| < (2H)^{-1}.$$

Then

$$\sum_{u=1}^{H} \min\left(N, \|\zeta u + \mu\|^{-1}\right) \ll (\log H) \min\left(\frac{NH}{t}, \frac{H}{\|\mu t\|}, \frac{1}{\|\zeta t\|}\right).$$

Proof. Write $u = tz + j$ $(j = 1, \ldots, t)$. The sum in question is

$$\leqq \sum_{z=0}^{[H/t]} \sum_{j=1}^{t} \min\left(N, \|\zeta j + \zeta tz + \mu\|^{-1}\right)$$

and is

$$\ll \sum_{z=0}^{[H/t]} \left\{\min\left(N, t\,\|\zeta t^2 z + \mu t\|^{-1}\right) + t \log t\right\}$$

by Lemma 3.6. Writing $\lambda = \zeta t - r$, we obtain

$$\ll H \log N + \sum_{z=0}^{[H/t]} \min\left(N, t\,\|\lambda t z + \mu t\|^{-1}\right). \qquad (3.13)$$

It is clear that this is

$$\ll H \log N + (HN)/t \ll (\log H)(NH/t),$$

which is the first of the desired estimates.

To get the other estimates we now apply Lemma 3.5 to the sum in (3.13), with $R = [H/t] + 1$, $\rho = \lambda t$, $\sigma = \mu t - \lambda t$. (This is permissible, since

$$R |\rho| \leq 2(H/t) \cdot |\lambda t| = 2H \|\zeta t\| < 1.)$$

We obtain the estimate

$$\ll H \log N + \min (N, t\theta^{-1}) + t\Delta^{-1}R \log R \ll \Delta^{-1}H \log H.$$

Here Δ is the maximum of $\|\lambda tz + \mu t\|$ for $z = 0, \ldots, [H/t]$. Clearly $\Delta \geq \|\mu t\|$. But since $H |\lambda| < \frac{1}{2}$, we have also

$$|\lambda t[H/t]| = \|\lambda t[H/t]\| \leq \|\lambda t[H/t] + \mu t\| + \|\mu t\| \leq 2\Delta,$$

whence $\Delta \geq \frac{1}{2} |\lambda| t[H/t] \geq \frac{1}{4}\lambda H = \frac{1}{4}H \|\zeta t\|$. We therefore obtain

$$\Delta^{-1}H \log H \ll \log H \min (H \|\mu t\|^{-1}, \|\zeta t\|^{-1})$$

and the lemma follows.

3.3. Schmidt's extension of Heilbronn's theorem

Theorem 3.2. *Let* $N > C_3(\varepsilon)$. *Let* α *and* β *be real. Then there is a natural number* n *with*

$$n \leq N, \qquad \|\alpha n^2 + \beta n\| < N^{-1/2+\varepsilon}. \tag{3.14}$$

Proof. Suppose if possible no such n exists. Let $M = [N^{(1/2)-\varepsilon}] + 1$. It was shown in Section 3.1 that the lower bound (3.9) holds, and in Lemma 3.4 that there are coprime integers a, q,

$$1 \leq q \leq M^2N^\varepsilon, \qquad |\alpha q - a| < MN^{\varepsilon-2}. \tag{3.15}$$

Let d be a divisor of q. Let Σ_d be the double sum in (3.9), but restricted to summands with $(m, q) = d$. Writing $q = dq_1$, $m = dm_1$,

$$\Sigma_d = \sum_{\substack{m_1=1 \\ (m_1,q_1)=1}}^{[2M/d]} \sum_{u=1}^{2N} \min (N, \|\alpha dm_1u + \beta dm_1\|^{-1}).$$

Since the number of divisors of q is $\ll N^\delta$, there is a d with

$$\Sigma_d \gg N^{2-\delta}M^{-1}. \tag{3.16}$$

We now consider the inner sum in the definition of Σ_d. It is the type of sum considered in Lemma 3.7, with $H = 2N$, $\zeta = \alpha dm_1$ and $\mu = \beta dm_1$.

With $t = q_1$, $r = am_1$ we have $(t, r) = 1$ and

$$t \leqq q < N$$

by (3.15). Further,

$$|\zeta t - r| = |\alpha d m_1 t - a m_1| = |\alpha q - a| m_1 < N^{-2+\varepsilon} M \cdot 2M < (4N)^{-1}$$

by (3.15). The hypotheses of Lemma 3.7 are satisfied, and the inner sum in the definition of Σ_d is

$$\ll (\log N) \min \left(\frac{N^2}{q_1}, \frac{N}{\|\beta q m_1\|}, \frac{1}{\|\alpha q m_1\|} \right).$$

In view of $\|\alpha q m_1\| = \|\alpha q\| m_1$, we can deduce from (3.16) that

$$\sum_{m_1=1}^{[2M/d]} \min \left(\frac{N^2}{q_1}, \frac{N}{\|\beta q m_1\|}, \frac{1}{\|\alpha q\| m_1} \right) \gg N^{2-2\delta} M^{-1}. \qquad (3.17)$$

Now we apply Dirichlet's theorem (for a second time—recall Lemma 3.3). There are coprime integers u and w with

$$1 \leqq u \leqq 4M \quad \text{and} \quad \|\beta q u\| = |\beta q u - w| < (4M)^{-1}. \qquad (3.18)$$

It is important to observe that for those m_1 counted by (3.17) for which $u \nmid m_1$, we have

$$\|\beta q m_1\| \geqq (2u)^{-1}. \qquad (3.19)$$

For otherwise, we would have

$$|\beta q m_1 - e| < (2u)^{-1} \qquad \text{for some integer } e,$$
$$|\beta q u - w| < (4M)^{-1}.$$

This leads to

$$|eu - m_1 w| < \tfrac{1}{2} + m_1/4M \leqq 1,$$

so that $eu - m_1 w = 0$, and u is a divisor of $m_1 w$; hence of m_1, which is absurd.

Consequently, we have

$$\sum_{\substack{m_1=1 \\ u \nmid m_1}}^{[2M/d]} \frac{1}{\|\beta q m_1\|} \ll \sum_{m_1=1}^{[2M/d]} \min \left(2u, \frac{1}{\|\beta q m_1\|} \right)$$

$$\ll \left(\frac{M}{du} + 1 \right)(2u + u \log u)$$

$$\ll M \log N \qquad (3.20)$$

by Lemma 3.2, and (3.18).

Since $NM \log N$ is of smaller order of magnitude than $N^{2-2\delta} M^{-1}$, it follows from (3.17) and (3.20) that

$$N^{2-2\delta}M^{-1} \ll \sum_{\substack{m_1=1 \\ u|m_1}}^{[2M/d]} \min\left(\frac{N^2}{q_1}, \frac{1}{\|\alpha q\| m_1}\right)$$

$$= \sum_{v=1}^{[2M/du]} \min\left(\frac{N^2}{q_1}, \frac{1}{\|\alpha q\| uv}\right).$$

In particular,

$$N^{2-2\delta}M^{-1} \ll 2M(du)^{-1}(N^2/q_1) = 2MN^2/qu.$$

Putting $n = qu$, we have $n \ll M^2 N^{2\delta}$, whence

$$n \leqq N.$$

Furthermore,

$$N^{2-2\delta}M^{-1} \ll (\|\alpha q\| u)^{-1} \log N$$

and

$$\|\alpha q\| u \leqq MN^{3\delta-2}. \tag{3.21}$$

So

$$\|\alpha n^2 + \beta n\| \leqq nu \|\alpha q\| + \|\beta qu\| < NMN^{3\delta-2} + (4M)^{-1} < M^{-1}$$

from (3.18), (3.21). This contradicts our hypothesis, and the theorem is proved.

3.4 A simultaneous approximation result

Schmidt's technique can be adapted to prove the following theorem (Baker (1981*d*); earlier, Baker and Gajraj (1976*b*) gave a weaker result). The exponent $-\frac{1}{4} + \varepsilon$ is perhaps disappointing, but it seems difficult to improve.

Theorem 3.3. *Let $N > C_4(\varepsilon)$. Let α and β be real numbers. Then there is a natural number $n \leqq N$ satisfying*

$$\max(\|\alpha n^2\|, \|\beta n\|) < N^{-(1/4)+\varepsilon}. \tag{3.22}$$

Proof. Suppose that no such n exists. Let $M = [N^{1/4-\varepsilon}] + 1$. Let $\psi(x)$ be the non-negative function in Lemma 2.1, with $\Delta = 1/(2M)$, $r = [3/\delta] + 1$, so that

$$\psi(x) = 0 \quad \text{for } \|x\| \geqq M^{-1}, \quad \hat{\psi}(0) = M^{-1}. \tag{3.23}$$

By considering $\hat{\psi}(m)$ with $|m| \leqq M$ and $|m| > M$ we easily obtain

$$\sum_{m=-\infty}^{\infty} |\hat{\psi}(m)| \ll 1. \tag{3.24}$$

Moreover,

$$\sum_{|m|>M^{1+\delta}} |\hat{\psi}(m)| \ll M^{-r\delta} \ll M^{-3}. \tag{3.25}$$

We observe that, from (3.23) and the hypothesis,

$$\sum_{n=1}^{N} \psi(\alpha n^2)\psi(\beta n) = 0,$$

and therefore

$$\sum_{(m_1, m_2) \neq (0,0)} \hat{\psi}(m_1)\hat{\psi}(m_2) \sum_{n=1}^{N} e(m_1 \alpha n^2 + m_2 \beta n) = -N\hat{\psi}(0)^2 = -NM^{-2}. \tag{3.26}$$

The contribution to the left-hand side of (3.26) from $|m_1| \geqq M^{1+\delta}$ has modulus at most

$$N \sum_{m_2=-\infty}^{\infty} |\hat{\psi}(m_2)| \sum_{|m_1| \geqq M^{1+\delta}} |\hat{\psi}(m_1)| \ll NM^{-3}$$

from (3.24), (3.25). This is of smaller order than NM^{-2}; similarly for the contribution from $|m_2| \geqq M^{1+\delta}$. Since

$$|\hat{\psi}(m_1)\hat{\psi}(m_2)| \leqq \hat{\psi}(0)^2,$$

we infer from (3.26) that

$$\sum_{\substack{|m_1|<M^{1+\delta} \\ (m_1, m_2) \neq (0,0)}} \sum_{|m_2|<M^{1+\delta}} \left| \sum_{n=1}^{N} e(\alpha m_1 n^2 + \beta m_2 n) \right| \geqq \frac{N}{4}. \tag{3.27}$$

There are two cases to consider.

Case I. The contribution to the left-hand side of (3.27) from pairs (m_1, m_2) with $m_1 = 0$ is at least $N/8$:

$$\sum_{1 \leqq m \leqq M^{1+\delta}} \left| \sum_{n=1}^{N} e(m\beta n) \right| \geqq \frac{N}{8}.$$

Then

$$B = \sum_{1 \leqq m \leqq M^{1+\delta}} \|m\beta\|^{-1} \gg N \tag{3.28}$$

from (3.4). There is a natural number $r \leqq N^{1-\delta}$ such that

$$|r\beta - a| < N^{-1+\delta}.$$

If $r > 2M^{1+\delta}$, then

$$\|m\beta\| \geqq \|ma/r\| - mN^{-1+\delta}r^{-1} \geqq r^{-1} - (2r)^{-1} = (2r)^{-1}$$

for $m = 1, \ldots, [M^{1+\delta}]$. The argument at the beginning of the proof of Lemma 3.2 then yields

$$B \ll r \log r,$$

which contradicts (3.28). Thus $r \leq 2M^{1+\delta}$. Now, by Heilbronn's theorem, there is an integer $t \leq N^{1/2}$ such that

$$\|r^2 t^2 \alpha\| < N^{-1/4+\varepsilon}.$$

Let $n = rt$, then $n \leq 2N^{1/2}M^{1+\delta} \leq N$,

$$\|n\beta\| \leq t \|r\beta\| \leq N^{1/2}N^{-1+\delta} < M^{-1},$$

so that the initial hypothesis on $(n^2\alpha, n\beta)$ is wrong. We conclude that Case I does not occur. We must have:

Case II. The contribution to the sum in (3.27) from pairs with $m_1 \neq 0$ is at least $N/8$. In this case, we see that

$$\sum_{|m_2|<M^{1+\delta}} \sum_{m_1=1}^{[M^{1+\delta}]} \left| \sum_{n=1}^{N} e(\alpha m_1 n^2 + \beta m_2 n) \right|^2 \gg N^2 M^{-2-2\delta} \qquad (3.29)$$

after an application of Cauchy's inequality. Next, just as in (3.5), (3.7),

$$\sum_{m_1=1}^{[M^{1+\delta}]} \left| \sum_{n=1}^{N} e(m_1 \alpha n^2 + m_2 \beta n) \right|^2 \ll \sum_{m_1=1}^{[M^{1+\delta}]} \sum_{v=1}^{N} \min(N, \|2m_1 v\alpha\|^{-1})$$

$$\ll N^{\delta} \sum_{z=1}^{[2M^{1+\delta}N]} \min(N, \|z\alpha\|^{-1}). \qquad (3.30)$$

Combining (3.29) and (3.30),

$$\sum_{z=1}^{[2M^{1+\delta}N]} \min(N, \|z\alpha\|^{-1}) \gg N^{2-2\delta}M^{-3}. \qquad (3.31)$$

We apply Lemma 3.3. The right-hand side of (3.31) is at least $N^{1+\delta} + R^{1+\delta}$, in the notation of that lemma. For

$$R^{1+\delta} \leq (2M^{1+\delta}N)^{1+\delta} \leq N^{2-3\delta}M^{-3}.$$

Hence that lemma yields coprime integers q, a with

$$1 \leq q \ll NR(N^{2-2\delta}M^{-3})^{-1}$$

$$\ll M^4 N^{3\delta},$$

$$|\alpha q - a| \ll N^{3\delta-2}M^3.$$

We also have available the inequality (3.8), which we now apply with $m = 1$ and with αm_1 in place of α and βm_2 in place of β:

$$\left| \sum_{n=1}^{N} e(\alpha m_1 n^2 + \beta m_2 n) \right|^2 \ll \sum_{u=1}^{2N} \min(N, \|2(\alpha m_1 u + \beta m_2)\|^{-1}).$$

In conjunction with (3.29), then,

$$\sum_{|m_2|<M^{1+\delta}} \sum_{m_1=1}^{[M^{1+\delta}]} \sum_{u=1}^{2N} \min{(N, \|2(\alpha m_1 u + \beta m_2)\|^{-1})} \gg N^2 M^{-2-2\delta}.$$

Indeed, we can use the same divisor argument as in Section 3.1, and obtain

$$\sum_{|y|<2M^{1+\delta}} \sum_{x=1}^{[4M^{1+\delta}N]} \min{(N, \|\alpha x + \beta y\|^{-1})} \gg N^{2-\delta}M^{-2}.$$

We apply Lemma 3.7 to the sum over x, with $H = [4M^{1+\delta}N]$, $t = q$, $\zeta = \alpha$. This is permissible since $\|q\alpha\| \ll N^{3\delta-2}M^3 < H^{-1}N^{-\delta}$. Thus

$$N^{2-2\delta}M^{-2} \ll \sum_{|y|<2M^{1+\delta}} \min{\left(\frac{M^{1+\delta}N^2}{q}, \frac{M^{1+\delta}N}{\|\beta qy\|}, \frac{1}{\|\alpha q\|}\right)}. \tag{3.32}$$

This is rather like (3.17) and we proceed, as we did there, to apply Dirichlet's theorem to βq. There is a natural number $t \leq 4M^{1+\delta}$ satisfying

$$|\beta q t - z| < (4M^{1+\delta})^{-1} \tag{3.33}$$

where $(z, t) = 1$. It is not difficult to show, arguing as in Section 3.3, that

$$\|\beta qy\| > (2t)^{-1}$$

whenever $|y| < 2M^{1+\delta}$ and $t \nmid y$. The contribution of these integers y to the sum in (3.32) is thus

$$\leq M^{1+\delta}N \sum_{0<|y|<2M^{1+\delta}} \min{(2t, \|\beta qy\|^{-1})}$$

$$\ll M^{1+\delta}N\left(\frac{M^{1+\delta}}{t} + 1\right)(2t + t\log t)$$

$$\ll M^2 N^{1+\delta} \tag{3.34}$$

by Lemma 3.2. Since $M^2 N^{1+\delta} < N^{-\delta}(N^{2-2\delta}M^{-2})$ we learn from (3.32) and (3.34) that

$$N^{2-2\delta}M^{-2} \ll \sum_{\substack{|y|<2M^{1+\delta} \\ t|y}} \min{\left(\frac{M^{1+\delta}N^2}{q}, \frac{1}{\|\alpha q\|}\right)}.$$

As $t \leq 4M^{1+\delta}$, the number of terms in the last sum is $\ll M^{1+\delta}/t$. It follows that

$$N^{2-2\delta}M^{-2} \ll \frac{M^{1+\delta}}{t} \min{\left(\frac{M^{1+\delta}N^2}{q}, \frac{1}{\|\alpha q\|}\right)},$$

so that $qt \ll M^4 N^{4\delta}$.
     Thus $qt \leq N$, and

$$\|\alpha q\| t \leq M^3 N^{-2+3\delta} < M^{-1}. \tag{3.35}$$

Now we get a contradiction from (3.33) and (3.35), since $n = qt$ satisfies (3.22). This completes the proof of Theorem 3.3.

3.5 Weyl's inequality

The method of 'squaring the modulus' used in Section 3.1 can be applied repeatedly to a polynomial of degree k. For any function $\phi(x)$ on the natural numbers, put $\Delta_y \phi(x) = \phi(x + y) - \phi(x)$. Define the difference operator $\Delta_{y_1 \cdots y_t}$ inductively by

$$\Delta_{y_1 \cdots y_t} \phi(x) = \Delta_{y_t} (\Delta_{y_1 \cdots y_{t-1}} \phi(x)).$$

This operator is easily seen to be independent of the ordering $y_1, \ldots, y_t$. In the following lemma, as usual, an empty sum denotes zero, and $Y_j = y_1 + \cdots + y_j$.

Lemma 3.8. (Weyl) *Let $\phi(x)$ be a real-valued function defined for $x = 1, 2, \ldots$. Let h be an integer ≥ 2, and let $H = 2^{h-1}$. Then the sum*

$$S(\phi) = \sum_{n=1}^{N} e(\phi(n))$$

satisfies

$$|S(\phi)|^H \ll N^{H-1} + N^{H-h} \left| \sum_{y_1=1}^{N} \cdots \sum_{y_{h-1}=1}^{N} \sum_{1 \leq n < n + Y_{h-1} \leq N} e(\Delta_{y_1 \cdots y_{h-1}}(\phi(n))) \right|.$$

The implied constant depends only on h.

Proof. We have

$$|S(\phi)|^2 = \sum_{n_1=1}^{N} \sum_{n_2=1}^{N} e(\phi(n_2) - \phi(n_1))$$

$$= N + 2 \operatorname{Re} \left(\sum_{1 \leq n_1 < n_2 \leq N} e(\phi(n_2) - \phi(n_1)) \right)$$

$$= N + 2 \operatorname{Re} \left(\sum_{y_1=1}^{N} \sum_{1 \leq n < n + y_1 \leq N} e(\Delta_{y_1}(\phi(n))) \right). \qquad (3.36)$$

The case $h = 2$ of the lemma follows at once.

Now assume the truth of the lemma for $h - 1$, and square the inequality of the lemma for $h - 1$. Using Cauchy's inequality, and the observation that $(a + b)^2 \ll a^2 + b^2$, we obtain

$$|S(\phi)|^H \ll N^{H-2} + N^{H-2(h-1)} N^{h-2}$$

$$\times \sum_{y_1=1}^{N} \cdots \sum_{y_{h-2}=1}^{N} \left| \sum_{1 \leq n < n + Y_{h-2} \leq N} e(\Delta_{y_1 \cdots y_{h-2}}(\phi(n))) \right|^2.$$

Now we apply (3.36), with $\Delta_{y_1 \cdots y_{h-2}}\phi$ in place of ϕ and $N - (y_1 + \cdots + y_{h-2})$ in place of N. We obtain

$$|S(\phi)|^H \ll N^{H-2} + N^{H-h} \sum_{y_1=1}^{N} \cdots \sum_{y_{h-2}=1}^{N}$$

$$\times \left\{ N + 2\operatorname{Re} \sum_{y_{h-1}=1}^{N} \sum_{1 \le n < n + Y_{h-1} \le N} e(\Delta_{y_1 \cdots y_{h-1}}(\phi(n))) \right\}.$$

The lemma follows at once.

Lemma 3.9. *For a polynomial* $f(x) = \alpha x^k + \beta x^{k-1} + \cdots$ *of degree* $k \ge 2$,

$$\Delta_{y_1 \cdots y_{k-}} f(x) = y_1 \cdots y_{k-1}(\tfrac{1}{2}k!\,\alpha(2x + y_1 + \cdots + y_{k-1}) + (k-1)!\,\beta).$$

Proof. The case $k = 2$ is straightforward. In the step from $k - 1$ to k, we note that

$$\Delta_{y_1 \cdots y_{k-1}}(\alpha x^k + \beta x^{k-1} + \cdots) = \Delta_{y_1 \cdots y_{k-1}}(\alpha x^k + \beta x^{k-1})$$

since the Δ_y operator reduces the degree in x by one. Thus

$$\Delta_{y_1 \cdots y_{k-1}}(\alpha x^k + \beta x^{k-1} + \cdots)$$

$$= \Delta_{y_2 \cdots y_{k-1}}(\alpha((x + y_1)^k - x^k) + \beta((x + y_1)^{k-1} - x^{k-1}))$$

$$= \Delta_{y_2 \cdots y_{k-1}}\left(\alpha k y_1 x^{k-1} + \alpha \binom{k}{2} y_1^2 x^{k-2} + \beta(k - 1)y_1 x^{k-2}\right)$$

$$= y_2 \cdots y_{k-1}(\tfrac{1}{2}\alpha k y_1(k - 1)!\,(2x + y_2 + \cdots + y_{k-1}) + \alpha\binom{k}{2}y_1^2(k - 2)!$$

$$+ \beta(k - 1)y_1(k - 2)!),$$

which is the desired result.

Lemma 3.10. *Suppose* $L \le N^k$ *and* $f(x) = \alpha x^k + \beta x^{k-1} + \cdots$
If

$$\sum_{m=1}^{L} |S_m(f)|^K \gg A, \qquad (3.37)$$

where

$$A \gg N^{K-1+\eta}L, \qquad (3.38)$$

then there exist coprime integers $r \le LN^{K+\eta}A^{-1}$ *and* s *with*

$$|\alpha r - s| \le N^{K-k+\eta}A^{-1}.$$

Proof. Combining Lemmas 3.8, 3.9, we see that

$$|S_m(f)|^K \ll N^{K-1} + N^{K-k} \sum_{y_1=1}^{N} \cdots \sum_{y_{k-1}=1}^{N} \left| \sum_x e(k!\, m y_1 \cdots y_{k-1} \alpha x + g) \right|$$

where $g = g(y_1, \ldots, y_{k-1}, m)$ is independent of x. The sum over x is taken over less than N consecutive integers, so that by (3.4) the inner sum is

$$\ll \min\,(N, \|k!\, my_1 \cdots y_{k-1}\alpha\|^{-1})$$

and

$$\sum_{m=1}^{L} |S_m(f)|^K \ll N^{K-1}L + N^{K-k} \sum_{y_1=1}^{N} \cdots \sum_{y_{k-1}=1}^{N}$$

$$\times \sum_{m=1}^{L} \min\,(N, \|k!\, my_1 \cdots y_{k-1}\alpha\|^{-1})$$

$$\ll N^{K-k+\delta} \sum_{z=1}^{k!N^{k-1}L} \min\,(N, \|z\alpha\|^{-1}), \tag{3.39}$$

since for $z \leqq k!\, N^{k-1}L$ there are at most N^δ ways of writing z as a product $k!\, my_1 \cdots y_{k-1}$.

Combining (3.37) and (3.39),

$$AN^{-K+k-\delta} \ll \sum_{z=1}^{k!N^{k-1}L} \min\,(N, \|z\alpha\|^{-1}).$$

We may apply Lemma 3.3, since in the notation of that lemma,

$$N^{1+\delta} + R^{1+\delta} \ll N^{k-1+\eta}L \ll AN^{-K+k-\delta} = B.$$

Thus there exist coprime integers r and t with

$$r \ll NRB^{-1} \ll LN^{K+\delta}A^{-1}$$

and

$$|\alpha r - t| < B^{-1}N^\delta \ll A^{-1}N^{K-k+\eta}.$$

This proves the lemma.

Weyl's argument here does not give information about the coefficients of f other than α, the leading coefficient. For the second highest coefficient, this was remedied by Schmidt (1977*b*), Lemma 11A, adapting his method given here in the proof of Theorem 3.2. We shall omit the details, since we obtain a good result for *all* the coefficients in Chapter 5.

However, we do pause to prove

Theorem 3.4 (*Danicic*).   *Suppose that $N > C_5(k, \varepsilon)$. Let α be real, then there is a natural number $n \leqq N$ satisfying*

$$\|\alpha n^k\| < N^{\varepsilon - 1/K}.$$

Proof.   Suppose that no such n exists. By Theorem 2.2 with $M =$

$[N^{1/K-\varepsilon}] + 1,$

$$\sum_{m=1}^{M} |S_m(f)| > \frac{N}{6}.$$

Here $S_m(f)$ is as in Lemma 3.10 with $f(x) = \alpha x^k$. An application of Hölder's inequality yields

$$\sum_{m=1}^{M} |S_m(f)|^K \gg N^K M^{-K+1}.$$

It is clear that (3.38) holds with $L = M$, $A = N^K M^{-K+1}$. Therefore there exist coprime integers r, t,

$$1 \leq r \leq L N^{K+\varepsilon} A^{-1} \ll M^K N^{\varepsilon},$$
$$|\alpha r - t| \leq N^{K-k+\varepsilon} A^{-1} \ll M^{K-1} N^{-k+\varepsilon}.$$

It is clear that $r \leq N$ and that

$$\|\alpha r^k\| \leq N^{k-1} \|\alpha r\| < M^{-1}.$$

This contradicts the hypothesis. The theorem is proved.

We observe that Liu (1970a) and Cook (1976a) have published slightly sharper versions of the Heilbronn–Danicic theorem, based on the observation that

$$d(n) < n^{1/(\log \log n)}$$

for large n (see Hardy and Wright (1979), Theorem 317). At present one cannot prove anything as strong as

$$\min_{1 \leq n \leq N} \|\alpha n^k\| < (\log N)^A N^{-1/K}$$

for some $A > 0$ and $N > C_6(A, k)$.

It is an interesting challenge to attempt to prove Theorem 3.4 without the use of exponential sums. For weaker results in this direction see van der Corput (1939) and van der Corput and Pisot (1939a, 1939b).

4
Vinogradov's mean value method

4.1 Approximations to the coefficients $\alpha_2, \ldots, \alpha_k$

The theorems obtained in this chapter are quite similar to results in Vinogradov (1971), the second edition of his book. (See Vinogradov (1950), (1951), (1957) for the development of the method.) For the convenience of the reader, however, we have arranged the account so that it follows on naturally from a reading of Chapter 5 of Vaughan (1981).

We recall some of Vaughan's results. For a given $X > 1$ write

$$F(\boldsymbol{\theta}) = \sum_{0 < x \leq X} e(\theta_1 x + \theta_2 x^2 + \cdots + \theta_k x^k).$$

Vaughan gives the following estimate for the integral

$$J_s^{(k)}(X) = \int_{U_k} |F(\boldsymbol{\theta})|^{2s} \, d\boldsymbol{\theta}.$$

Theorem 4.1 (*Vinogradov's mean value theorem*). *For each pair of natural numbers k, l,*

$$J_{lk}^{(k)}(X) \leq C_1(l, k) X^{2lk - k(k+1)/2 + \theta}$$

where $\theta = \frac{1}{2}k^2(1 - 1/k)^l$.

Note that $J_s^{(k)}(X)$ is the number of solutions of

$$\sum_{r=1}^{s} (x_r^j - y_r^j) = 0 \qquad (1 \leq j \leq k) \tag{4.1}$$

with $0 < x_r, y_r \leq X$.

Vinogradov discovered that one can pass from estimates of the mean value $J_s^{(k-1)}(X)$ to bounds for sums

$$S(f) = \sum_{n=1}^{N} e(\alpha_1 x + \cdots + \alpha_k x^k),$$

given a rational approximation to one of the coefficients of $f(x) = \alpha_1 x + \cdots + \alpha_k x^k$, other than α_1. Vaughan gives a new method of doing this, and proves

Theorem 4.2. *Suppose that there exist j, a, q with $2 \leqq j \leqq k$, $|\alpha_j - a/q| \leqq q^{-2}$, $(a, q) = 1$, $q \leqq N^j$. Then*

$$S(f) \ll (J_s^{(k-1)}(2N)N^{k(k-1)/2}(qN^{-j} + N^{-1} + q^{-1}))^{1/2s} \log 2N.$$

We need to recall some of the details of the proof of Theorem 4.2. The basic tool is the following l-dimensional version of the large sieve inequality (Vaughan's Lemma 5.3):

Lemma 4.1. *Suppose that $\delta_j > 0$ $(j = 1, \ldots, l)$ and that Γ is a set of points in $\mathbb{R}^l$ such that the sets*

$$\mathscr{R}(\gamma) = \{\beta : \|\beta_j - \gamma_j\| < \delta_j, \ 0 \leqq \beta_j < 1\} \qquad (\gamma \in \Gamma)$$

are pairwise disjoint. Let $N_1, \ldots, N_l$ denote natural numbers and $\mathcal{N}$ denote the set of integer l-tuples $\mathbf{n} = (n_1, \ldots, n_l)$ with $1 \leqq n_j \leqq N_j$. Then the sum

$$T(\beta) = \sum_{\mathbf{n} \in \mathcal{N}} a(\mathbf{n})e(\mathbf{n}\beta),$$

where the $a(\mathbf{n})$ are complex numbers, satisfies

$$\sum_{\gamma \in \Gamma} |T(\gamma)|^2 \leqq C_2(l) \sum_{\mathbf{n} \in \mathcal{N}} |a(n)|^2 \prod_{j=1}^{l} (N_j + \delta_j^{-1}).$$

Let us write $\mathbf{v}^{(k)}(x) = (x, x^2, \ldots, x^k)$ and let $\mathcal{M}$ be an arbitrary nonempty set of integers in $[1, N]$, with $|\mathcal{M}| = M$. The initial inequality involved in Vaughan's proof of Theorem 4.2 is

$$S(f) \ll M^{-1}(\log 2N) \sup_{0 \leqq \beta \leqq 1} \sum_{m \in \mathcal{M}} |g(m, \beta)| \qquad (4.2)$$

where

$$g(m, \beta) = \sum_{x=1}^{2N} e(\mathbf{v}^{(k)}(x - m)\mathbf{\alpha} + x\beta) \qquad (4.3)$$

(see Vaughan (1981), equation (5.23)); here $\mathbf{\alpha} = (\alpha_1, \ldots, \alpha_k)$.
If we write

$$\gamma_j(m) = \sum_{h=j}^{k} \alpha_h \binom{h}{j}(-m)^{h-j} \qquad (1 \leqq j \leqq k - 1)$$

and $\mathbf{\gamma}(m) = (\gamma_1(m), \ldots, \gamma_{k-1}(m))$, then a Taylor expansion gives

$$\mathbf{v}^{(k)}(x - m)\mathbf{\alpha} = \mathbf{v}^{(k-1)}(x)\mathbf{\gamma}(m) + x^k \alpha_k + \sum_{j=1}^{k} \alpha_j(-m)^j. \qquad (4.4)$$

Vaughan (1981; equation (5.33)) shows that

$$\|(k!)^k \alpha_j(x - y)\| \ll \sum_{h=j-1}^{k-1} \|\gamma_h(x) - \gamma_h(y)\| N^{h-j+1} \qquad (2 \leqq j \leqq k) \quad (4.5)$$

for $1 \leqq x, y \leqq N$.

We are now in a position to prove

Theorem 4.3.   *Let $k \geqq 2$, let l be a natural number and*

$$\theta = \tfrac{1}{2}(k-1)^2 \left(\frac{k-2}{k-1}\right)^l.$$

Suppose that, for some $N > C_3(k, l, \varepsilon)$, we have

$$|S(f)| \gg A, \tag{4.6}$$

where

$$(NA^{-1})^{2(k-1)l} \ll N^{1-\theta-\varepsilon}. \tag{4.7}$$

Then there are coprime pairs of integers q_j, a_j $(j = 2, \ldots, k)$ such that

$$q_j \geqq 1, \qquad |q_j\alpha_j - a_j| \leqq N^{-j+\theta+\varepsilon}(NA^{-1})^{2(k-1)l} \tag{4.8}$$

for $j = 2, \ldots, k$, and the least common multiple q_0 of $q_2, \ldots, q_k$ satisfies

$$q_0 \leqq (NA^{-1})^{2(k-1)l}N^{\theta+\varepsilon}. \tag{4.9}$$

Proof.   By Dirichlet's theorem there are coprime pairs of integers q_j, a_j satisfying

$$q_j \leqq N^{j-\theta-\varepsilon}(NA^{-1})^{-2(k-1)l} \tag{4.10}$$

and (4.8), for $j = 2, \ldots, k$. Let q_0 denote the least common multiple of $q_2, \ldots, q_k$. We apply Theorem 4.1 in conjunction with the case $s = (k-1)l$ of Theorem 4.2. This gives

$$S(f) \ll N(N^\theta(q_j N^{-j} + N^{-1} + q_j^{-1}))^{1/2(k-1)l} \log 2N$$

for $j = 2, \ldots, k$. Comparing with (4.6), we see that

$$N^{-\theta}(NA^{-1})^{-2(k-1)l}(\log N)^{-2(k-1)l} \ll q_j N^{-j} + N^{-1} + q_j^{-1}. \tag{4.11}$$

Now $q_j N^{-j} + N^{-1}$ is of smaller order of magnitude than the left-hand side of (4.11), by (4.10) and (4.7). We deduce that

$$q_j \ll N^\theta(NA^{-1})^{2(k-1)l}(\log N)^{2(k-1)l} \tag{4.12}$$

for $j = 2, \ldots, k$. This is a first step towards (4.9).

Let $1 \leqq x \leqq N$ be given and suppose that y, $1 \leqq y \leqq N$, satisfies

$$\|(k!)^k \alpha_j(x - y)\| \leqq N^{1-j} \qquad (j = 2, \ldots, k). \tag{4.13}$$

Then

$$\|(k!)^k a_j(x - y)/q_j\| \leqq N^{1-j} + (k!)^k N^{-j+\theta+\varepsilon}(NA^{-1})^{2(k-1)l}q_j^{-1} \tag{4.14}$$

by (4.8). The right-hand side of (4.14) is smaller than q_j^{-1}. After all,

$$q_j < \tfrac{1}{2}N \leq \tfrac{1}{2}N^{j-1}$$

from (4.12) and (4.7), while

$$(k!)^k N^{-j+\theta+\varepsilon}(NA^{-1})^{2(k-1)l} < \tfrac{1}{2}$$

from (4.7). We conclude that

$$q_j \mid (k!)^k a_j(x-y) \qquad (j=2, \ldots, k)$$

and, in fact, that $q_0 \mid (k!)^k(x-y)$. The number of possibilities for y is at most

$$R = \frac{(k!)^k N}{q_0} + 1.$$

We deduce that there is a set $\mathcal{M}$ of integers x in $[1, N]$ such that $M = |\mathcal{M}|$ satisfies

$$M \geq N/(R+1), \tag{4.15}$$

and such that, for each pair x, y with $x \in \mathcal{M}$, $y \in \mathcal{M}$, $x \neq y$,

$$\max_{2 \leq j \leq k} \|(k!)^k \alpha_j(x-y)\| \, N^{1-j} > 1.$$

By (4.5), for every such pair x, y we have

$$\|\gamma_h(x) - \gamma_h(y)\| \gg N^{-h} \tag{4.16}$$

for some h, $1 \leq h \leq k-1$.

Let us write $s = l(k-1)$. We are now ready to apply Lemma 4.1 to estimate the sum

$$\sum_{m \in \mathcal{M}} |g(m, \beta)|^{2s}.$$

In fact, from (4.3), (4.4),

$$g(m, \beta)^s = \left\{ \sum_{x=1}^{2N} e(\mathbf{v}^{(k-1)}(x)\boldsymbol{\gamma}(m) + x^k \alpha_k + x\beta + \sum_{j=1}^{k} \alpha_j(-m)^j) \right\}^s$$

$$= e\left(s \sum_{j=1}^{k} \alpha_j(-m)^j\right) \sum_{n_1=1}^{2Ns} \cdots \sum_{n_{k-1}=1}^{(2N)^{k-1}s} a(\mathbf{n})e(\mathbf{n}\boldsymbol{\gamma}(m))$$

where

$$a(\mathbf{n}) = \sum_{x_1, \ldots, x_s}' e(x_1^k + \cdots + x_s^k)\alpha_k + (x_1 + \cdots + x_s)\beta).$$

Here the sum Σ' is restricted to the solutions $x_1, \ldots, x_s$ of the simultaneous equations

$$x_1^h + \cdots + x_s^h = n_h \qquad (1 \leq h \leq k-1)$$

with $1 \leq x_r \leq 2N$.

Thus, in view of (4.16),

$$\sum_{m \in \mathcal{M}} |g(m, \beta)|^{2s} \le \sum_{\gamma \in \Gamma} \left| \sum_{\mathbf{n}} a(\mathbf{n}) e(\mathbf{n}\gamma) \right|^2 \qquad (4.17)$$

where $\Gamma = \{\gamma(m): m \in \mathcal{M}\}$. We apply Lemma 4.1 to the right-hand side of (4.17), with $l = k - 1$ and $N_j = s(2N)^j$. We may take $\delta_j \gg N^{-j}$ because of (4.16). Moreover,

$$\sum_{\mathbf{n} \in \mathcal{N}} |a(\mathbf{n})|^2 \le \sum_{\mathbf{n} \in \mathcal{N}} \sideset{}{''}\sum_{x_1, \ldots, x_s, y_1, \ldots, y_s} 1$$

where the inner sum is restricted to x_r, y_r in $[1, 2N]$ with

$$x_1^h + \cdots + x_s^h = y_1^h + \cdots + y_s^h = n_h \qquad (1 \le h \le k - 1),$$

so that

$$\sum_{\mathbf{n} \in \mathcal{N}} |a(n)|^2 \le J_s^{(k-1)}(2N). \qquad (4.18)$$

Combining (4.17) with Lemma 4.1 and (4.18), we obtain

$$\sum_{m \in \mathcal{M}} |g(m, \beta)|^{2s} \ll J_s^{(k-1)}(2N) N^{k(k-1)/2}.$$

Hence, by (4.2) and Hölder's inequality,

$$|S(f)|^{2s} \ll M^{-1} (\log N)^{2s} J_s^{(k-1)}(2N) N^{k(k-1)/2}$$

$$\ll (q_0^{-1} + N^{-1}) (\log N)^{2s} J_s^{(k-1)}(2N) N^{k(k-1)/2}$$

in view of (4.15). If we now insert the lower bound (4.6) and the upper bound for $J_{l(k-1)}^{(k-1)}(2N)$ from Theorem 4.1, we obtain

$$A^{2l(k-1)} \ll (q_0^{-1} + N^{-1}) (\log N)^{2l(k-1)} N^{2l(k-1)+\theta}$$

with θ as in the statement of the present theorem. We know that

$$N^{-1} (\log N)^{2l(k-1)} N^{2l(k-1)+\theta}$$

is of smaller order of magnitude than $A^{2l(k-1)}$, from (4.7). We conclude that

$$q_0 \ll (\log 2N)^{2l(k-1)} (NA^{-1})^{2l(k-1)} N^{\theta}$$

which implies (4.9), and completes the proof of Theorem 4.3.

4.2 The final coefficient lemma

The approximation to $\alpha_2, \ldots, \alpha_k$ derived in Theorem 4.3 can be extended to the full set of coefficients $\alpha_1, \ldots, \alpha_k$, provided that A is large enough to ensure that q and $N^2 \|q\alpha_2\|, \ldots, N^k \|q\alpha_k\|$ are smaller than $N^{1-\varepsilon}$. This principle was used by Vinogradov (1971) and Baker (1981*a*).

Lemma 4.2 (*Hua*).   Let $G(x) = u_k x^k + \cdots + u_1 x$ be a polynomial with integer coefficients. Let q be an integer and write d for the greatest common divisor,

$$d = (q, u_2, \ldots, u_k).$$

Then when $1 \leqq m \leqq q$, we have

$$\sum_{x=1}^{m} e_q(G(x)) = O(q^{1-(1/k)+\varepsilon} d^{1/k}).$$

Proof.   This is Theorem 2 of Hua (1965); most of it is also in Vaughan (1981), Chapter 7. For the development of the ideas see Hua (1940), and for refinements see Hua (1957), Chen (1977). For stronger estimates when more is known about the zeros of G, see Loxton and Vaughan (to appear).

Lemma 4.3.   Suppose that $X < Y$, F'' exists and is continuous on $[X, Y]$ and F' is monotonic on $[X, Y]$. Let H_1, H_2 denote integers such that $H_1 \leqq F'(y) \leqq H_2$ for every y in $[X, Y]$. Then

$$\sum_{X < x \leqq Y} e(F(x)) = \sum_{h=H_1}^{H_2} \int_X^Y e(F(\alpha) - \alpha h) \, d\alpha + O(\log(2 + H)).$$

Proof.   This is Lemma 4.2 of Vaughan (1981).

Lemma 4.4.   Let $k \geqq 2$. Let $f(x) = \alpha_k x^k + \cdots + \alpha_1 x$ and suppose that $q, u_1, \ldots, u_k$ are integers,

$$d = (q, u_2, \ldots, u_k), \tag{4.19}$$

and we have

$$|q\alpha_j - u_j| \leqq (2k^2)^{-1} N^{1-j} \qquad (1 \leqq j \leqq k). \tag{4.20}$$

Writing

$$\beta_j = \alpha_j - (u_j/q) \quad (j = 1, \ldots, k), \qquad g(x) = \sum_{j=1}^{k} \beta_j x^j,$$

$$G(x) = \sum_{j=1}^{k} u_j x^j, \qquad S(q) = \sum_{v=1}^{q} e_q(G(v)),$$

we have

$$\sum_{n=1}^{T} e(f(n)) = q^{-1} S(q) \int_0^T e(g(y)) \, dy + O(q^{1-(1/k)+\varepsilon/2} d^{1/k})$$

for all integers T, $1 \leqq T \leqq N$.

Proof. Write $S = \sum_{n=1}^{T} e(f(n))$. Then

$$S = \sum_{n=1}^{T} \sum_{\substack{v=1 \\ v \equiv n \pmod q}}^{q} e\left(g(n) + \frac{G(v)}{q}\right)$$

$$= q^{-1} \sum_{b=1}^{q} \sum_{n=1}^{T} e\left(g(n) - \frac{bn}{q}\right) \sum_{v=1}^{q} e\left(\frac{G(v) + bv}{q}\right). \quad (4.21)$$

After all,

$$q^{-1} \sum_{b=1}^{q} e\left(\frac{b(v - n)}{q}\right)$$

is 1 if $v \equiv n \pmod q$ and 0 if not.

Let $1 \le b \le q$. Now $[0, T]$ may be dissected into $O(1)$ intervals I, in each of which $g_b(x) = g(x) - (bx/q)$ is monotonic. The derivative $g_b'(x) = g'(x) - (b/q)$ lies between $-(b + \frac{1}{2})/q$ and $-(b - \frac{1}{2})/q$ on $[0, T]$, by (4.20). Thus Lemma 4.3 can be applied with $H_1 = -2$, $H_2 = 0$. We obtain

$$\sum_{n \in I} e\left(g(n) - \frac{bn}{q}\right) = \sum_{h=-2}^{0} \int_{I} e\left(g(y) - \frac{by}{q} - yh\right) dy + O(1),$$

and on summing over I,

$$\sum_{n=1}^{T} e\left(g(n) - \frac{bn}{q}\right) = \sum_{h=-2}^{0} \int_{0}^{T} e\left(g(y) - \frac{by}{q} - yh\right) dy + O(1). \quad (4.22)$$

When $1 \le b \le q - 1$ and $0 \le y \le T$, one has

$$\left| g'(y) - \frac{b}{q} - h \right| \ge \left\| g'(y) - \frac{b}{q} \right\| \ge \tfrac{1}{2} \left\| \frac{b}{q} \right\|.$$

By integration by parts of $e(g_b(y) - hy)$, we find that

$$\int_{0}^{T} e\left(g(y) - \frac{by}{q} - yh\right) \ll \|b/q\|^{-1}.$$

In (4.22) this gives

$$\sum_{n=1}^{T} e\left(g(n) - \frac{b}{q}\right) \ll \left\| \frac{b}{q} \right\|^{-1} \quad (b = 1, \ldots, q - 1).$$

Using this bound in (4.21), and taking Lemma 4.2 into account, we have

$$S - q^{-1} S(q) \sum_{n=1}^{T} e(g(n)) \ll q^{-1} \sum_{b=1}^{q-1} \left\| \frac{b}{q} \right\|^{-1} \left| \sum_{v=1}^{q} e\left(\frac{G(v) + bv}{q}\right) \right|$$

$$\ll q^{-1} \sum_{b=1}^{q-1} \left\| \frac{b}{q} \right\|^{-1} q^{1-(1/k)+\varepsilon/4} d^{1/k}$$

$$\ll q^{1-(1/k)+\varepsilon/2} d^{1/k}. \quad (4.23)$$

We now use (4.22) again with $b = q$. We have

$$|g'(y) \pm 1| \geq \tfrac{1}{2} \qquad (0 \leq y \leq T).$$

Integration by parts gives

$$\int_0^T e(g(y) \pm y) \, dy = O(1),$$

and we find that

$$\sum_{n=1}^T e(g(n)) = \int_0^T e(g(y)) \, dy + O(1).$$

In (4.23) this yields

$$S - q^{-1} S(q) \int_0^T e(g(y)) \, dy \ll q^{1-(1/k)+\varepsilon/2} d^{1/k}$$

and Lemma 4.4 is proved.

In an earlier version of this lemma (Baker (1981*a*)) the hypotheses are a little stronger. Euler's sum formula was used (cf. Estermann (1936)) rather than Lemma 4.3, which is a truncated Poisson summation formula. The present argument closely follows Vaughan (1981), Theorem 4.1. For further discussion of the case $f(x) = \alpha x^k$ see Vaughan (1984).

Lemma 4.5. *Let* $g(x) = e(\beta_k x^k + \cdots + \beta_1 x)$. *Then for* $0 < T \leq N$ *we have*

$$\int_0^T e(g(x)) \, dx \ll N Z^{-1/k}$$

where $Z = \max (1, N |\beta_1|, \ldots, N^k |\beta_k|)$.

Proof. See, for example, Vaughan (1981), Theorem 7.3. The theorem seems to appear first in Vinogradov (1951).

Lemma 4.6. *Let* $f(x) = \alpha_k x^k + \cdots + \alpha_1 x$, *and suppose there are integers* $N > C_6(k, \varepsilon)$, $r, v_2, \ldots, v_k$ *such that*

$$|\alpha_j r - v_j| \leq N^{1-j}/4k^4 \qquad (4.24)$$

$(j = 2, \ldots, k)$. *Suppose further that for some* T, $1 \leq T \leq N$, *we have*

$$\left| \sum_{n=1}^T e(f(n)) \right| \geq H \geq r^{1-1/k} N^\varepsilon. \qquad (4.25)$$

Write d for the greatest common divisor, $d = (r, v_2, \ldots, v_k)$. *Then there is a natural number* $t \leqq 2k^2$ *such that*

$$trd^{-1} \leqq (NH^{-1})^k N^{3k\varepsilon},$$

$$t |\alpha_j r - v_j| d^{-1} \leqq (NH^{-1})^k N^{-j+3k\varepsilon} \qquad (j = 2, \ldots, k)$$

$$\|trd^{-1}\alpha_1\| \leqq (NH^{-1})^k N^{-1+3k\varepsilon}.$$

Proof. Write $x = rd^{-1}$, $w_j = v_j d^{-1}$ $(j = 2, \ldots, k)$. By Dirichlet's theorem there are coprime integers t, u, such that

$$1 \leqq t \leqq 2k^2, \qquad |tx\alpha_1 - u| \leqq (2k^2)^{-1}.$$

Write $q = xt$, $u_j = w_j t$ $(j = 2, \ldots, k)$. Then

$$(q, u_2, \ldots, u_k) = t(x, w_2, \ldots, w_k) = t \leqq 2k^2, \qquad (4.26)$$

and, in view of (4.24),

$$|q\alpha_j - u_j| = td^{-1} |r\alpha_j - v_j| \leqq N^{1-j}/(2k^2) \qquad (j = 2, \ldots, k).$$

We may therefore apply Lemma 4.4, with t in place of d. Now because of (4.25), (4.26), the quantity $O(q^{1-1/k+\varepsilon/2}d^{1/k})$ is smaller than $\frac{1}{4}H$. It follows that

$$\left| q^{-1}S(q) \int_0^T e(g(y)) \, dy \right| > \frac{1}{2}H,$$

where $S(q)$ and $g(y)$ are as in Lemma 4.4.

We now use the estimate

$$q^{-1}S(q) \ll q^{-1/k}N^{2\varepsilon}$$

which follows from (4.26) and Lemma 4.2. In the notation of Lemma 4.5, then, we see that

$$H \ll q^{-1/k}N^{1+2\varepsilon}Z^{-1/k},$$

or

$$qZ = \max(q, N \|q\alpha_1\|, \ldots, N^k \|q\alpha_k\|) \ll N^{k+2k\varepsilon}H^{-k}.$$

Since $N > C_6(k, \varepsilon)$, this proves Lemma 4.6.

4.3 Application to fractional parts

We can now prove a version of the result mentioned in the last chapter: large Weyl sums yield a good simultaneous rational approximation to all the coefficients.

Theorem 4.4. *Let* $J = 8k^2(\log k + \frac{1}{2} \log \log k + 2)$, *where* $k \geqq 4$. *Let* $N >$

$C_7(k)$. Let $f(x) = \alpha_k x^k + \cdots + \alpha_1 x$ and suppose that

$$\sum_{m=1}^{M} |S_m(f)| \geq P \tag{4.27}$$

where $P \ll N$,

$$(MNP^{-1})^J \leq N. \tag{4.28}$$

Then there are integers $y, u_1, \ldots, u_k$ such that

$$1 \leq y \leq M(MNP^{-1})^k N^\varepsilon, \tag{4.29}$$

$$|y\alpha_j - u_j| \leq (MNP^{-1})^k N^{\varepsilon-j} \quad (j = 1, \ldots, k). \tag{4.30}$$

Proof. We apply Theorem 4.3 with

$$l = [k \log (4k^2 \log k)] + 1,$$

so that

$$\theta = \tfrac{1}{2}(k-1)^2 \left(\frac{k-2}{k-1}\right)^l < \tfrac{1}{2}(k-1)^2 e^{-l/k} < \frac{1}{8 \log k}.$$

Now

$$J \geq 4k^2(2 \log k + \log \log k + 1.7)\left(1 + \frac{1}{3 \log k}\right)$$

$$\geq 4k^2(2 \log k + \log \log k + 1.7)(1 - 2\theta)^{-1}$$

$$> 4kl(1 - 2\theta)^{-1}.$$

We suppose (as we may) that ε is small as a function of k. Now there is a natural number $m \leq M$ such that

$$|S_m(f)| \gg A = PM^{-1}.$$

We have

$$(NA^{-1})^{4(k-1)l} \ll (MNP^{-1})^{4(k-1)l}$$

$$\ll N^{4(k-1)l/J} \leq N^{1-2\theta-4\varepsilon} \tag{4.31}$$

from (4.28). The hypotheses of Theorem 4.3 are satisfied with mf in place of f. Thus there is an integer q_0, satisfying

$$1 \leq q_0 \leq (NA^{-1})^{2(k-1)l} N^{\theta+\varepsilon} \leq N^{1-2\varepsilon},$$

and

$$\|mq_0\alpha_j\| \leq q_0 N^{-j+\theta+\varepsilon}(NA^{-1})^{2(k-1)l}$$

$$\leq (NA^{-1})^{4(k-1)l} N^{-j+2\theta+2\varepsilon} \leq N^{-j+1-2\varepsilon}$$

from (4.8), (4.9) and (4.31).

Now it is permissible to apply Lemma 4.6 with mf in place of f, $\varepsilon/(4k)$

in place of ε, $A = H$ and $r = q_0$. It is clear that (4.25) is satisfied, since

$$A \gg N^{1-(1/J)} \geqq N^{1-(1/k)+2\varepsilon} \geqq q_0^{1-1/k} N^{2\varepsilon}.$$

Let t be the integer provided by Lemma 4.6 and let $y = trd^{-1}m$. Then y satisfies (4.29) and (4.30). This proves Theorem 4.4.

Theorem 4.5.   *Let $k \geqq 4$ and let J be as in Theorem 4.4. Then for any real polynomial $f(x) = \alpha_k x^k + \cdots + \alpha_1 x$ and $N > C_8(k)$ there is a natural number $n \leqq N$ satisfying*

$$\|f(n)\| < N^{-1/J}.$$

Proof.   We suppose the contrary. With $M = [N^{1/J}] + 1$, we have

$$\sum_{m=1}^{M} |S_m(f)| > N/6$$

from Theorem 2.2. By Theorem 4.4, there is a natural number y such that

$$y \ll M^{k+1} N^\varepsilon \ll N^{(k+1)/J+\varepsilon}.$$

$$\|y\alpha_j\| \ll N^{k/J-j+\varepsilon} \qquad (j = 1, \ldots, k).$$

It is clear that $y \leqq N$ and that

$$\|f(y)\| \leqq \sum_{j=1}^{k} N^{j-1} \|y\alpha_j\| < N^{-1/J}.$$

This contradicts the hypothesis, and the theorem is proved.

5
The solution of Davenport's problem

5.1 More about difference operators

In this chapter our objective is the inequality

$$\min_{1 \leq n \leq N} \|f(n)\| < N^{-1/K + \varepsilon} \tag{5.1}$$

mentioned in the introduction, where $f(x) = \alpha_k x^k + \cdots + \alpha_1 x$ and $N > C_1(k, \varepsilon)$. As a short calculation shows, (5.1) is stronger than Theorem 4.5 for $k \leq 13$.

The difference operator $\Delta_{y_1 \cdots y_t}$ was defined in Chapter 3.

Lemma 5.1. *Let* $2 \leq h < k$. *Then*

$$\Delta_{y_1 \cdots y_{h-1}}(mf(x)) = my_1 \cdots y_{h-1} \sum_{j=0}^{k-h+1} \sum_{s=j+h-1}^{k} \alpha_s b(s, h, j) x^j,$$

where
(i) $b(s, h, j)$ *is a homogeneous polynomial in* $y_1, \ldots, y_{h-1}$ *of degree* $s - h - j + 1$, *with integer coefficients* $\ll 1$, *and*
(ii) *we have*

$$b(j + h - 1, h, j) = (j + h - 1)(j + h - 2) \cdots (j + 1). \tag{5.2}$$

Proof. It is clear that, for $1 \leq s \leq k$, $\Delta_{y_1 \cdots y_{h-1}}(x^s)$ is a homogeneous polynomial in $x, y_1, \ldots, y_{h-1}$ of degree s, divisible by $y_1, \ldots, y_{h-1}$. It follows that

$$\Delta_{y_1 \cdots y_{h-1}}(x^s) = y_1 \cdots y_{h-1} \sum_{j=0}^{s-h+1} b(s, h, j) x^j$$

where $b(s, h, j)$ has the properties given in (i). As for property (ii), it is easy to pick out the leading term from $\Delta_{y_1 \cdots y_{h-1}}(x^{j+h-1})$. By linearity, we have

$$\Delta_{y_1 \cdots y_{h-1}}\left(\sum_{s=1}^{k} m\alpha_s x^s\right) = \sum_{s=1}^{k} m\alpha_s y_1 \cdots y_{h-1} \sum_{j=0}^{s-h+1} b(s, h, j) x^j$$

$$= my_1 \cdots y_{h-1} \sum_{j=0}^{k-h+1} \sum_{s=j+h-1}^{k} \alpha_s b(s, h, j) x^j.$$

Our second lemma is due to Birch and Davenport (1958a).

Lemma 5.2. *Suppose θ is real, and suppose there exist R distinct integer pairs x, z satisfying*

$$|\theta x - z| < \zeta, \tag{5.3}$$

$$0 < |x| < X, \tag{5.4}$$

where $R \geq 24\zeta X > 0$. Then all integer pairs x, z satisfying (5.3), (5.4) have the same ratio z/x.

Proof. The hypothesis implies that $X > 1$, since otherwise $R = 0$. It also implies that $\zeta < \frac{1}{2}$. For if $\zeta \geq \frac{1}{2}$ we have the crude bound

$$R \leq (2X + 1)(2\zeta + 1) \leq 12X\zeta.$$

By Dirichlet's theorem, there exist coprime integers p, q such that

$$0 < q \leq 2X, \qquad |q\theta - p| < (2X)^{-1}.$$

If x and z satisfy (5.3) and (5.4), then

$$\begin{aligned}|xp - zq| &\leq |x(p - q\theta)| + |q(x\theta - z)| \\ &< X(2X)^{-1} + q\zeta = \tfrac{1}{2} + q\zeta.\end{aligned} \tag{5.5}$$

Suppose if possible that $q\zeta > \frac{1}{2}$. Then the number of possible residue classes for $x \pmod q$ is less than $2(\frac{1}{2} + q\zeta) + 1 < 6q\zeta$ and consequently the number of possibilities for x is less than

$$6q\zeta(2Xq^{-1} + 1) < 12\zeta X + 12\zeta X.$$

Since x determines z with at most one possibility by (5.3), we obtain $R < 24\zeta X$, a contradiction.

Thus $q\zeta \leq \frac{1}{2}$. Now (5.5) implies that $z/x = p/q$ whenever x and z satisfy (5.3), (5.4), and Lemma 5.2 is proved.

Note the similarity to the proof of (3.19).

Lemma 5.3. *Let $1 \leq h < k$, and let $H = 2^{h-1}$. Then*

$$2H(k - h) \leq K \tag{5.6}$$

and

$$H + 2H(k - h)\left(1 - \frac{1}{k - h + 1}\right) \leq K. \tag{5.7}$$

Proof. Writing $r = k - h$, the inequalities reduce to $2r \leq 2^r$ (which is rather obvious) and

$$1 + (2r^2/(r + 1)) \leq 2^r \qquad (r = 1, 2, \dots) \tag{5.8}$$

The inequality (5.8) is seen to hold for $r = 1, 2$. For $r \geq 3$ it is a consequence of

$$1 + 2r \leqq 2^r,$$

which is easy to prove by induction.

5.2   Finding out about one more coefficient

Theorem 5.1. *Let* $N > C_2(k, \varepsilon)$. *Let* M *be a natural number and* P *a positive number such that*

$$(MNP^{-1})^{K+\varepsilon} \leqq N. \tag{5.9}$$

Suppose that

$$\sum_{m=1}^{M} |S_m(f)| \geqq P. \tag{5.10}$$

Then there exists a natural number q *and integers* $v_1, \ldots, v_k$ *such that*

$$q < (MNP^{-1})^k N^\varepsilon, \ (q, v_1, \ldots, v_k) = 1, \ (q, v_2, \ldots, v_k) \leqq MN^\varepsilon, \tag{5.11}$$

$$|\alpha_j q - v_j| = \|\alpha_j q\| < M^{-1}(MNP^{-1})^k N^{\varepsilon-j} \quad (j = 1, 2, \ldots, k). \tag{5.12}$$

The reader should be able to deduce (5.1). Nevertheless, we shall give the details in the next section.

We now state a proposition which implies Theorem 5.1.

Proposition 5.1. *Let* $1 \leqq h < k$, $H = 2^{h-1}$. *Given all the hypotheses of Theorem* 5.1, *suppose further that there is a natural number* r *satisfying*

$$r < (MNP^{-1})^{2H(k-h)} N^\eta, \tag{5.13}$$

$$\|\alpha_j r\| < (MNP^{-1})^{2H(k-h)} M^{-1} N^{\eta-j} \quad (j = h+1, \ldots, k). \tag{5.14}$$

Then there is a natural number w *satisfying*

$$w < (MNP^{-1})^{H(k-h+1)} N^\eta \tag{5.15}$$

$$\|w\alpha_j\| < (MNP^{-1})^{H(k-h+1)} M^{-1} N^{\eta-j} \quad (j = h, \ldots, k). \tag{5.16}$$

In case $h = 1$, $\|w\alpha_j\| = |w\alpha_j - v_j|$ *where* $(w, v_1, \ldots, v_k) = 1$, $(w, v_2, \ldots, v_k) \ll MN^\delta$.

We note that

$$2H(k - h) = 2^{(h+1)-1}(k - (h + 1) + 1).$$

Thus the proposition says that if we have some information about simultaneous approximation to $\alpha_k, \ldots, \alpha_{h+1}$ then we can get better information about simultaneous approximation to $\alpha_k, \ldots, \alpha_h$. (Naturally, we need some large Weyl sums to do so.)

Proof of Theorem 5.1. Suppose for a moment that the proposition has been proved. We proceed to deduce Theorem 5.1 by constructing a sequence of natural numbers $r_k, r_{k-1}, \ldots, r_1$ such that $w = r_h$ satisfies (5.15) and (5.16). First of all, by Hölder's inequality, the hypothesis (5.10) implies

$$\sum_{m=1}^{M} |S_m(f)|^K \geqq P^K M^{-K+1}.$$

We apply Lemma 3.10 with $L = M$,

$$A = P^K M^{-K+1} \gg N^{K-1+\eta} M$$

by (5.9). Thus there exists a natural number r_k,

$$r_k \leqq M N^{K+\eta} A^{-1} \leqq (MNP^{-1})^K N^\eta,$$

$$\|\alpha_k r_k\| \leqq N^{K-k+\eta} A^{-1} \leqq (MNP^{-1})^K M^{-1} N^{-k+\eta}.$$

Then $r = r_k$ satisfies (5.13) and (5.14) with $h = k - 1$. Let r_{k-1} be the natural number w supplied by the Proposition in the case $h = k - 1$. Generally, once we have found r_{h+1}, where $1 \leqq h < k$, we can take $r = r_{h+1}$ in the Proposition and obtain a natural number $w = r_h$.

After $k - 1$ steps we obtain a natural number r_1 which satisfies (5.11) and (5.12) with $q = r_1$, and Theorem 5.1 is proved.

Proof of Proposition 5.1. Suppose first that $h > 1$. From (5.10) we find that

$$P^H M^{-H+1} \leqq \sum_{m=1}^{M} |S_m(f)|^H$$

using Hölder's inequality. Applying Weyl's inequality (Lemma 3.8) we have

$$P^H M^{-H+1} \ll M N^{H-1} + N^{H-h} \sum_{m=1}^{M} \sum_{y_1=1}^{N} \cdots \sum_{y_h=1}^{N} \left| \sum_{n=1}^{N-Y} e(\Delta_{y_1 \cdots y_{h-1}}(mf(n))) \right| \tag{5.17}$$

where Y denotes $y_1 + \cdots + y_{h-1}$. The term $M N^{H-1}$ is of smaller order of magnitude than $P^H M^{-H+1}$. For, in view of (5.9),

$$M^H N^{H-1} P^{-H} = (MNP^{-1})^H N^{-1}$$

$$\leqq (MNP^{-1})^K N^{-1} \leqq N^{-\delta}.$$

Thus

$$N^h (MNP^{-1})^{-H} M \ll \sum_{m=1}^{M} \sum_{y_1=1}^{N} \cdots \sum_{y_{h-1}=1}^{N} \left| \sum_{n=1}^{N-Y} e(\Delta_{y_1 \cdots y_{h-1}}(mf(n))) \right|.$$

Let us write $\mathbf{y} = (m, y_1, \ldots, y_{h-1})$. We subdivide the last sum over $\mathbf{y}$

into $\ll \log N$ sums Σ_U. Here Σ_U is a sum over $\mathbf{y}$ satisfying

$$1 \leq m \leq M, \quad 1 \leq y_i \leq N \ (i = 1, \ldots, h-1), \quad U \leq my_1 \cdots y_{h-1} < 2U. \tag{5.18}$$

Each value of U satisfies

$$1 \leq U \leq N^{h-1}M. \tag{5.19}$$

We select the sum Σ_U having largest modulus and further subdivide it into at most N^δ sums $\Sigma_{U,D}$. Here $\Sigma_{U,D}$ is a sum over those $\mathbf{y}$ satisfying (5.18) and

$$(r, my_1 \cdots y_{h-1}) = D. \tag{5.20}$$

Each value of D is a divisor of r. We can choose D such that, in an obvious notation,

$$N^{h-\delta}(MNP^{-1})^{-H}M(\log N)^{-1} \ll \sum_{\substack{\mathbf{y} \\ (5.18),(5.20)}} T(\mathbf{y}) \tag{5.21}$$

where

$$T(\mathbf{y}) = \left| \sum_{n=1}^{N-Y} e(\Delta_{y_1 \cdots y_{h-1}}(mf(n))) \right|.$$

Those $\mathbf{y}$ for which $T(\mathbf{y}) \leq N^{1-2\delta}(MNP^{-1})^{-H}$ give a contribution to the term in (5.21) which is of smaller order of magnitude than the left-hand side. Let us cover the interval $(N^{1-2\delta}(MNP^{-1})^{-H}, N]$ by $\ll \log N$ subintervals of the type $(B, 2B]$. We see that there is a

$$B \geq N^{1-2\delta}(MNP^{-1})^{-H} \tag{5.22}$$

such that the set $\mathcal{B}$ of $\mathbf{y}$ with (5.18), (5.20) and

$$B < T(\mathbf{y}) \leq 2B \tag{5.23}$$

has

$$\sum_{\mathbf{y} \in \mathcal{B}} T(\mathbf{y}) \gg N^{h-\delta}(MNP^{-1})^{-H}M(\log N)^{-2}.$$

Consequently,

$$|\mathcal{B}| \geq N^{h-2\delta}(MNP^{-1})^{-H}MB^{-1}. \tag{5.24}$$

Now pick any $\mathbf{y}$ in $\mathcal{B}$. According to Lemma 5.1,

$$\Delta_{y_1 \cdots y_{h-1}}(mf(x)) = my_1 \cdots y_{h-1}(A_{k-h+1}x^{k-h+1} + \cdots + A_1x + A_0)$$

where for $j = 1, \ldots, k-h+1$ we have

$$A_j = \sum_{s=j+h-1}^{k} \alpha_s b(s, h, j). \tag{5.25}$$

We may suppose without loss of generality that, in addition to (5.13) and

(5.14), r satisfies $\|r\alpha_s\| = |r\alpha_s - u_s|$ with

$$(r, u_{h+1}, \ldots, u_k) = 1.$$

We note that, from Lemma 5.3 and (5.9),

$$r < (MNP^{-1})^K N^\eta \leq N^{1-2\delta}, \tag{5.26}$$

$$|r\alpha_s - u_s| < (MNP^{-1})^K M^{-1} N^{\eta-s} < M^{-1} N^{-s+1-2\delta} (h+1 \leq s \leq k). \tag{5.27}$$

Let us see what this tells us about $\|r(my_1 \cdots y_{h-1} A_j)\| \ (2 \leq j \leq k-h+1)$. From (5.27),

$$rA_j = \sum_{s=j+h-1}^{k} (u_s + O(M^{-1} N^{-s+1-2\delta})) b(s, h, j).$$

Let

$$V_j = \sum_{s=j+h-1}^{k} u_s b(s, h, j). \tag{5.28}$$

Then

$$|rA_j - V_j| \ll \sum_{s=j+h-1}^{k} M^{-1} N^{-s+1-2\delta} |b(s, h, j)|$$

$$\ll M^{-1} \sum_{s=j+h-1}^{k} N^{-s+1-2\delta} N^{s-h-j+1}$$

$$\ll M^{-1} N^{-(j+h)+2-2\delta}$$

in view of Lemma 5.1 (i). Plainly we have

$$|r(my_1 \cdots y_{h-1} A_j) - my_1 \cdots y_{h-1} V_j| < N^{-j+1-2\delta} \qquad (j = 2, \ldots, k-h+1) \tag{5.29}$$

for all $\mathbf{y}$ in $\mathcal{B}$.

Because of (5.26) and (5.29), we may reasonably hope to apply the final coefficient lemma (Lemma 4.6) to

$$g(x) = \Delta_{y_1 \cdots y_{h-1}} (mf(x))$$

with δ in place of ε and $k-h+1$ in place of k. After all, (5.23) gives the first inequality in (4.25), with $T = N - Y$ and $H = B$. But we do have to verify the second inequality in (4.25):

$$r^{1-1/(k-h+1)} N^{2\delta} B^{-1}$$

$$\leq (MNP^{-1})^{(1-1/(k-h+1))2H(k-h)+H} N^{-1+\eta}$$

$$\leq (MNP^{-1})^K N^{-1+\eta} \leq 1. \tag{5.30}$$

Here we used in turn (5.13), (5.22), (5.7) and (5.9).

We deduce that there is a natural number $t = t(\mathbf{y}) \leq 2k^2$ such that

$$trd^{-1} \leq (NB^{-1})^{k-h+1} N^\eta, \tag{5.31}$$

$$td^{-1} |r(my_1 \cdots y_{h-1} A_j) - my_1 \cdots y_{h-1} V_j| \leq (NB^{-1})^{k-h+1} N^{-j+\eta} \tag{5.32}$$

for $j = 2, \ldots, k - h + 1$, and

$$\|trd^{-1}(my_1 \cdots y_{h-1}A_1)\| \leqq (NB^{-1})^{k-h+1}N^{-1+\eta}. \tag{5.33}$$

Here $d = d(\mathbf{y})$ is the greatest common divisor of r and $my_1 \cdots y_{h-1}V_2, \ldots, my_1 \cdots y_{h-1}V_{k-h+1}$. To proceed further we must show that d is essentially the same as D.

Let $d_0 = (r, V_2, \ldots, V_{k-h+1})$ and write

$$b_j = (j + h - 1)(j + h - 2) \cdots (j + 1).$$

We may show inductively that, for $j = k - h, k - h - 1, \ldots, 1$,

$$d_0 | b_{k-h+1} \cdots b_{j+1} u_{j+h}. \tag{5.34}$$

To start this process, we note that

$$b_{k-h+1}u_k = V_{k-h+1}$$

from (5.28) and (5.2), and that $d_0 | V_{k-h+1}$. This gives (5.34) for $j = k - h$. Suppose now that (5.34) holds for some j, $1 < j \leqq k - h$. Now

$$b_{k-h+1} \cdots b_j u_{j+h-1} = b_{k-h+1} \cdots b_{j+1}\left(V_j - \sum_{s=j+h}^{k} u_s b(s, h, j)\right)$$

from (5.28) and (5.2). Each of

$$V_j, b_{k-h+1} \cdots b_{j+1} u_s \qquad (j + h \leqq s \leqq k)$$

is divisible by d_0, and (5.34) follows with $j - 1$ in place of j. We conclude that

$$d_0 | b_{k-h+1} \cdots b_2(r, u_{h+1}, \ldots, u_k) = b_{k-h+1} \cdots b_2.$$

Moreover, from the definition of d,

$$D = (r, my_1 \cdots y_{h-1}) | d \leqq Dd_0 \ll D. \tag{5.35}$$

Let $\theta = rA_1$. For any $\mathbf{y}$ in $\mathcal{B}$ write $x = x(\mathbf{y}) = tmy_1 \cdots y_{h-1}D^{-1}$ and let $z = z(\mathbf{y})$ be the nearest integer to $x(\mathbf{y})\theta$. Then in view of (5.33), (5.35) we have

$$|x\theta - z| \leqq dD^{-1} \|(trd^{-1})A_1 my_1 \cdots y_{h-1}\|$$
$$< (NB^{-1})^{k-h+1}N^{-1+\eta} = \zeta, \tag{5.36}$$

say. Moreover, it is clear that

$$0 < x < X = MN^{h-1+\delta}. \tag{5.37}$$

The number of divisors of x is $< N^\delta$, so we obtain (say) R distinct products x as $\mathbf{y}$ runs through $\mathscr{B}$, where

$$
\begin{aligned}
R \geqq |\mathscr{B}| \, N^{-\eta} &\geqq N^{h-\eta}(MNP^{-1})^{-H}MB^{-1} \\
&\geqq 24(NB^{-1})^{k-h+1}N^{-1+\eta}MN^{h-1+\delta} \\
&= 24\zeta X.
\end{aligned}
\tag{5.38}
$$

(The second and third inequalities follow from (5.24), (5.22), (5.6) and (5.9)).

By Lemma 5.2, there are integers u and v such that $v > 0$, $(u, v) = 1$ and $z/x = u/v$ for all $\mathbf{y}$ in $\mathscr{B}$. Since all the distinct products $tmy_1 \cdots y_{h-1}$ lie in $[U, 2UN^\delta]$ and Dv evidently divides each one, there must be

$$
\ll UN^\delta/Dv
$$

such products. That is

$$
R \ll UN^\delta/Dv.
\tag{5.39}
$$

Moreover, for any choice of $\mathbf{y}$ in $\mathscr{B}$,

$$
|vrA_1 - u| = |v\theta - u| = vx^{-1}|\theta x - z| < v(UD^{-1})^{-1}\zeta.
\tag{5.40}
$$

Let $w = rvb_{k-h+1} \cdots b_2 b_1$. We shall show that w satisfies (5.15) and (5.16). First of all,

$$
w \ll rv \ll rD^{-1} \cdot Dv \ll (NB^{-1})^{k-h+1}N^\eta \cdot UN^\delta R^{-1}
\tag{5.41}
$$

from (5.31), (5.35) and (5.39). The inequality (5.15) follows after applying (5.19), the second inequality of (5.38), and (5.22).

From (5.32) it follows that, for $2 \leqq j \leqq k - h + 1$,

$$
\begin{aligned}
\|rA_j\| &\leqq U^{-1}d(NB^{-1})^{k-h+1}N^{-j+\eta} \\
&\ll U^{-1}D(NB^{-1})^{k-h+1}N^{-j+\eta}.
\end{aligned}
$$

In view of (5.39), (5.40) and (5.36), we have

$$
\begin{aligned}
\|vrA_j\| &\ll vU^{-1}D(NB^{-1})^{k-h+1}N^{-j+\eta} \\
&\ll R^{-1}(NB^{-1})^{k-h+1}N^{-j+\eta}.
\end{aligned}
\tag{5.42}
$$

for all the suffices $j = 1, 2, \ldots, k - h + 1$.

It remains to pass from the set of coefficients $A_1, \ldots, A_{k-h+1}$ to the set $\alpha_h, \alpha_{h+1}, \ldots, \alpha_k$. We have

$$
\|w\alpha_k\| \ll \|b_{k-h+1}vr\alpha_k\| = \|vrA_{k-h+1}\|
$$

from (5.25). For $1 \leqq j < k - h + 1$, we have

$$
\begin{aligned}
&\|b_{k-h+1} \cdots b_j vr\alpha_{j+h-1}\| \\
&= \left\| b_{k-h+1} \cdots b_{j+1}vr\left(A_j - \sum_{s=j+h}^{k} \alpha_s b(s, h, j)\right) \right\| \\
&\ll \|vrA_j\| + \sum_{s=j+h}^{k} N^{s-h-j+1}\|b_{k-h+1} \cdots b_{j+1}vr\alpha_s\|
\end{aligned}
$$

from (5.25). We can thus deduce inductively from (5.42) that

$$\|w\alpha_{j+h-1}\| \ll \|b_{k-h+1} \cdots b_j v r \alpha_{j+h-1}\|$$
$$\ll R^{-1}(NB^{-1})^{k-h+1} N^{-j+\eta}$$

for $j = k - h + 1, k - h, \ldots, 1$. As in the case of (5.41), we can deduce (5.16), and the Proposition is proved in the case when $h > 1$.

Now suppose that $h = 1$. The argument is similar but simpler. Using (5.10) we can find a positive U, $1 \leq U \leq M$, a divisor D of r, and a set $\mathcal{B}$ of natural numbers m such that

$$U \leq m < 2U, \qquad (m, r) = D, \qquad |S_m(f)| \geq B$$

for all $m \in \mathcal{B}$, where B and $\mathcal{B}$ satisfy (5.22) and (5.24) with $h = 1$. Just as above, we have

$$r \leq N^{1-\delta}, \qquad (r, mu_2, \ldots, mu_k) = D$$
$$|rm\alpha_s - mu_s| < N^{-s+1-2\delta} \qquad (2 \leq s \leq k)$$

and (5.30). An application of the final coefficient lemma yields a natural number $t = t(m) < N^\delta$ such that

$$trD^{-1} \leq (NB^{-1})^k N^\eta,$$
$$tD^{-1}|rm\alpha_s - mu_s| \leq (NB^{-1})^k N^{-s+\eta} \qquad (2 \leq s \leq k),$$

and

$$\|trD^{-1}m\alpha_1\| < (NB^{-1})^k N^{-1+\eta} = \zeta.$$

For any m in $\mathcal{B}$ let $x = x(m) = tmD^{-1}$. With $\theta = r\alpha_1$, we have

$$|\theta x - z| < \zeta, \qquad 0 < x < X = MN^\delta$$

for R distinct products x, where R satisfies (5.38). For some v, dividing each x, and $w = rv$, we can deduce all the estimates (5.39)–(5.42) with α_j in place of A_j. The inequalities (5.15) and (5.16) follow.

We still have to consider $(w, v_k, \ldots, v_2)$ and $(w, v_k, \ldots, v_1)$ where $\|w\alpha_j\| = |w\alpha_j - v_j|$. In fact, it is clear from the above that, with u as in (5.40),

$$v_1 = u, \quad v_j = vu_j \qquad (j \geq 2).$$

Since $(v, u) = 1$ and $(r, u_k, \ldots, u_2) = 1$ we have

$$(w, v_k, \ldots, v_1) = (rv, vu_k, \ldots, vu_2, u) = 1$$

while

$$(w, v_k, \ldots, v_2) = v(r, u_k, \ldots, u_2) = v \ll MN^\delta$$

from (5.39). This completes the proof in the case $h = 1$, and the proposition is completely proved.

5.3 Small values of $f(n)$ modulo one

Theorem 5.2. *Let* $f(x) = \alpha_k x^k + \cdots + \alpha_1 x$. *Let* $N > C_2(k, \varepsilon)$. *Then there is a natural number* $n \leq N$ *such that*

$$\|f(n)\| < N^{-(1/K)+\varepsilon}.$$

Proof. Suppose that no such n exists. Let $M = [N^{(1/K)-\varepsilon}] + 1$. By Theorem 2.2,

$$\sum_{m=1}^{M} |S_m(f)| > \frac{N}{6}.$$

We may apply Theorem 5.1 with $P = N/6$. The hypothesis (5.9) is evidently satisfied. Thus there exists q satisfying

$$q \leq M^k N^{2\varepsilon} \leq N,$$
$$\|\alpha_j q\| \ll M^{k-1} N^{\varepsilon-j} \qquad (j = 1, \ldots, k).$$

Accordingly,

$$\|f(q)\| \ll \sum_{j=1}^{k} N^{j-1} M^{k-1} N^{\varepsilon-j} \ll k M^{k-1} N^{\varepsilon-1} \ll M^{-1} N^{-\varepsilon},$$

and $\|f(q)\| < M^{-1}$. This contradicts the hypothesis, and the theorem is proved.

It is convenient for our later work to state a slight strengthening of Theorem 4.4.

Theorem 5.3. *Let* $k \geq 4$. *The analogue of Theorem 4.4 is true with* (4.29) *and* (4.30) *replaced by the stronger inequalities*

$$1 \leq y \leq (MNP^{-1})^k N^\varepsilon, \quad |y\alpha_j - u_j| \leq M^{-1}(MNP^{-1})^k N^{\varepsilon-j} \qquad (j = 1, \ldots, k),$$

and with the additional conditions $(y, u_1, \ldots, u_k) = 1$,

$$(y, u_2, \ldots, u_k) \ll MN^\delta.$$

Proof. We use the case $h = 1$ of the proposition. The hypotheses of Theorem 4.4 yield a natural number y satisfying

$$y \leq M(MNP^{-1})^k N^\delta$$
$$\|y\alpha_j\| \leq (MNP^{-1})^k N^{\delta-j} \qquad (j = 1, \ldots, k).$$

(One needs to alter the constant $C_7(k)$.)
Now

$$(MNP^{-1})^k \ll (MNP^{-1})^{2(k-1)} M^{-1}$$

because $P \ll N$ and $k + 1 \leq 2(k - 1)$. Thus Proposition 5.1 is applicable with $h = 1$, $r = y$. The outcome is a natural number w and integers $v_1, \ldots, v_k$, which do all that is required for Theorem 5.3.

It is rather curious that for $k \geq 3$ the information about approximation to $\alpha_1, \ldots, \alpha_k$ gained from a hypothesis

$$\min_{1 \leq n \leq N} \|f(n)\| > N^{-1/(\min(J, K + \varepsilon))}$$

is stronger than we need to get a contradiction. There seems to be no way to take advantage of this to strengthen Theorem 4.5 or Theorem 5.2. We need Weyl sums larger than $N^{1-1/(\min(J, K + \varepsilon))}$ to 'set the ball rolling' with a nontrivial rational approximation to one coefficient. However, the 'stronger information' is useful when we are discussing simultaneous approximation questions, as we shall see in Chapter 8.

6
The method of sums of kth powers

6.1 Counting solutions of an equation

In this chapter we are going to improve Danicic's Theorem 3.4 for $k \geqq 9$. We recall the following idea, applied to Waring's problem by Hardy and Littlewood (see Vaughan (1981)). Let $1 \geqq \lambda_1 \geqq \lambda_2 \geqq \cdots \geqq \lambda_h$ be a suitably chosen sequence of positive numbers. Then the set of natural numbers

$$x_1^k + \cdots + x_h^k$$

with

$$P^{\lambda_j} < x_j < 2P^{\lambda_j} \qquad (j = 1, \ldots, h)$$

has cardinality of the order of magnitude $\gg P^{\lambda_1 + \cdots + \lambda_h}$ (that is, as large as it possibly could be). In the present chapter we use quite closely related ideas.

Definition 6.1. *Let $1 \geqq \lambda_1 \geqq \cdots \geqq \lambda_h$ be positive numbers. We say that $\lambda_1, \ldots, \lambda_h$ are* permissible exponents *if, whenever $P \geqq M \geqq 1$, and $m_1, \ldots, m_h$ are nonzero integers in $[-M, M]$, the number H of solutions of*

$$m_1 x_1^k + \cdots + m_h x_h^k = m y_1^k + \cdots + m_h y_h^k \qquad (6.1)$$

with

$$P^{\lambda_j} < x_j, \ y_j < 2P^{\lambda_j} \qquad (j = 1, \ldots, h) \qquad (6.2)$$

satisfies

$$H \ll P^{\lambda_1 + \cdots + \lambda_h + \varepsilon} M^h / |m_1 \cdots m_h|.$$

Implied constants in Chapters 6, 7, 8 and 10 depend at most on k, ε, and h.

A very similar notion (*admissible* exponents) can be found, for example, in Davenport (1939) and Davenport and Erdös (1939). For a given k, h, the most useful permissible exponents are those for which

$$\Lambda = \lambda_1 + \cdots + \lambda_h - (k - 1)$$

has the largest value. It is essential that $\Lambda > 0$.

Actually, it is fairly obvious from the definition that

$$\Lambda \leq 1. \tag{6.3}$$

To see this, take $m_1 = \cdots = m_h = 1$ in Definition 6.1, and let Q be the number of distinct values taken by $x_1^k + \cdots + x_h^k$ with $P^{\lambda_j} < x_j < 2P^{\lambda_j}$. Then, with $Y = h(2P^{\lambda_1})^k$,

$$Q = \sum_{\substack{m \leq Y \\ F(m) > 0}} 1 \ll P^k,$$

where $F(m)$ is the number of solutions of

$$x_1^k + \cdots + x_h^k = m \qquad (P^{\lambda_j} < x_j < 2P^{\lambda_j}).$$

Observe that

$$\sum_{m \leq Y} F(m)^2 = H$$

in the terminology of Definition 6.1, while

$$P^{\lambda_1 + \cdots + \lambda_h} \ll \sum_{m \leq Y} F(m) \leq \left\{ \sum_{m \leq Y} F(m)^2 \right\}^{1/2} \left\{ \sum_{\substack{m \leq Y \\ F(m) > 0}} 1 \right\}^{1/2}$$

$$\leq H^{1/2} Q^{1/2} \ll P^{(\lambda_1 + \cdots + \lambda_h + \varepsilon)/2} P^{k/2}$$

giving $\lambda_1 + \cdots + \lambda_h \leq k + \varepsilon$, when $P \to \infty$. Since ε is an arbitrary positive number, (6.3) follows.

Throughout this chapter we write $\theta = 1 - (1/k)$. It is quite easy to show that

$$\lambda_j = \theta^{j-1} \qquad (j = 1, \ldots, h)$$

are permissible exponents, using the following lemma.

Lemma 6.1. *Let $\mu_1, \ldots, \mu_r$ be positive numbers $(2 \leq r \ll 1)$ with $\mu_j \leq \theta \mu_{j-1}$ for all $j > 1$. Let $P > 1$. Then for given z, the number of solutions of*

$$m_1 y_1^k + \cdots + m_r y_r^k = z \tag{6.4}$$

(with given nonzero $m_1, \ldots, m_r$ in $[-M, M]$) satisfying

$$P^{\mu_j} < y_j < 2P^{\mu_j}$$

is

$$\ll M^{r-1}/|m_1 \cdots m_{r-1}|.$$

Proof. Let

$$Y_1, \ldots, Y_r$$

be any solution of (6.4). Then for any other solution $y_1, \ldots, y_r$ of (6.4), we have

$$\lambda = |m_1(y_1^k - Y_1^k)| \geq m_1 |y_1 - Y_1| k P^{\mu_1(k-1)}.$$

We note that λ may also be written

$$\lambda = |m_2 y_2^k + \cdots + m_r y_r^k - (m_2 Y_2^k + \cdots + m_r Y_r^k)| \ll MP^{\mu_2 k}.$$

Thus

$$y_1 - Y_1 \ll P^{\mu_2 k - \mu_1(k-1)} M |m_1|^{-1}$$
$$\ll M |m_1|^{-1}.$$

There are thus $\ll M |m_1|^{-1}$ possibilities for y_1. Once y_1 has been specified, we repeat the argument; there are at most

$$\ll \frac{M}{|m_2|}$$

possibilities for y_2, and so on. The proof is easily completed.

Corollary. $\lambda_j = \theta^{j-1}$ $(j = 1, \ldots, h)$ *are permissible exponents.*

Proof. There are

$$\ll P^{\lambda_1 + \cdots + \lambda_h}$$

possibilities for $x_1, \ldots, x_h$ in the solutions of (6.1) satisfying (6.2). Once $x_1, \ldots, x_h$ have been specified, the number of possible $y_1, \ldots, y_h$ is

$$\ll M^h / |m_1 \cdots m_h|$$

by Lemma 6.1, and the result follows.

A rather more sophisticated argument gives a slightly larger Λ. It is adapted from Vaughan (1977); see also Vaughan (1981), Theorem 6.1.

Lemma 6.2. *Let* $h \geq 3$, $\lambda_1 = 1$,

$$\lambda_2 = \frac{k^2 - \theta^{h-3}}{k^2 + k - k\theta^{h-3}}, \qquad \lambda_j = \frac{k^2 - k - 1}{k^2 + k - k\theta^{h-3}} \theta^{j-3} \quad (3 \leq j \leq h).$$

Then $\lambda_1, \ldots, \lambda_h$ *are permissible exponents.*

Proof. Let $P \geq M \geq 1$. Let H_s denote the number of solutions of

$$m_1 x_1^k + \cdots + m_s x_s^k = m_1 y_1^k + \cdots + m_s y_s^k \tag{6.5}$$

with $P^{\lambda_j} < x_j$, $y_j < 2P^{\lambda_j}$, and $x_s \neq y_s$. (Here $m_1, \ldots, m_h$ are given nonzero integers in $[-M, M]$.) It is clear that, since $H_1 = 0$,

$$H \ll \sum_{s=2}^{h} H_s P^{\lambda_{s+1} + \cdots + \lambda_h} + P^{\lambda_1 + \cdots + \lambda_h}, \tag{6.6}$$

where H is defined in Definition 6.1.

Now H_2 is the number of solutions of

$$m_1(x_1^k - y_1^k) = -m_2(x_2^k - y_2^k)$$

with $x_2 \neq y_2$ and $P^{\lambda_j} < x_j, y_j < 2P^{\lambda_j}$. For each given pair x_2, y_2 with $x_2 \neq y_2$, the number of possible choices for x_1, y_1 is $\ll P^\varepsilon$. (After all,

$$x_1 - y_1, \quad x_1^{k-1} + x_1^{k-2}y_1 + \cdots + y_1^{k-1}$$

are divisors of $m_2(x_2^k - y_2^k)$, and once they have been specified it is clear that there is at most one possible pair x_1, y_1.) Hence

$$H_2 \ll P^{2\lambda_2 + \varepsilon} \ll P^{\lambda_1 + \lambda_2} \tag{6.7}$$

(we may suppose that ε is small).

For $s \geq 3$,

$$H_s = H_s' + 2H_s'' \tag{6.8}$$

where H_s' is the number of solutions of (6.5) with the additional constraint $x_1 = y_1$, and H_s'' is the number with $x_1 > y_1$. Thus

$$H_s' \ll P^{\lambda_1} L_s, \tag{6.9}$$

where L_s is the number of solutions of

$$m_2 x_2^k + \cdots + m_s x_s^k = m_2 y_2^k + \cdots + m_s y_s^k.$$

Now we may apply Lemma 6.1 to obtain

$$L_s \ll P^{\lambda_2 + \cdots + \lambda_s} \frac{M^{s-1}}{|m_2 \cdots m_s|} \tag{6.10}$$

since $\lambda_3 < \theta \lambda_2$.

Now we turn to H_s''. The number of choices for x_2, y_2 is $\ll P^{2\lambda_2}$. For any such choice, (6.5) becomes

$$m_1(x_1^k - y_1^k) + A + \sum_{j=3}^{s} m_j(x_j^k - y_j^k) = 0 \tag{6.11}$$

where A is fixed. Let $l = x_1 - y_1$. Then $x_1^k - y_1^k > lP^{k-1}$. Also

$$A + \sum_{j=3}^{s} m_j(x_j^k - y_j^k) \ll MP^{k\lambda_2}. \tag{6.12}$$

Hence $0 < l \ll P^{k\lambda_2 - k + 1} M / |m_1|$.

Now (6.11) implies

$$A + m_1(y_1 + l)^k - m_1 y_1^k \ll MP^{k\lambda_3}. \tag{6.13}$$

For a given l, let y and $y + j$ be two possible values of y_1 for which (6.13) holds. Then

$$(y + j + l)^k - (y + j)^k - (y + l)^k + y^k \ll MP^{k\lambda_3} |m_1|^{-1},$$

whence

$$ljP^{k-2} \ll MP^{k\lambda_3}|m_1|^{-1}.$$

Thus the number of possible choices for y_1 is

$$\ll 1 + MP^{k\lambda_3-k+2}l^{-1}|m_1|^{-1}. \tag{6.14}$$

For given x_1, y_1, (6.11) becomes

$$A_1 + \sum_{j=3}^{s} m_j(x_j^k - y_j^k) = 0 \tag{6.15}$$

where A_1 is fixed. The number of possible choices for $y_3, \ldots, y_{s-1}$ is $\ll P^{\lambda_3+\cdots+\lambda_{s-1}}$ and, for any such choice, the number of choices for $x_3, \ldots, x_{s-1}$ is

$$\ll \frac{M^{s-3}}{|m_3 \cdots m_{s-1}|}.$$

(This is proved just like Lemma 6.1.)

Given $y_3, \ldots, y_{s-1}, x_3, \ldots, x_{s-1}$, (6.15) becomes

$$A_2 + m_s(x_s^k - y_s^k) = 0$$

where A_2 is fixed, and since $x_s \neq y_s$ the number of choices for x_s, y_s is $\ll P^{\varepsilon/2}$. Therefore, by (6.14),

$$H_s'' \ll \frac{M^{s-3}P^{2\lambda_2}}{|m_3 \cdots m_{s-1}|} \sum_{0 < l \ll MP^{k\lambda_2-k+1}/|m_1|} \left(1 + \frac{M}{|m_1|}P^{k\lambda_3-k+2}l^{-1}\right)P^{\lambda_3+\cdots+\lambda_{s-1}+\varepsilon/2}$$

Thus by (6.8)–(6.10),

$$H_s \ll \frac{M^s}{|m_1 \cdots m_s|}\{P^{\lambda_1+\cdots+\lambda_s} + P^{2\lambda_2}(P^{k\lambda_2-k+1} + P^{k\lambda_3-k+2})P^{\lambda_3+\cdots+\lambda_{s-1}+\varepsilon}\}.$$

It is easy to check that, for $s = 3, \ldots, h$,

$$(k+1)\lambda_2 - k \leq \lambda_s; \qquad \lambda_2 + k\lambda_3 - k + 1 \leq \lambda_s.$$

Hence

$$H_s \ll \frac{M^s}{|m_1 \cdots m_s|}P^{\lambda_1+\cdots+\lambda_s+\varepsilon}. \tag{6.16}$$

The lemma follows from (6.6) and (6.16).

The other ingredient of our analysis is a lemma of Vinogradov, which Vaughan ((1981), Lemma 5.4) proves by means of the large sieve: the letter p denotes a prime variable.

Lemma 6.3. *Let*

$$V(\alpha) = \sum_{X/2 < p \leq X} \sum_{|y| \leq Y} b_y e(\alpha p^k y)$$

where the b_y are arbitrary complex numbers. Suppose that a, q are coprime integers with

$$|q\alpha - a| \leq (2X^k)^{-1}, \quad X < q \leq 2X^k \ll Y. \tag{6.17}$$

Then

$$V(\alpha) \ll \left(XY^{1+\delta} \sum_{|y| \leq Y} |b_y|^2 \right)^{1/2}.$$

6.2   A theorem about αn^k

The following theorem is due to Baker (1981*b*) although it is closely related to work of Vinogradov (1954), Chapter 5.

Theorem 6.1. *Let $\lambda_1, \ldots, \lambda_h$ be permissible exponents with $\Lambda = \lambda_1 + \cdots + \lambda_h - (k-1) > 0$. Then for $\varepsilon > 0$, $N > C_1(k, h, \varepsilon)$ and any real α, there is a natural number $n \leq N$ satisfying*

$$\min_{1 \leq n \leq N} \|\alpha n^k\| < N^{\varepsilon - \Lambda/(4h + 2\theta + \Lambda k^{-1})}.$$

Proof. We may suppose that ε is small as a function of $k, \lambda_1, \ldots, \lambda_h$. Let $N > C_1(k, h, \varepsilon)$. We define Δ by

$$2\Delta = N^{\varepsilon - \Lambda/(4h + 2\theta + \Lambda k^{-1})} > N^{-1/4h}$$

(the last inequality by (6.3)).
   Suppose if possible that

$$\|\alpha n^k\| \geq 2\Delta \quad (n = 1, \ldots, N).$$

We write $M = [\Delta^{-(1+\delta)}]$ and

$$R = N^{1/2} M^{1/2k}, \quad P = N^{1/2} M^{-1/2k}.$$

Then $R > P > 1$. Let $r = [2h/\delta] + 1$.
   Let $\psi(x)$ be the function defined (using r, Δ) in Lemma 2.1 and define

$$\chi(z) = 2\Delta - \psi(z).$$

Then $\chi(z) = 2\Delta$ for $\|z\| \geq 2\Delta$ and

$$\chi(z) = \sum_{\substack{m=-\infty \\ m \neq 0}}^{\infty} b(m) e(mx),$$

where

$$|b(m)| \ll \min(\Delta, |m|^{-r-1} \Delta^{-r}), \tag{6.18}$$

so that

$$\sum_{|m|>M} |b(m)| \ll (\Delta M)^{-r} \ll \Delta^{2h}. \qquad (6.19)$$

Also, just as in (3.24),

$$\sum_{m=-\infty}^{\infty} |b(m)| \ll 1. \qquad (6.20)$$

Since $RP = N$, we have

$$\chi(\alpha p^k x_j^k) = 2\Delta$$

whenever

$$\tfrac{1}{4}R < p < \tfrac{1}{2}R, \qquad P^{\lambda_j} < x_j < 2P^{\lambda_j}.$$

Consequently,

$$S_j(p) > \Delta P^{\lambda_j}, \quad \text{for} \quad \tfrac{1}{4}R < p < \tfrac{1}{2}R, \qquad j = 1, \ldots, h,$$

where

$$S_j(p) = \sum_{P^{\lambda_j} < x_j < 2P^{\lambda_j}} \chi(\alpha p^k x_j^k).$$

Let

$$L = \sum_{\frac{1}{4}R < p < \frac{1}{2}R} \prod_{j=1}^{h} S_j(p).$$

Since R is large, we have

$$L > \Delta^h P^{\lambda_1 + \cdots + \lambda_h} R^{1-\delta}. \qquad (6.21)$$

On the other hand, we have

$$S_j(p) = \sum_{m_j \neq 0} b(m_j) \sum_{P^{\lambda_j} < x_j < 2P^{\lambda_j}} e(p^k m_j x_j^k),$$

so that

$$L = \sum_{m_1 \neq 0} \cdots \sum_{m_h \neq 0} b(m_1) \cdots b(m_h) T(m_1, \ldots, m_h), \qquad (6.22)$$

where

$$T(m_1, \ldots, m_h) = \sum_{\frac{1}{4}R < p < \frac{1}{2}R} \sum_{P^{\lambda_j} < x_j < 2P^{\lambda_j}} e(\alpha p^k (m_1 x_1^k + \cdots + m_h x_h^k)).$$

It is clear that

$$|T(m_1, \ldots, m_h)| < RP^{\lambda_1 + \cdots + \lambda_h}. \qquad (6.23)$$

From (6.19), (6.20), (6.23), the contribution to the sum in (6.22) from those $m_1, \ldots, m_h$, for which *any* of $|m_1|, \ldots, |m_h|$ exceeds M, has modulus

$$\ll \Delta^{2h} RP^{\lambda_1 + \cdots + \lambda_h}.$$

We deduce from (6.18), (6.21) that

$$\sum_{0<|m_1|\leq M} \cdots \sum_{0<|m_h|\leq M} |T(m_1,\ldots,m_h)| \gg P^{\lambda_1+\cdots+\lambda_h}R^{1-\delta}. \qquad (6.24)$$

By Dirichlet's theorem, there is a natural number q such that

$$q \leq 2^{1-k}R^k, \qquad |q\alpha - a| \leq 2^{k-1}R^{-k}, \qquad \text{where } (a, q) = 1.$$

If $q \leq R$, then

$$\|\alpha q^k\| \leq q^{k-1}\|\alpha q\| \leq 2^k R^{(k-1)-k} \leq \Delta,$$

which is contrary to hypothesis. Thus we must have $q > R$.

We rewrite $T(m_1,\ldots,m_h)$ in the form

$$T(m_1,\ldots,m_h) = \sum_{\frac{1}{4}R<p<\frac{1}{2}R} \sum_{|n|\leq Y} \varphi_n e(\alpha p^k n)$$

where $Y = hM(2P)^k$; φ_n is the number of sets $x_1,\ldots,x_h$ satisfying

$$m_1 x_1^k + \cdots + m_h x_h^k = n, \qquad P^{\lambda_j} < x_j < 2P^{\lambda_j}.$$

Thus $\sum_n \varphi_n^2$ is the number of solutions of (6.1) satisfying (6.2). Since $\lambda_1,\ldots,\lambda_h$ are permissible exponents and N is large, we have

$$\sum_{|n|\leq Y} \varphi_n^2 \leq P^{\lambda_1+\cdots+\lambda_h+\delta}M^h/|m_1\cdots m_h|.$$

We apply Lemma 6.3 with $X = R/2$, so that (6.17) is satisfied. After all,

$$2X^k < R^k = MP^k \leq Y.$$

We obtain

$$T(m_1,\ldots,m_h) \ll \left(RY^{1+\delta}\sum_{|n|\leq Y}\varphi_n^2\right)^{1/2}$$
$$\ll R^{(1/2)+\eta}M^{(h+1)/2}P^{(k+\lambda_1+\cdots+\lambda_h)/2}|m_1\cdots m_h|^{-1/2}.$$

Summing over $m_1,\ldots,m_h$, we obtain

$$\sum_{0<|m_1|\leq M} \cdots \sum_{0<|m_h|\leq M} |T(m_1,\ldots,m_h)|$$
$$\ll M^{(h+1)/2}R^{(1/2)+\eta}P^{(k+\lambda_1+\cdots+\lambda_h)/2}\left(\sum_{0<|m|\leq M}m^{-1/2}\right)^h$$
$$\ll M^{h+(1/2)}R^{(1/2)+\eta}P^{(k+\lambda_1+\cdots+\lambda_h)/2}.$$

Comparing this with (6.24),

$$P^{\lambda_1+\cdots+\lambda_h}R^{1-\delta} \ll M^{h+(1/2)}R^{(1/2)+\eta}P^{(k+\lambda_1+\cdots+\lambda_h)/2}. \qquad (6.25)$$

We might pause at this point to observe that we have been unharmed by the factor $M^h/|m_1 \cdots m_h|$. We'd have obtained the same bound (6.25) if $M^h/|m_1 \cdots m_h|$ had been replaced by 1.

Rearranging (6.25), we find that

$$\Delta^{h+(1/2)-(2-\Lambda)/4k} \ll N^{-(\Lambda/4)+\eta}.$$

This contradicts the definition of Δ, and the theorem is proved.

6.3 Conclusion

Theorem 6.2. *Let $k \geq 10$ and let $N > C_2(k)$. Then for any real α there is a natural number $n \leq N$ such that*

$$\|\alpha n^k\| < N^{-1/U(k)},$$

where $U(k) = 4k (\log k + \log \log k + 3)$.

Proof. We apply Theorem 6.1, with $\lambda_j = \theta^{j-1}$ $(j = 1, \ldots, h)$ for simplicity. We have

$$\Lambda = 1 - k\theta^h. \tag{6.26}$$

We take

$$h = [\log (4k \log k)] + 1.$$

Then

$$k\theta^h < ke^{-h/k} < 1/(4 \log k),$$

while

$$4h + 2\theta + \Lambda k^{-1} < 4k \log (4k \log k) + 6$$
$$< 4k(\log k + \log \log k + 1\cdot55),$$

$$\frac{4h + 2\theta + \Lambda k^{-1}}{1 - k\theta^h} < 4k(\log k + \log \log k + 1\cdot55)\left(1 + \frac{1}{3 \log k}\right)$$

$$< 4k(\log k + \log \log k + 3)$$

as required.

It is of interest to be more precise for small k. If we take $\lambda_1, \ldots, \lambda_h$ as in Lemma 6.2 we have

$$\Lambda = 1 - k\left(\frac{k^3 - 3k^2 + k + 2}{k^3 + k^2 - k^2\theta^{h-3}}\right)\theta^{h-3}$$

instead of (6.26). For example, we obtain $\min_{1 \leq n \leq N} \|\alpha n^9\| < N^{-1/159}$ for $N > C_3$, by taking $h = 31$.

It does not seem to be possible to reach the exponent $-1/128$ for $k = 8$ by the methods of this chapter. However, there is scope for further work. Lemma 6.3 has been slightly sharpened by Thanigasalam (1982), while ideas in a recent paper of Vaughan (1985) should lead to a sharpening of Lemma 6.2.

7
Schmidt's lattice method

7.1 Introduction

In this chapter we prove

Theorem 7.1 *Let* $\alpha_1, \ldots, \alpha_h, \beta_1, \ldots, \beta_h$ *be real numbers. Let* $N > C_1(h, \varepsilon)$. *Then there is a natural number* $n \leq N$ *such that*

$$\|n^2\alpha_i + n\beta_i\| < N^{\varepsilon - 1/(h^2+h)} \qquad (i = 1, \ldots, h).$$

Schmidt (1977*b*) proved the case $\beta_1 = \cdots = \beta_h = 0$, which contains all the key ideas, and gave a weaker version of Theorem 7.1 with the exponent $\varepsilon - 1/(2h^2)$. Baker (1978), (1980) proved Theorem 7.1.

In the next chapter we consider inequalities of the form

$$1 \leq n \leq N, \|f_i(n)\| < N^{\varepsilon - \sigma(h)} \qquad (i = 1, \ldots, h)$$

where $f_1, \ldots, f_h$ are polynomials of degree $\leq k$ without constant term. We shall use the lemmas proved in the present chapter, and generalize some of the procedures. In fact we refer back to the present chapter throughout the rest of our studies of Diophantine inequalities modulo one.

We recall some notions from the geometry of numbers. A convenient reference is Cassels (1959). Let E be a subspace of Euclidean space $\mathbb{R}^h$ having dimension t, $1 \leq t \leq h$. It is convenient to specify an orthonormal basis

$$\mathbf{x}_1, \ldots, \mathbf{x}_t$$

in E. The definitions given below are easily seen to be independent of the choice of $\mathbf{x}_1, \ldots, \mathbf{x}_t$.

The *determinant* $\det(\mathbf{y}_1, \ldots, \mathbf{y}_t)$ of t points $\mathbf{y}_1, \ldots, \mathbf{y}_t$ in E is defined to be $|\det A|$, where A is the linear mapping satisfying

$$A\mathbf{x}_i = \mathbf{y}_i \qquad (i = 1, \ldots, t).$$

A *lattice* in E is a subgroup Λ of E of the form

$$\Lambda = \{m_1\mathbf{z}_1 + \cdots + m_t\mathbf{z}_t : m_i \in \mathbb{Z}\} \tag{7.1}$$

where $\mathbf{z}_1, \ldots, \mathbf{z}_t$ are given linearly independent points of E. When the

relation (7.1) holds we say that $\mathbf{z}_1, \ldots, \mathbf{z}_t$ is a *basis* of Λ or *generates* Λ.
The *determinant* of Λ is

$$d(\Lambda) = \det(\mathbf{z}_1, \ldots, \mathbf{z}_t)$$

(it is independent of the choice of basis of Λ).

Let $\mathbf{w}_1, \ldots, \mathbf{w}_t$ be linearly independent points in Λ. The lattice Γ
generated by $\mathbf{w}_1, \ldots, \mathbf{w}_t$ is a subgroup of Λ, whose index in Λ is

$$v = [\Lambda:\Gamma] = d(\Gamma)/d(\Lambda).$$

It follows that

$$v\Lambda = \{v\mathbf{l}:\mathbf{l} \in \Lambda\} \subset \Gamma. \tag{7.2}$$

We observe here that, as a consequence of the last paragraph,

$$\det(\mathbf{w}_1, \ldots, \mathbf{w}_t) \geqq d(\Lambda) \tag{7.3}$$

for any linearly independent points $\mathbf{w}_1, \ldots, \mathbf{w}_t$ in Γ.

The open unit ball in $\mathbb{R}^h$ will be denoted by K_0 and its closure by $\bar{K}_0$.

Theorem 7.2   *Let $N > C_2(h, \varepsilon)$. Let Λ be a lattice in $\mathbb{R}^h$ such that*

$$K_0 \cap \Lambda = \{\mathbf{0}\} \tag{7.4}$$

and

$$d(\Lambda)^{h+1+\varepsilon} \leqq N. \tag{7.5}$$

Then for any $\mathbf{a}_1, \mathbf{a}_2$ in R^h there is a natural number n satisfying

$$n \leqq N, \qquad n^2\mathbf{a}_2 + n\mathbf{a}_1 \in K_0 + \Lambda. \tag{7.6}$$

The rest of the chapter will be devoted to proving Theorem 7.2.

Corollary.   *Let $N > C_2(h, \varepsilon)$. Let reals $\alpha_1, \ldots, \alpha_h, \beta_1, \ldots, \beta_h$ be given.
Let $0 < \psi_1, \ldots, \psi_h < 1$ with*

$$\psi_1 \cdots \psi_h \geqq N^{-1/(h+1+\varepsilon)}.$$

Then there is a natural number $n \leqq N$ satisfying

$$\|n^2\alpha_i + n\beta_i\| < \psi_i \qquad (i = 1, \ldots, h). \tag{7.7}$$

Proof.   The lattice Λ in $\mathbb{R}^h$ consisting of

$$(\psi_1^{-1}m_1, \ldots, \psi_h^{-1}m_h) \qquad (m_i \in \mathbb{Z})$$

has determinant $(\psi_1 \cdots \psi_h)^{-1}$. By Theorem 7.2 there is an $n \leqq N$ with

$$n^2(\psi_1^{-1}\alpha_1, \ldots, \psi_h^{-1}\alpha_h) + n(\psi_1^{-1}\beta_1, \ldots, \psi_h^{-1}\beta_h) \in K_0 + \Lambda$$

and (7.7) follows at once.

Theorem 7.1 is, of course, an immediate consequence of the corollary.

I cannot see how to prove Theorem 7.1 without the use of general lattices.

7.2 Successive minima

Let Λ be a lattice in E where dim $E = t$.

A lattice point $\mathbf{l}$ in Λ is called *primitive* if it is part of a basis of Λ. Every lattice point $\mathbf{l}$ is a multiple of a primitive point:

$$\mathbf{l} = m\mathbf{l}^* \qquad (m \text{ a natural number, } \mathbf{l}^* \text{ primitive}).$$

(Cassels (1959), Corollary 3, p. 14). The set of primitive points of Λ is denoted by Λ^*.

A *fundamental parallelepiped* of Λ is a set

$$\mathscr{F} = \{x_1\mathbf{l}_1 + \cdots + x_t\mathbf{l}_t : 0 \leq x_1, \ldots, x_t < 1\}$$

where $\mathbf{l}_1, \ldots, \mathbf{l}_t$ is a basis of Λ.

For each integer j with $1 \leq j \leq h$, let λ_j be the infimum of all $\lambda \geq 0$ such that $\lambda \bar{K}_0$ contains j linearly independent points of Λ. Clearly each λ_j is actually a minimum, and

$$0 < \lambda_1 \leq \lambda_2 \leq \cdots \leq \lambda_t < \infty.$$

We call $\lambda_1, \ldots, \lambda_t$ the successive minima of Λ (with respect to $\bar{K}_0$). We constantly use Minkowski's inequalities

$$d(\Lambda) \ll \lambda_1 \cdots \lambda_t \ll d(\Lambda) \qquad (7.8)$$

(Cassels (1959), Chapter VIII. The implied constants depend only on t).

The following definition applies only when Λ is a lattice in $\mathbb{R}^h$; that is, $t = h$. The *polar lattice* to Λ is the lattice Π of points $\mathbf{p}$ having $\mathbf{pl} \in \mathbb{Z}$ for every $\mathbf{l} \in \Lambda$. We have the relation

$$d(\Pi) = d(\Lambda)^{-1} \qquad (7.9)$$

(Cassels (1959), p. 24). We also need Mahler's inequalities

$$1 \ll \lambda_i \pi_{h+1-i} \ll 1 \qquad (7.10)$$

for the successive minima $\pi_1, \ldots, \pi_h$ of Π (Cassels (1959), Chapter VIII).

For the remainder of this chapter and throughout Chapter 8, Λ will denote a lattice satisfying (7.4), and we write $\Delta = d(\Lambda)$ and define Π as above.

Lemma 7.1. *The number of $\mathbf{p}$ in Π with $0 < |\mathbf{p}| \leq B$ is $< C_3(h) \Delta \max(B, B^h)$.*

Proof. The hypothesis (7.4) implies that $\lambda_i \geq 1$ $(i = 1, \ldots, h)$, from which we obtain

$$\pi_j \ll 1 \qquad (j = 1, \ldots, h) \tag{7.11}$$

in view of (7.10).

If $B < \pi_1$, then there is no $\mathbf{p}$ with $0 < |\mathbf{p}| \leq B$. Let $B \geq \pi_1$, and let j be the greatest integer for which $\pi_j \leq B$. From the definition of the successive minima we find that there are linearly independent points $\mathbf{p}_1, \ldots, \mathbf{p}_h$ in Π with $|\mathbf{p}_i| = \pi_i$. Let S_j be the subspace spanned by $\mathbf{p}_1, \ldots, \mathbf{p}_j$. Then

$$\Pi_j = \Pi \cap S_j$$

is a lattice in S_j. Now $|\mathbf{p}| \leq B$, $\mathbf{p} \in \Pi$ implies that $\mathbf{p} \in \Pi_j$. (This follows from $B < \pi_{j+1}$ if $j < h$ and is trivial if $j = h$.) Thus $\pi_1, \ldots, \pi_j$ are the successive minima of Π_j. By Minkowski's inequalities,

$$\pi_1 \cdots \pi_j \ll d(\Pi_j) \ll \pi_1 \cdots \pi_j. \tag{7.12}$$

Moreover, Π_j has a fundamental parallelepiped $\mathcal{F}_j$ of diameter $\ll \pi_j$. (This is a rather easy deduction from Cassels (1959), Corollary 2, p. 13.)

The number of lattice points of length $\leq B$ is at most the number of translates of $\mathcal{F}_j$ by points of Π_j which intersect the ball $|\mathbf{p}| \leq B$. This number is

$$\ll d(\Pi_j)^{-1}(B + \mathrm{diam}\, \mathcal{F}_j)^j$$
$$\ll (\pi_1 \cdots \pi_j)^{-1}(B^j + \pi_j^j) \ll (\pi_1 \cdots \pi_h)^{-1}B^j$$
$$\ll d(\Pi)^{-1}B^j = \Delta B^j \leq \Delta \max(B, B^h)$$

in view of (7.12), (7.11), Minkowski's inequalities, and (7.9). This proves Lemma 7.1.

7.3   An auxiliary function

We begin with a few simple remarks about Fourier analysis on a lattice. For a complex-valued function g in $L^1(U_h)$ of period one in each variable, one writes

$$\hat{g}(\mathbf{n}) = \int_{U_h} g(\mathbf{y})e(-\mathbf{n}\mathbf{y})\, \mathrm{d}\mathbf{y} \qquad (\mathbf{n} \in \mathbb{Z}^h).$$

From Parseval's relation (Zygmund (1968), Chapter 17, (1.7)) we know that $\hat{g} = 0$ implies $g = 0$ almost everywhere, at least for $g \in L^2(U_h)$. By applying this to

$$g(\mathbf{x}) = h(\mathbf{x}) - \sum_{\mathbf{n} \in \mathbb{Z}^h} \hat{h}(\mathbf{n})e(\mathbf{n}\mathbf{x}),$$

we easily prove that

$$h(\mathbf{x}) = \sum_{\mathbf{n} \in \mathbb{Z}^h} \hat{h}(\mathbf{n})e(\mathbf{n}\mathbf{x}) \qquad (\mathbf{x} \in \mathbb{R}^h) \tag{7.13}$$

whenever h is continuous, of period one in each variable, and $\sum_{\mathbf{n}} |\hat{h}(\mathbf{n})| < \infty$.

If we now apply a linear transformation to the variable $\mathbf{x}$ in (7.13) we come up with the following result.

Lemma 7.2. *Let f be a continuous function on $\mathbb{R}^h$ periodic with respect to Λ, that is*

$$f(\mathbf{y} + \mathbf{l}) = f(\mathbf{y}) \qquad (\mathbf{y} \in \mathbb{R}^h, \mathbf{l} \in \Lambda).$$

Let $\mathscr{F}$ be a fundamental parallelepiped of Λ. Let us write

$$c_{\mathbf{p}} = \Delta^{-1} \int_{\mathscr{F}} f(\mathbf{y}) e(-\mathbf{p}\mathbf{y}) \, d\mathbf{y} \qquad (\mathbf{p} \in \Pi).$$

If $\sum_{\mathbf{p} \in \Pi} |c_{\mathbf{p}}| < \infty$, then

$$f(\mathbf{x}) = \sum_{\mathbf{p} \in \Pi} c_{\mathbf{p}} e(\mathbf{p}\mathbf{x}) \qquad (\mathbf{x} \in \mathbb{R}^h).$$

We now derive a lattice analogue of the construction in Section 2.1.

Lemma 7.3. *Let $r \geq 2$ be a natural number. There exists a real valued function $\psi(\mathbf{x})$ with*

(i) $\psi(\mathbf{x} + \mathbf{l}) = \psi(\mathbf{x})$ $(\mathbf{l} \in \Lambda, \mathbf{x} \in \mathbb{R}^h)$;
(ii) $\psi(\mathbf{x}) = 0$ *unless* $\mathbf{x} \in K_0 + \Lambda$;
(iii) *for all $\mathbf{x}$ in $\mathbb{R}^h$,*

$$\psi(\mathbf{x}) = 2^{-h} \Delta^{-1} m(K_0) + \sum_{\substack{\mathbf{p} \in \Pi \\ \mathbf{p} \neq 0}} c_{\mathbf{p}} e(\mathbf{p}\mathbf{x}),$$

where

$$|c_{\mathbf{p}}| < C_4(h, r) \Delta^{-1} \min(1, |\mathbf{p}|^{-r}). \tag{7.14}$$

Proof. Let $\sigma = 1/(2r)$. Define

$$\psi_0(\mathbf{x}) = \begin{cases} 1 & \text{if } \mathbf{x} \in \frac{1}{2} K_0 + \Lambda \\ 0 & \text{otherwise,} \end{cases}$$

and define $\psi_1, \ldots, \psi_r$ inductively by

$$\psi_t(\mathbf{x}) = (m(K_0)\sigma^h)^{-1} \int_{|\mathbf{z}| \leq \sigma} \psi_{t-1}(\mathbf{x} + \mathbf{z}) \, d\mathbf{z}$$

$(t = 1, \ldots, r)$. Then $\psi_0(\mathbf{x})$, and hence each $\psi_t(\mathbf{x})$, is periodic with respect to Λ and has $0 \leq \psi_t(\mathbf{x}) \leq 1$. We see that $\psi_t(\mathbf{x}) = 0$ for $\mathbf{x} \notin (\frac{1}{2} + t\sigma) K_0 + \Lambda$, and hence $\psi = \psi_r$ satisfies (i) and (ii).

Define $c_{\mathbf{p}}^{(t)}$ for $t = 0, 1, \ldots, r$ and $\mathbf{p} \in \Pi$ by

$$c_{\mathbf{p}}^{(t)} = \Delta^{-1} \int_{\mathscr{F}} \psi_t(\mathbf{x}) e(-\mathbf{p}\mathbf{x}) \, d\mathbf{x}$$

($\mathscr{F}$ as in Lemma 7.2). Since $K_0 \cap \Lambda = \{0\}$, the intersection of $\frac{1}{2}K_0 + \Lambda$ with $\mathscr{F}$ has measure $m(\frac{1}{2}K_0) = 2^{-h}m(K_0)$, so that

$$c_0^{(0)} = \Delta^{-1} \int_{\mathscr{F}} \psi_0(\mathbf{x}) \, d\mathbf{x} = 2^{-h} \, \Delta^{-1} m(K_0).$$

Now for $t = 1, 2, \ldots,$

$$c_{\mathbf{p}}^{(t)} = \Delta^{-1} \int_{\mathscr{F}} \psi_t(\mathbf{x}) e(-\mathbf{px}) \, d\mathbf{x}$$

$$= \Delta^{-1} \int_{\mathscr{F}} d\mathbf{x} \, e(-\mathbf{px}) (m(K_0)\sigma^h)^{-1} \int_{|z| \leq \sigma} \psi_{t-1}(\mathbf{x} + \mathbf{z}) \, d\mathbf{z}$$

$$\ll \Delta^{-1} \int_{\mathscr{F}} d\mathbf{x} \, \psi_{t-1}(\mathbf{x}) \int_{|z| \leq \sigma} e(-\mathbf{p}(\mathbf{x} - \mathbf{z})) \, d\mathbf{z}$$

$$= c_{\mathbf{p}}^{(t-1)} \int_{|z| \leq \sigma} e(\mathbf{pz}) \, d\mathbf{z}.$$

Now if, say, p_1 is the coordinate of $\mathbf{p}$ of maximum modulus, we integrate over z_1 first in $\int e(\mathbf{pz}) \, d\mathbf{z}$. This integral over z_1 is $\ll |p_1|^{-1} \ll |\mathbf{p}|^{-1}$. We get the same estimate for $\int_{|z| \leq \sigma} e(\mathbf{pz}) \, d\mathbf{z}$. Thus

$$|c_{\mathbf{p}}^{(t)}| \ll |c_{\mathbf{p}}^{(t-1)}| \min (1, |\mathbf{p}|^{-1}).$$

Repeated application gives

$$|c_{\mathbf{p}}^{(r)}| \ll |c_{\mathbf{p}}^{(0)}| \min (1, |\mathbf{p}|^{-r}).$$
$$\ll c_0^{(0)} \min (1, |\mathbf{p}|^{-r}) \ll \Delta^{-1} \min (1, |\mathbf{p}|^{-r}).$$

Thus $\sum_{\mathbf{p}} |c_{\mathbf{p}}^{(r)}| < \infty \ (r \geq 2)$. By Lemma 7.2,

$$\psi(\mathbf{x}) = \psi_r(\mathbf{x}) = \sum_{\mathbf{p} \in \Pi} c_{\mathbf{p}}^{(r)} e(\mathbf{px}) \qquad (\mathbf{x} \in \mathbb{R}^h)$$

and ψ has all the desired properties.

Lemma 7.4. *Let $N > C_5(h, k, \varepsilon)$. Let $\mathbf{x}_1, \ldots, \mathbf{x}_N$ be a sequence in $\mathbb{R}^h$ and suppose that*

$$\mathbf{x}_n \notin K_0 + \Lambda \quad (n = 1, \ldots, N).$$

Then, provided that $\Delta \leq N^k$,

$$\sum_{\substack{\mathbf{p} \in \Pi \\ 0 < |\mathbf{p}| < N^\delta}} \left| \sum_{n=1}^{N} e(\mathbf{px}_n) \right| \gg N.$$

Proof. Let ψ be as in Lemma 7.3 with $r = h + 2 + [2k\delta^{-1}]$. We know that

$$\sum_{n=1}^{N} \psi(x_n) = 0.$$

Using the Fourier expansion of ψ,

$$N \Delta^{-1} 2^{-h} m(K_0) + \sum_{\mathbf{p} \in \Pi, \mathbf{p} \neq 0} c_{\mathbf{p}} \sum_{n=1}^{N} e(\mathbf{p} \mathbf{x}_n) = 0,$$

and consequently

$$\sum_{\mathbf{p} \in \Pi, \mathbf{p} \neq 0} |c_{\mathbf{p}}| \left| \sum_{n=1}^{N} e(\mathbf{p} \mathbf{x}_n) \right| \gg \Delta^{-1} N. \tag{7.15}$$

Now, in view of (7.14),

$$\sum_{\mathbf{p} \in \Pi; |\mathbf{p}| \geq N^{\delta}} |c_{\mathbf{p}}| \left| \sum_{n=1}^{N} e(\mathbf{p} \mathbf{x}_n) \right| \leq N \sum_{\mathbf{p} \in \Pi; |\mathbf{p}| \geq N^{\delta}} |c_{\mathbf{p}}|$$

$$\ll N \Delta^{-1} \sum_{\mathbf{p} \in \Pi; |\mathbf{p}| \geq N^{\delta}} |\mathbf{p}|^{-r}$$

$$\leq N \Delta^{-1} \sum_{j=0}^{\infty} \sum_{\mathbf{p} \in \Pi; 2^j N^{\delta} \leq |\mathbf{p}| < 2^{j+1} N^{\delta}} (2^j N^{\delta})^{-r}.$$

According to Lemma 7.1, the number of summands in the inner sum is $\ll \Delta 2^{(j+1)h} N^{\delta h}$. Thus

$$\sum_{\mathbf{p} \in \Pi; |\mathbf{p}| \geq N^{\delta}} |c_{\mathbf{p}}| \left| \sum_{n=1}^{N} e(\mathbf{p} \mathbf{x}_n) \right| \ll N^{1-\delta(r-h)} \sum_{j=0}^{\infty} 2^{-j(r-h)}$$

$$\ll N^{1-\delta(r-h)} \ll N^{-k}. \tag{7.16}$$

Now N^{-k} is of smaller order of magnitude than $N \Delta^{-1}$. Thus

$$\sum_{\substack{\mathbf{p} \in \Pi \\ 0 < |\mathbf{p}| < N^{\delta}}} |c_{\mathbf{p}}| \left| \sum_{n=1}^{N} e(\mathbf{p} \mathbf{x}_n) \right| \gg \Delta^{-1} N$$

from (7.15) and (7.16). Since $|c_{\mathbf{p}}| \ll \Delta^{-1}$, the desired result follows.

7.4 Rational approximation to inner products

The result in this section will not be needed in full generality until Chapter 8.

Lemma 7.5. *Let* $N > C_6(h, k, \varepsilon)$. *Suppose that*

$$\Delta^{K+\varepsilon} \leq N. \tag{7.17}$$

Let $\mathbf{a}_1, \ldots, \mathbf{a}_k$ *be points of* $\mathbb{R}^h$. *Suppose that there is no integer* n *satisfying*

$$1 \leq n \leq N, \, n^k \mathbf{a}_k + \cdots + n \mathbf{a}_1 \in K_0 + \Lambda. \tag{7.18}$$

Write

$$S_{\mathbf{p}} = \sum_{n=1}^{N} e(n^k (\mathbf{a}_k \mathbf{p}) + \cdots + n(\mathbf{a}_1 \mathbf{p})) \qquad (\mathbf{p} \in \Pi).$$

Then there are numbers a and B such that

$$\Delta^{-1} \ll a < N^\delta, \tag{7.19}$$

$$B \geqq N^{1-h\delta} \Delta^{-1} a^{-1}, \tag{7.20}$$

and a set $\mathcal{B}$ of primitive points of Π with the properties

$$a \leqq |\mathbf{p}| < 2a, \quad \sum_{t=1}^{[N^\delta a^{-1}]} |S_{t\mathbf{p}}| \geqq B \quad \text{for all } \mathbf{p} \in \mathcal{B}, \tag{7.21}$$

$$|\mathcal{B}| \gg N B^{-1} (\log N)^{-2}. \tag{7.22}$$

Moreover, for each $\mathbf{p}$ in $\mathcal{B}$ there are integers $q = q(\mathbf{p})$, $v_1 = v_1(\mathbf{p}), \ldots, v_k = v_k(\mathbf{p})$ satisfying

$$1 \leqq q < a^{-k} B^{-k} N^{k+2k\delta}, \tag{7.23}$$

$$(q, v_1, \ldots, v_k) = 1, \quad (q, v_2, \ldots, v_k) \leqq N^{2\delta} a^{-1}, \tag{7.24}$$

$$|q\mathbf{a}_j \mathbf{p} - v_j| < a^{-k+1} B^{-k} N^{k-j+2k\delta} \quad (j = 1, \ldots, k). \tag{7.25}$$

Proof.   By Lemma 7.4 we have

$$\sum_{0 < |\mathbf{p}| < N^\delta} |S_\mathbf{p}| \gg N.$$

It follows that

$$\sum_{\substack{\mathbf{p} \in \Pi^*; \\ 0 < |\mathbf{p}| < N^\delta}} S(\mathbf{p}) \gg N, \tag{7.26}$$

where

$$S(\mathbf{p}) = \sum_{t=1}^{[N^\delta/|\mathbf{p}|]} |S_{t\mathbf{p}}|.$$

In the sum (7.26) we have $|\mathbf{p}| \geqq \pi_1$, where π_1 is the first minimum of Π. Because of (7.11) and Minkowski's inequalities, we have

$$|\mathbf{p}| \geqq \pi_1 \gg d(\Pi)(\pi_2 \cdots \pi_k)^{-1} \gg d(\Pi) = \Delta^{-1}. \tag{7.27}$$

The interval $\Delta^{-1} \ll x < N^\delta$ is the union of $\ll \log N$ subintervals of the type $a \leqq x < 2a$, by (7.17). Thus there is an a satisfying (7.19) such that

$$\sum_{\substack{\mathbf{p} \in \Pi^*; \\ a \leqq |\mathbf{p}| < 2a}} S(\mathbf{p}) \gg N(\log N)^{-1}. \tag{7.28}$$

The sum here has $\ll \Delta \max(a, a^h) \leqq \Delta a N^{(h-1)\delta}$ summands, by Lemma 7.1. The summands with $S(\mathbf{p}) \leqq N^{1-h\delta} \Delta^{-1} a^{-1}$ give a contribution which is of small order of magnitude than the right-hand side of (7.28). Now $S(\mathbf{p}) < N^{1+\delta} |\mathbf{p}|^{-1} \leqq N^2$ by (7.27), (7.17). Dividing the interval

$$N^{1-h\delta} \Delta^{-1} a^{-1} \leqq x < N^2$$

into $\ll \log N$ subintervals of the type $B \leqq x < 2B$, we see that there is a B

satisfying (7.20) such that the set $\mathcal{B}$ of $\mathbf{p}$ in Π^* with

$$a \leq |\mathbf{p}| < 2a, \qquad B \leq S(\mathbf{p}) < 2B$$

has

$$\sum_{\mathbf{p} \in \mathcal{B}} S(\mathbf{p}) \gg N/(\log N)^2.$$

We immediately obtain (7.21) and (7.22).

It remains to deduce the rational approximations (7.23)–(7.25), using Theorem 5.1. In applying this theorem we replace ε by δ, take $M = [N^\delta a^{-1}]$ and $P = B$. The condition (5.9) is satisfied because

$$(MNP^{-1})^{K+\delta} \ll (N^{1+\delta} a^{-1} B^{-1})^{K+\delta} \ll N^{1-\delta}$$

by (7.20), (7.17). Thus for each $\mathbf{p}$ in $\mathcal{B}$ there exists a natural number $q = q(\mathbf{p})$ and integers $v_1(\mathbf{p}), \ldots, v_k(\mathbf{p})$ satisfying (7.24) and

$$q < (MNP^{-1})^k N^\delta < a^{-k} B^{-k} N^{k+2k\delta},$$
$$|q(\mathbf{a}_j\mathbf{p}) - v_j| < (MNP^{-1})^k M^{-1} N^{\delta-j} \leq a^{-k+1} B^{-k} N^{k-j+2k\delta}.$$

This completes the proof of Lemma 7.5.

7.5 A determinant argument

The hypotheses of Lemma 7.5 yield integers $q(\mathbf{p})(\mathbf{p} \in \mathcal{B})$ such that, modulo Λ, $q(\mathbf{p})\mathbf{a}_i$ is nearly orthogonal to $\mathbf{p}$ ($i = 1, \ldots, k$). It would be very helpful if $q(\mathbf{p}_j)$ took the same value, $q(\mathbf{p}_j) = q$ say, for a set of linearly independent vectors $\mathbf{p}_1, \ldots, \mathbf{p}_d$ with $d \geq 2$. We would be able to infer that, after a small shift in position, $q^k \mathbf{a}_k, \ldots, q\mathbf{a}_1$ lie in the $(h - d)$-dimensional orthogonal complement of $\mathbf{p}_1, \ldots, \mathbf{p}_d$. (We disregard a shift by lattice points). Now by an induction on dimension, we may be able to make $x^k q^k \mathbf{a}_k + \cdots + xq\mathbf{a}_1$ lie in $K_0 + \Lambda$ with $x \leq N/q$, and get a contradiction.

Our tool for ensuring $q(\mathbf{p})$ has a constant value on certain subsets of $\mathcal{B}$ is the following lemma.

Lemma 7.6. *Let $N > C_7(h, k, \varepsilon)$. Let $\mathcal{A}$ be a set of vectors in Π with $|\mathbf{p}| \leq N$ for all $\mathbf{p} \in \mathcal{A}$. Suppose that the linear span S of $\mathcal{A}$ has dimension t and that any set of t vectors in $\mathcal{A}$ have determinant $\leq Z$. Let $\mathbf{e} \in \mathbb{R}^h$. Let U, V be positive numbers such that for each $\mathbf{p}$ in $\mathcal{A}$ there are coprime integers $l(\mathbf{p}), w(\mathbf{p})$ having*

$$1 \leq l(\mathbf{p}) \leq U \leq N, \qquad |l(\mathbf{p})\mathbf{e}\mathbf{p} - w(\mathbf{p})| \leq V. \tag{7.29}$$

Suppose further that $\Delta \leq N^k$,

$$ZU^tV \, \Delta N^\delta \leq 1. \tag{7.30}$$

Then there is an integer l and a subset $\mathscr{C}$ of $\mathscr{A}$ with $|\mathscr{C}| \geq |\mathscr{A}| N^{-\delta}$, $l(\mathbf{p}) = l$ for all $\mathbf{p}$ in $\mathscr{C}$.

Proof. Let $\Pi_S = \Pi \cap S$; then Π_S is a lattice in S. Since the successive minima of Π are $\pi_1, \ldots, \pi_h \ll 1$, and the successive minima of Π_S are (say)

$$\mu_1 \geq \pi_1, \ldots, \mu_t \geq \pi_t,$$

we have

$$d(\Pi_S) \gg \mu_1 \cdots \mu_t \geq \pi_1 \cdots \pi_t \gg \pi_1 \cdots \pi_h \gg d(\Pi) = \Delta^{-1}.$$

Here we have used Minkowski's inequalities and (7.9).

Let Π_S' be the $(t+1)$-dimensional lattice of points $(\mathbf{p}, u)$ with $\mathbf{p} \in \Pi_S$ and $u \in \mathbb{Z}$; then

$$d(\Pi_S') = d(\Pi_S) \gg \Delta^{-1}. \tag{7.31}$$

Choose and fix linearly independent points $\mathbf{p}_1, \ldots, \mathbf{p}_t$ of $\mathscr{A}$ and let $\mathbf{p}_0$ be any point of $\mathscr{A}$. Write $l_i = l(\mathbf{p}_i)$, $w_i = w(\mathbf{p}_i)$. Then the points

$$(l_i\mathbf{p}_i, w_i) \qquad (i = 0, 1, \ldots, t)$$

lie in Π_S'. Their determinant D is the same as the determinant with rows

$$(l_i\mathbf{p}_i, el_i\mathbf{p}_i - w_i) \qquad (i = 0, 1, \ldots, t),$$

so that

$$D \ll ZU^t V \ll \Delta^{-1} N^{-\delta} \ll d(\Pi_S') N^{-\delta}$$

in view of the hypotheses of the lemma, (7.29), (7.30) and (7.31). Thus $D < d(\Pi_S')$, and indeed $D = 0$.

Given a basis $\mathbf{x}_1, \ldots, \mathbf{x}_t$ of Π_S, write

$$\mathbf{p}_i = c_{i1}\mathbf{x}_1 + \cdots + c_{it}\mathbf{x}_t \qquad (i = 0, 1, \ldots, t).$$

The determinant with rows

$$(l_i c_{i1}, \ldots, l_i c_{it}, w_i) \qquad (i = 0, 1, \ldots, t)$$

is also zero. It follows that $l_0 \mid w_0 Q$, where Q is the modulus of the $t \times t$ determinant with rows

$$(l_i c_{i1}, \ldots, l_i c_{it}) \qquad (i = 1, \ldots, t).$$

Indeed, $l_0 = l(\mathbf{p}_0)$ divides Q.

Now

$$|\det(c_{ij})| \det(\mathbf{x}_1, \ldots, \mathbf{x}_t) = \det(\mathbf{p}_1, \ldots, \mathbf{p}_t) \leq N^t$$

from the hypotheses. Hence

$$Q \ll UN^{2t} \det(\mathbf{x}_1, \ldots, \mathbf{x}_t)^{-1} \ll N^{2t+1} \Delta \ll N^{2t+k+1}.$$

Thus Q has fewer than N^δ divisors. Therefore there is a set $\mathscr{C}$ of $\mathbf{p}_0$ having at least $|\mathscr{A}| N^{-\delta}$ members and an integer l such that $l(\mathbf{p}_0) = l$ for all $\mathbf{p}_0$ in $\mathscr{C}$. This completes the proof of Lemma 7.6.

Lemma 7.7. *Let $h \geqq 2$, $N \geqq C_8(h, \varepsilon)$. Suppose that*

$$\Delta^{h+1+\varepsilon} \leqq N. \tag{7.32}$$

Let $\mathbf{a}_1, \mathbf{a}_2 \in \mathbb{R}^h$ and suppose (7.6) is insoluble. Then either
(i) *there is a primitive $\mathbf{p} \in \Pi$ with $|\mathbf{p}| \leqq N^\delta$ and a natural number q with*

$$q \leqq N^\eta |\mathbf{p}|^{-2}, \tag{7.33}$$

$$\|q\mathbf{a}_j\mathbf{p}\| < N^{\eta-j} |\mathbf{p}|^{-1} \qquad (j = 1, 2). \tag{7.34}$$

Or
(ii) *there are two linearly independent points $\mathbf{p}_1, \mathbf{p}_2$ in Π with $|\mathbf{p}_i| \leqq N^\delta$ $(i = 1, 2)$ and a natural number q with*

$$q \leqq N^\eta \Delta^2, \ \|q\mathbf{a}_j\mathbf{p}_i\| \leqq N^{\eta-j} \Delta^2 \qquad (i, j = 1, 2). \tag{7.35}$$

Proof. We apply Lemma 7.5 with $k = 2$. There is a set $\mathscr{B}$ of primitive points $\mathbf{p}$, and there are integers $q = q(\mathbf{p})$, $v_1 = v_1(\mathbf{p})$, $v_2 = v_2(\mathbf{p})$ such that

$$(q, v_1, v_2) = 1, \qquad (q, v_2) \leqq N^{2\delta} a^{-1}, \tag{7.36}$$

$$q < a^{-2} B^{-2} N^{2+4\delta}, \tag{7.37}$$

$$|q\mathbf{a}_j\mathbf{p} - v_j| < a^{-1} B^{-2} N^{2-j+4\delta} \qquad (j = 1, 2). \tag{7.38}$$

We also have the inequalities (7.19)–(7.22).

Let us write $s = s(\mathbf{p}) = (q, v_2)$, $r = r(\mathbf{p}) = qs^{-1}$, $v = v(\mathbf{p}) = v_2 s^{-1}$. Then we may rewrite the above relations as

$$1 \leqq rs < a^{-2} B^{-2} N^{2+4\delta} \tag{7.39}$$

$$s |r(\mathbf{a}_2\mathbf{p}) - v| < a^{-1} B^{-2} N^{4\delta} \qquad (r, v) = 1, \tag{7.40}$$

$$|s(r\mathbf{a}_1\mathbf{p}) - v_1| < a^{-1} B^{-2} N^{1+4\delta}, \qquad (s, v_1) = 1, \tag{7.41}$$

$$1 \leqq s \leqq N^{2\delta} a^{-1}. \tag{7.42}$$

There are two cases to consider. Suppose first that

$$B \geqq N^{1-2h\delta}. \tag{7.43}$$

Take any $\mathbf{p}$ in $\mathscr{B}$. Then alternative (i) holds with this choice of $\mathbf{p}$, because

$$q < a^{-2} B^{-2} N^{2+4\delta} \leqq a^{-2} N^{4h\delta} < |\mathbf{p}|^{-2} N^{5h\delta}$$

by (7.37), (7.43), (7.21), while similarly

$$\|q\mathbf{a}_j\mathbf{p}\| < a^{-1} B^{-2} N^{2-j+4\delta} < a^{-1} N^{4h\delta-j} < |\mathbf{p}|^{-1} N^{5h\delta-j}.$$

Now suppose that (7.43) is false. Write b_t for the maximum number of elements of $\mathscr{B}$ which lie in a subspace of dimension t. We have $1 \leqq b_1 \leqq 2$,

$$b_1 \leqq b_2 \leqq \cdots \leqq b_h = |\mathscr{B}|,$$

while

$$|\mathscr{B}| \gg NB^{-1}(\log N)^{-2} \gg N^{2h\delta}(\log N)^{-2} \tag{7.44}$$

because (7.43) is violated.

Suppose for a moment that for some t, $2 \leqq t \leqq h-1$, we have

$$b_t \geqq N^{2\delta}b_{t-1}.$$

Let S_t be a subspace with $b_t = |\mathscr{B} \cap S_t|$. Now we apply Lemma 7.6 with $\mathscr{A} = \mathscr{B} \cap S_t$, $\mathbf{e} = \mathbf{a}_2$, $l(\mathbf{p}) = r(\mathbf{p})$, $w(\mathbf{p}) = v(\mathbf{p})$. Evidently the linear span of $\mathscr{A}$ is S_t, for otherwise $|\mathscr{A}| \leqq b_{t-1}$. Moreover, we may take $Z = (2a)^t$ in applying the lemma. As for U and V, we may take

$$U = a^{-2}B^{-2}N^{2+4\delta}, \qquad V = a^{-1}B^{-2}N^{4\delta}$$

in view of (7.39) and (7.40). The condition (7.30) is satisfied, because

$$ZU^tV \Delta N^\delta \leqq (2a)^t(a^{-2}B^{-2}N^{2+4\delta})^t a^{-1}B^{-2}N^{5\delta} \, \Delta$$
$$\ll (a^{-2}B^{-2})^{t+1}N^{2t+\eta}\Delta$$
$$\ll \Delta^{2t+3}N^{-2+\eta} \ll \Delta^{2h+1}N^{-2+\eta} \ll 1.$$

Here we used in turn (7.19), (7.20) and (7.32).

We deduce that there is a subset $\mathscr{C}$ of $\mathscr{B} \cap S_t$ having $|\mathscr{C}| > b_{t-1}N^\delta$, and a natural number r such that $r(\mathbf{p}) = r$ for all $\mathbf{p}$ in $\mathscr{C}$.

We can get a similar conclusion if $b_t < N^{2\delta}b_{t-1}(t = 2, \ldots, h-1)$. For then we have

$$b_{h-1} \leqq b_1 N^{2(h-2)\delta} \leqq 2N^{2(h-2)\delta}.$$

The (x_1, x_2) plane may be covered by $\ll N^{1-2h\delta}B^{-1}$ angular sections σ of centre $\mathbf{0}$ and of angle $BN^{2h\delta-1}$. There is a section σ such that the set $\mathscr{B}_\sigma$ of $\mathbf{p}$ in $\mathscr{B}$, whose projection into the (x_1, x_2)-plane lies in σ, has

$$|\mathscr{B}_\sigma| \gg |\mathscr{B}| \, BN^{2h\delta-1} \gg N^{2h\delta}(\log N)^{-2}$$

in view of (7.22). Thus $|\mathscr{B}_\sigma| > N^{3\delta}b_{h-1}$, and we may apply Lemma 7.6 with $\mathscr{A} = \mathscr{B}_\sigma$. Here $\mathbf{e}$, $l(\mathbf{p})$, $w(\mathbf{p})$, U, V are as before, but now we can take

$$Z \ll a^h BN^{2h\delta-1}$$

because all points of $\mathscr{B}_\sigma$ lie in σ. We have

$$ZU^hV \Delta N^\delta \ll a^h BN^{2h\delta-1}(a^{-2}B^{-2}N^{2+4\delta})^h a^{-1}B^{-2}N^{5\delta}\Delta$$
$$\ll (a^{-1}B^{-1})^{2h+1}N^{2h-1+\eta}\Delta$$
$$\ll \Delta^{2h+2}N^{-2+\eta} \ll 1,$$

again using (7.19), (7.20) and (7.32). Thus $\mathscr{C}$ exists as before, this time with $t = h$.

We have 'fixed r' in (7.39)–(7.42), but we still want to 'fix s' (we could not 'fix q' at one shot, because we may have $(q, v_2) > 1$, which spoils the condition $(l(\mathbf{p}), w(\mathbf{p})) = 1$ in Lemma 7.6). This time we don't need to distinguish the cases $t = h$, $t < h$. We apply Lemma 7.6 again with $\mathscr{A} = \mathscr{C}$, $\mathbf{e} = r\mathbf{a}_1$, $l(\mathbf{p}) = s(\mathbf{p})$, $w(\mathbf{p}) = v_1(\mathbf{p})$. In view of (7.41), (7.42) we may take

$$U = N^{2\delta}a^{-1}, \qquad V = a^{-1}B^{-2}N^{1+4\delta}.$$

We take $Z = (2a)^t$, and find that

$$ZU^tV\Delta N^\delta \leqq (2a)^t(N^{2\delta}a^{-1})^t a^{-1}B^{-2}N^{1+5\delta}\Delta$$
$$\ll a^{-2}B^{-2}N^{1+\eta}\Delta \ll \Delta^3 N^{-1+\eta} \ll 1.$$

Thus there is a subset $\mathscr{D}$ of $\mathscr{C}$ having $|\mathscr{D}| \geqq |\mathscr{C}| N^{-\delta} > b_{t-1}$ and $s(\mathbf{p}) = s$, say, for all $\mathbf{p}$ in $\mathscr{C}$. There are certainly two linearly independent vectors $\mathbf{p}_1$, $\mathbf{p}_2$ in $\mathscr{D}$, since $t \geqq 2$, and it is easy to check that alternative (ii) of Lemma 7.7 holds with $q = rs$. This completes the proof of Lemma 7.7.

7.6 Lattices in subspaces

We write $T^\perp$ for the orthogonal complement of a subspace T of R^h.

Lemma 7.8. *Let T be a subspace of $\mathbb{R}^h$ containing t linearly independent points $\mathbf{p}_1, \ldots, \mathbf{p}_t$ of Π where $t = \dim T$. Let $\Pi' = \Pi \cap T$ and let $\Lambda' = \Lambda \cap T^\perp$. Then Π' is a lattice in T, Λ' is a lattice in $T^\perp$ (unless $t = h$) and, for $t < h$,*

$$d(\Pi')\Delta \ll d(\Lambda') \ll d(\Pi')\Delta. \tag{7.45}$$

Proof. First of all Π' is a lattice in T, because T contains $t = \dim T$ linearly independent points of Π.

Let $\lambda_1, \ldots, \lambda_t$ be the successive minima of Π' and choose linearly independent points $\mathbf{q}_1, \ldots, \mathbf{q}_t$ in Π' such that

$$|\mathbf{q}_j| = \lambda_j.$$

By Minkowski's inequalities

$$d(\Pi') \leqq \det(\mathbf{q}_1, \ldots, \mathbf{q}_t) \leqq |\mathbf{q}_1| \cdots |\mathbf{q}_t| \ll d(\Pi'). \tag{7.46}$$

Thus, if we write $v = \det(\mathbf{q}_1, \ldots, \mathbf{q}_t)/d(\Pi')$, we have

$$v \ll 1; v = 1 \quad \text{if} \quad t = 1. \tag{7.47}$$

Let $\mathbf{z}_1, \ldots, \mathbf{z}_t$ be a basis of Π' and extent it to a basis $\mathbf{z}_1, \ldots, \mathbf{z}_t$, $\mathbf{q}_{t+1}, \ldots, \mathbf{q}_h$ of Π. (This is possible—see Cassels (1959), Corollary 3, p. 14). We know that $v\Pi'$ is contained in the lattice spanned by $\mathbf{q}_1, \ldots, \mathbf{q}_t$, so $v\Pi$ is contained in the lattice spanned by $\mathbf{q}_1, \ldots, \mathbf{q}_h$.

Let $l_1, \ldots, l_h$ be chosen to satisfy

$$l_i q_j = \begin{cases} 1 & \text{if } i = j \\ 0 & \text{if } i \neq j. \end{cases}$$

We have, for any $\mathbf{p}$ in Π,

$$vl_i\mathbf{p} = l_iv\mathbf{p} \in \mathbb{Z},$$

and so $vl_1, \ldots, vl_h$ belong to Λ. Moreover, $vl_{t+1}, \ldots, vl_h$ belong to $T^\perp$, so (unless $t = h$) $\Lambda' = \Lambda \cap T^\perp$ is a lattice in $T^\perp$.

To calculate $d(\Lambda')$, we begin by noting that

$$\det (l_1, \ldots, l_h) = 1/\det (\mathbf{q}_1, \ldots, \mathbf{q}_h) \leqq d(\Pi)^{-1} = \Delta.$$

Write T_j for the space spanned by $l_j, \ldots, l_h$ and decompose l_j as

$$l_j = l_j' + l_j''(l_j' \in T_{j+1}^\perp, l_j'' \in T_{j+1}).$$

Clearly

$$\begin{aligned}
\Delta \geqq \det (l_1, \ldots, l_h) &= |l_1'| \det (l_2, \ldots, l_h) = \cdots \\
&= |l_1'| \cdots |l_t'| \det (l_{t+1}, \ldots, l_h) \\
&\gg \det (l_1', \ldots, l_t') \det (vl_{t+1}, \ldots, vl_h) \\
&\geqq \det (l_1', \ldots, l_t') \, d(\Lambda').
\end{aligned} \tag{7.48}$$

Now $l_1', \ldots, l_t', \mathbf{q}_1, \ldots, \mathbf{q}_t$ all lie in $T_{t+1}^\perp$.
Also

$$l_i'\mathbf{q}_i = l_i'\mathbf{q}_i + l_i''\mathbf{q}_i = l_i\mathbf{q}_i = 1,$$

and, if $i > j$,

$$l_i'\mathbf{q}_j = l_i'\mathbf{q}_j + l_i''\mathbf{q}_j = l_i\mathbf{q}_j = 0.$$

It follows that $\det (l_i'\mathbf{q}_j) = 1$. Thus

$$\det (l_1', \ldots, l_t') \det (\mathbf{q}_1, \ldots, \mathbf{q}_t) = 1.$$

Combining this inequality with (7.46), and (7.48),

$$d(\Lambda') \leqq \det (\mathbf{q}_1, \ldots, \mathbf{q}_t)\Delta \ll d(\Pi')\Delta.$$

On the other hand, we may reverse the roles of Λ and Π to obtain

$$d(\Pi') \ll \Delta^{-1}d(\Lambda').$$

This completes the proof of the lemma.

Lemma 7.9. *Make all the hypotheses of Lemma 7.8. There is a natural number c,*

$$c \ll \det (\mathbf{p}_1, \ldots, \mathbf{p}_t)/d(\Pi'), \tag{7.49}$$

$$c = 1 \text{ if } t = 1 \text{ and } \mathbf{p}_1 \text{ is primitive,}$$

such that, given $\mathbf{a} \in \mathbb{R}^h$, $c\mathbf{a}$ *may be written in the form*

$$c\mathbf{a} = \mathbf{l} + \mathbf{s} + \mathbf{b} \tag{7.50}$$

where $\mathbf{l} \in \Lambda$, $\mathbf{s} \in T^{\perp}$ *and*

$$|\mathbf{b}| \ll d(\Pi')^{-1}(\max |\mathbf{p}_i|)^{t-1} \max_j \|\mathbf{p}_j \mathbf{a}\|. \tag{7.51}$$

For instance, if $\mathbf{p}_1, \ldots, \mathbf{p}_t$ is a basis of Π', and all $\|\mathbf{p}_j \mathbf{a}\|$ are small, we have $c \ll 1$ and a small multiple of $\mathbf{a}$ has been shown to lie 'nearly' in $T^{\perp}$, modulo Λ.

Proof. Write $A = \max |\mathbf{p}_i|$, $P = \max_i \|\mathbf{p}_i \mathbf{a}\|$. Let $\mathbf{q}_1, \ldots, \mathbf{q}_h, v, \mathbf{l}_1, \ldots, \mathbf{l}_h$ be as in the preceding proof. Let $\mathbf{w}_1, \ldots, \mathbf{w}_t$ be an orthonormal basis of T. We write

$$\mathbf{p}_j = p_{j1}\mathbf{w}_1 + \cdots + p_{jt}\mathbf{w}_t, \quad \mathbf{q}_j = q_{j1}\mathbf{w}_1 + \cdots + q_{jt}\mathbf{w}_t.$$

There are integers c_{ij} such that

$$v\mathbf{p}_j = c_{j1}\mathbf{q}_1 + \cdots + c_{jt}\mathbf{q}_t \quad (j = 1, \ldots, t).$$

Write c for the modulus of $\det(c_{ij})$ and C_{ij} for the cofactor of c_{ij} in this determinant. Obviously,

$$v^t \det(\mathbf{p}_1, \ldots, \mathbf{p}_t) = c \det(\mathbf{q}_1, \ldots, \mathbf{q}_t).$$

In view of (7.47), we obtain (7.49) immediately.

We require the inequality

$$d(\Pi')C_{ir} \ll A^{t-1}|\mathbf{q}_r| \quad (i = 1, \ldots, t, r = 1, \ldots, t). \tag{7.52}$$

To get this, we note that for a fixed j,

$$v p_{ji} = c_{j1}q_{1i} + \cdots + c_{jt}q_{ti} \quad (i = 1, \ldots, t).$$

Solving this set of t equations for c_{js} by Cramer's rule,

$$\det(\mathbf{q}_1, \ldots, \mathbf{q}_t)c_{js} \ll A \prod_{i \neq s} |\mathbf{q}_i|,$$

or

$$c_{js} \ll A |\mathbf{q}_s|^{-1}$$

because of (7.46). Now (7.52) follows at once because $\prod_{s \neq r}(A |\mathbf{q}_s|^{-1}) \leq A^{t-1}|\mathbf{q}_r| d(\Pi')^{-1}$.

We now observe that

$$v\mathbf{p}_j\mathbf{a} = vx_j + P_j \quad (j = 1, \ldots, t)$$

where $x_j \in \mathbb{Z}$ and $P_j \ll P$. That is,

$$c_{j1}\mathbf{q}_1\mathbf{a} + \cdots + c_{jt}\mathbf{q}_t\mathbf{a} = vx_j + P_j \quad (j = 1, \ldots, t).$$

For a fixed i, we multiply the jth equation by C_{ji} and add to get

$$c\mathbf{q}_i\mathbf{a} = vy_i + V_i,$$

where $y_i \in \mathbb{Z}$, and

$$V_i \ll A^{t-1}d(\Pi')^{-1}|\mathbf{q}_i|P \tag{7.53}$$

by (7.52).

Define $\mathbf{l} = v(y_1\mathbf{l}_1 + \cdots + y_t\mathbf{l}_t)$; then $\mathbf{l} \in \Lambda$ and

$$\mathbf{q}_i(c\mathbf{a} - \mathbf{l}) = vy_i + V_i - vy_i = V_i \qquad (i = 1, \ldots, t).$$

Write

$$c\mathbf{a} - \mathbf{l} = \mathbf{b} + \mathbf{s}$$

where $\mathbf{b} \in T$ and $\mathbf{s} \in T^{\perp}$. We have

$$\mathbf{q}_i\mathbf{b} = V_i \qquad (i = 1, \ldots, t)$$

because $\mathbf{q}_i \in T$. Writing

$$\mathbf{b} = b_1\mathbf{w}_1 + \cdots + b_t\mathbf{w}_t,$$

we can set up equations

$$q_{i1}b_1 + \cdots + q_{it}b_t = V_i \qquad (i = 1, \ldots, t)$$

for the coordinates of b. Solving by Cramer's rule,

$$\det(\mathbf{q}_1, \ldots, \mathbf{q}_t)b_j = \pm(Q_{1j}V_1 + \cdots + Q_{tj}V_t)$$

where Q_{ij} is the cofactor of q_{ij} in $\det(q_{rs})$. Now

$$|Q_{ij}| \ll \prod_{l \neq i}|\mathbf{q}_l|,$$

so that

$$|b_j| \ll d(\Pi')^{-1}\sum_{i=1}^{t}|Q_{ij}||V_i|$$

$$\ll d(\Pi')^{-1}\sum_{i=1}^{t}|\mathbf{q}_1|\cdots|\mathbf{q}_t|A^{t-1}d(\Pi')^{-1}P$$

from (7.53). Using (7.46) once more, we get (7.51) and the lemma is proved.

7.7 Proof of Theorem 7.2

We proceed by induction on h. The case $h = 1$ is just a rewording of Theorem 3.2. In the induction step let $N > C_2(h, \varepsilon)$ where $h \geqq 2$. Let $\mathbf{a}_1$, $\mathbf{a}_2 \in R^h$ and suppose if possible that (7.6) is insoluble; then the conclusion of Lemma 7.7 holds.

Suppose first that we are in alternative (i). We use Lemmas 7.8 and 7.9 in the case $t = 1$, $\mathbf{p}_1 = \mathbf{p}$, $T = $ linear span of $\mathbf{p}_1$. Since $\mathbf{p}$ is primitive, $d(\Pi') = |\mathbf{p}|$ and the lattice Λ' in $T^{\perp}$ has

$$d(\Lambda') \ll |\mathbf{p}|\,\Delta.$$

We take $\mathbf{a} = q^j \mathbf{a}_j$ in Lemma 7.9. We find that

$$q^j \mathbf{a}_j = \mathbf{l}_j + \mathbf{s}_j + \mathbf{b}_j,$$

where $\mathbf{l}_j \in \Lambda$, $\mathbf{s}_j \in T^{\perp}$ and

$$|\mathbf{b}_j| \ll |\mathbf{p}|^{-1}\,\|\mathbf{p}q^j \mathbf{a}_j\| \leqq |\mathbf{p}|^{-1} q^{j-1}\,\|q\mathbf{p}\mathbf{a}_j\|$$
$$\ll (|\mathbf{p}|^{-2})^{j-1} N^{\eta-j}\,|\mathbf{p}|^{-2}. \tag{7.54}$$

We now apply Theorem 7.2 in the $(h-1)$-dimensional space $T^{\perp}$, replacing ε by δ, Λ by $2\Lambda'$, $\mathbf{a}_j$ by $2\mathbf{s}_j$, and N by $[d(2\Lambda')^h N^\delta] \geqq d(\Lambda')^{h+\delta}$. Thus there is a natural number x,

$$x \leqq d(2\Lambda')^h N^\delta \ll |\mathbf{p}|^h \Delta^h N^\delta, \tag{7.55}$$

such that

$$2x^2 \mathbf{s}_2 + 2x\mathbf{s}_1 \in 2\Lambda' + K_0,$$

which implies

$$x^2 \mathbf{s}_2 + x\mathbf{s}_1 \in \Lambda + \tfrac{1}{2}K_0.$$

Now let $n = qx$. Then by (7.33), (7.55),

$$n \ll N^\eta\,|\mathbf{p}|^{-2} \cdot |\mathbf{p}|^h \Delta^h$$
$$\ll \Delta^h N^\eta \ll N^{1-\delta} \tag{7.56}$$

since $|\mathbf{p}| \leqq N^\delta$. Thus $n \leqq N$. Moreover,

$$n^2 \mathbf{a}_2 + n\mathbf{a}_1 = x^2(\mathbf{l}_2 + \mathbf{s}_2 + \mathbf{b}_2) + x(\mathbf{l}_1 + \mathbf{s}_1 + \mathbf{b}_1). \tag{7.57}$$

Of the terms on the right in (7.57),

$$x^2 \mathbf{s}_2 + x\mathbf{s}_1 + x^2 \mathbf{l}_2 + x\mathbf{l}_1 \in \tfrac{1}{2}K_0 + \Lambda,$$

while

$$|x^2 \mathbf{b}_2 + x\mathbf{b}_1| \ll \sum_{j=1}^{2} |\mathbf{p}|^{hj}\,\Delta^{hj} N^{\eta-j}\,|\mathbf{p}|^{-2j}$$

$$\ll \sum_{j=1}^{2} (\Delta^h N^{-1+\eta})^j \ll N^{-\delta}$$

by (7.54), (7.55). Thus $x^2 \mathbf{b}_2 + x\mathbf{b}_1 \in \tfrac{1}{2}K_0$, and $n^2 \mathbf{a}_2 + n\mathbf{a}_1 \in K_0 + \Lambda$. This is a contradiction, so alternative (ii) must hold.

We now apply Lemmas 7.8 and 7.9 with $t = 2$, $T = $ linear span of $\mathbf{p}_1$, $\mathbf{p}_2$. This time, Λ' is a lattice in $T^{\perp}$ with

$$d(\Lambda') \ll D\Delta$$

where we write $D = d(\Pi')$. Note that

$$N^{2\delta} \geqq D \gg \Delta^{-1}, \qquad c \ll N^{2\delta}D^{-1}. \qquad (7.58)$$

(We argue as in the proof of Lemma 7.6 to get the inequality $D \gg \Delta^{-1}$.)
For each $j \leqq 2$, $cq^j\mathbf{a}_j$ may be written in the form

$$cq^j\mathbf{a}_j = \mathbf{l}_j + \mathbf{s}_j + \mathbf{b}_j,$$

where $\mathbf{l}_j \in \Lambda$, $\mathbf{s}_j \in T^\perp$ and

$$|\mathbf{b}_j| \ll D^{-1}N^{(h-1)\delta}q^{j-1} \max_i \|q\mathbf{p}_i\mathbf{a}_j\|$$

$$\ll D^{-1}N^{n-j}\Delta^{2j} \qquad (7.59)$$

by (7.35).

It is best to distinguish two cases here. If $h > 2$, we apply Theorem 7.2
in the $(h-2)$-dimensional space $T^\perp$, replacing ε by δ, Λ by $2\Lambda'$, $\mathbf{a}_j$ by
$2\mathbf{s}_j$, and N by $[d(2\Lambda')^{h-1}N^\delta] \geqq d(2\Lambda')^{h-1+\delta}$. Thus there is a natural
number x,

$$x \leqq d(2\Lambda')^{h-1}N^\delta \ll D^{h-1}\Delta^{h-1}N^\delta, \qquad (7.60)$$

such that (as in alternative (i))

$$x^2 c\mathbf{s}_2 + x\mathbf{s}_1 \in \Lambda + \tfrac{1}{2}K_0.$$

Now

$$n = cqx \ll N^n D^{-1}\Delta^2 \cdot D^{h-1}\Delta^{h-1}$$

$$\ll N^n \Delta^{h+1} \ll N^{1-\delta}$$

by (7.58), (7.35) and (7.32). Thus $n \leqq N$. Moreover,

$$n^2\mathbf{a}_2 + n\mathbf{a}_1 = x^2 c(\mathbf{l}_2 + \mathbf{s}_2 + \mathbf{b}_2) + x(\mathbf{l}_1 + \mathbf{s}_1 + \mathbf{b}_1)$$

with

$$|x^2 c\mathbf{b}_2 + x\mathbf{b}_1| \ll \sum_{j=1}^{2} N^n (D^{h-1}\Delta^{h-1})^j (D^{-1})^{j-1}D^{-1}N^{-j}\Delta^{2j}$$

$$\ll \sum_{j=1}^{2} (D^{h-2}\Delta^{h+1}N^{n-1})^j \ll N^{-\delta}$$

by (7.60), (7.59), (7.58) and (7.32). Just as in alternative (i) we get a
contradiction.

On the other hand, if $h = 2$, then we define

$$n = cq \ll N^{2\delta}D^{-1}\Delta^2 N^n \ll \Delta^3 N^n \ll N^{1-\delta}$$

and note that $\mathbf{s}_1 = \mathbf{s}_2 = \mathbf{0}$. Thus

$$n^2\mathbf{a}_2 + n\mathbf{a}_1 = c(\mathbf{l}_2 + \mathbf{b}_2) + \mathbf{l}_1 + \mathbf{b}_1$$

with

$$|c\mathbf{b}_2 + \mathbf{b}_1| \ll \sum_{j=1}^{2} N^{\eta}(D^{-1})^{j-1}D^{-1}N^{-j}\Delta^{2j}$$

$$\ll \sum_{j=1}^{2} (D^{-1}\Delta^2 N^{\eta-1})^j \ll N^{-\delta}$$

by (7.59), (7.58) and (7.32). Again, we obtain a contradiction. Thus (7.6) is soluble after all. This completes the induction step, and Theorem 7.2 is proved.

8
The lattice method for polynomials of arbitrary degree

8.1 An inductive strategy

In this chapter we suppose that $k \geqq 3$. The results discussed here evolved in a number of papers. Cook (1975) was the first to calculate an exponent $\sigma(h, k)$ such that the inequalities

$$1 \leqq n \leqq N, \quad \|f_i(n)\| < N^{\varepsilon - \sigma(h,k)} \quad (i = 1, \ldots, h)$$

are soluble for arbitrary polynomials $f_1, \ldots, f_h$ of degree $\leqq k$ without constant term. Here $N > C_1(h, k, \varepsilon)$. Cook had $\sigma(h, k)$ of order of magnitude $8^{-h}(k^3 \log k)^{-1}$. (Hardy and Littlewood (1914) had conjectured that some such $\sigma(h, k) > 0$ existed.) This was improved somewhat by Gajraj (1976b). After Schmidt's work (1977b) on the quadratic case, much better results were found (Baker (1977), (1978), (1980a); Baker and Harman (1984b)). We follow the latter exposition here.

Theorem 8.1. *In the above notation, we may take*

$$\sigma(h, k) = 1/(Kh)$$

for pairs h, k such that $h \geqq 1$, $k \geqq 3$,

$$(h - 1)k \leqq K - 1. \tag{8.1}$$

We may take

$$\sigma(h, k) = 1/(h^2 + h\tau)$$

for $h \geqq 6K$, where $\tau = (k + 1)/2$.

These results remain valid with K replaced by $J = 8k^2(\log k + \frac{1}{2}\log\log k + 2)$ for $k \geqq 14$. The proofs can be repeated practically verbatim, using Theorem 5.3 in place of Theorem 5.1.

We shall deduce Theorem 8.1 from a lattice theorem.

Theorem 8.2. *Let $F(0) = 1$, $F(1) = K$ and for $h = 2, 3, \ldots$ let $F(h)$ be defined recursively as follows:*

$$F(h) = \max\left(K, h + \tau, \max_{2 \leqq m \leqq h} \min\left(F(h - m) + k, mk + 1\right)\right).$$

Let $N > C_1(h, k, \varepsilon)$. Let Λ be a lattice in $\mathbb{R}^h$ with $\Lambda \cap K_0 = \{0\}$ and

$$\Delta^{F(h)+\varepsilon} \leqq N. \tag{8.2}$$

Then for any $\mathbf{a}_1, \ldots, \mathbf{a}_k$ in $\mathbb{R}^h$ there is a natural number n satisfying

$$1 \leqq n \leqq N, \qquad n^k \mathbf{a}_k + \cdots + n \mathbf{a}_1 \in K_0 + \Lambda. \tag{8.3}$$

It is not very difficult to see that

$$F(h) = K \tag{8.4}$$

whenever (8.1) holds. A little harder is:

Lemma 8.1. *We have*

$$F(h) = h + \tau \qquad for \ h \geqq 6K.$$

Proof. We show by induction on h that

$$F(h) \leqq \max\left(h + \tau, \frac{h}{2} + 3K + \tau\right) \qquad (h \geqq 0). \tag{8.5}$$

(8.5) is evidently true for $h = 0, 1$. Now for any $h \geqq 2$, either

$$F(h) \leqq \max(h + \tau, K),$$

or

$$F(h) \leqq mk + 1 \qquad \text{for some } m \leqq \min(2k - 1, h) \tag{8.6}$$

or

$$F(h) \leqq F(h - m) + k \qquad \text{for some } m \geqq 2k. \tag{8.7}$$

We need only consider (8.6), (8.7). Since

$$(2k - 1)(k - (1/2)) \leqq 3K + \tau - 1$$

as one easily verifies, we have

$$mk + 1 \leqq \frac{m}{2} + 3K + \tau \leqq \frac{h}{2} + 3K + \tau$$

for $m \leqq \min(2k - 1, h)$, so that (8.6) implies (8.5).

If (8.7) holds, then by induction

$$F(h) \leqq \max\left(h - m + \tau, \frac{h - m}{2} + 3K + \tau\right) + k$$

$$\leqq \max\left(h + \tau, \frac{h}{2} + 3K + \tau\right).$$

This proves (8.5).

Clearly $F(h) \geqq h + \tau$. Suppose now $h \geqq 6K$. Then

$$F(h) \leqq \max \left(h + \tau, \frac{h}{2} + 3K + \tau \right) = h + \tau$$

This proves Lemma 8.1.

It is a simple matter to deduce Theorem 8.1 from Theorem 8.2 using (8.4) and Lemma 8.1. We argue as in the proof of the Corollary in Section 7.1.

8.2 Lemmas on exponential sums

To prove Theorem 8.2 we have to extract more information from the hypotheses of Theorem 5.1. Namely, we require more control over common factors between q and $v_1, \ldots, v_k$ in (5.11), (5.12).

Lemma 8.2. *Let* $2 \leqq j < k$. *Let* $R \geqq 1$, $Q \geqq 1$, $q = RQ$, $u_1, \ldots, u_k$ *be integers with* $Q \mid (u_{j+1}, \ldots, u_k)$ *and* $(Q, u_j, \ldots, u_2) = D$. *Let* $F(x) = u_k x^k + \cdots + u_1 x$. *Then*

$$\sum_{x=1}^{q} e_q(F(x)) \ll q^{1+\varepsilon} D^{1/j} Q_0^{-1/j} \tag{8.8}$$

where $Q_0 = Q/(Q, R^k)$.

Proof. Write $x = mR + t$ where $t = 1, \ldots, R$ and m runs over the values $0, 1, \ldots, Q - 1$. Then, writing S for the sum in (8.8),

$$S = \sum_{t=1}^{R} \sum_{m=0}^{Q-1} e \left(\frac{u_k(mR + t)^k + \cdots + u_1(mR + t)}{RQ} \right).$$

Now

$$S = \sum_{t=1}^{R} e \left(\frac{F(t)}{RQ} \right) \sum_{m=0}^{Q-1} e \left(\frac{u_j((mR + t)^j - t^j) + \cdots + u_1((mR + t) - t)}{RQ} \right) \tag{8.9}$$

since $QR \mid u_k((mR + t)^k - t^k) + \cdots + u_{j+1}((mR + t)^{j+1} - t^{j+1})$ for integer m, t.

Let t be fixed for a moment and write

$$g(m) = R^{-1}(u_j((mR + t)^j - t^j) + \cdots + u_1((mR + t) - t))$$

Then $g(m)$ has integer coefficients $a_j, a_{j-1}, \ldots, a_1$ say. Write $d = (Q, a_2, \ldots, a_j)$ and $d_0 = d/(d, R^k)$. It may be seen by examining $a_j, a_{j-1}, \ldots, a_2$ in turn that

$$d_0 \mid u_j, \; d_0 \mid u_{j-1}, \ldots, d_0 \mid u_2$$

Thus $d_0 \mid (Q, u_2, \ldots, u_j) = D$ and

$$d = d_0(d, R^k) \leqq D(Q, R^k). \tag{8.10}$$

We find by an application of Lemma 4.2 that

$$\sum_{m=0}^{Q-1} e_Q(g(m)) \ll Q^{1-(1/j)+\varepsilon} D^{1/j}(Q, R^k)^{1/j} \tag{8.11}$$

The left-hand side of (8.11) is the inner sum in (8.9). Thus

$$|S| \ll RQ^{1-(1/j)+\varepsilon} D^{1/j}(Q, R^k)^{1/j}$$
$$\ll q^{1+\varepsilon} D^{1/j}(Q/(Q, R^k))^{-1/j}$$

This completes the proof.

Lemma 8.3. *Let $k \geqq 3$ and $N > C_2(k, \varepsilon)$. Let M be a natural number and P a positive number such that*

$$P \ll N, \qquad (MNP^{-1})^{K+\varepsilon} \leqq N. \tag{8.12}$$

Suppose that

$$\sum_{m=1}^{M} |S_m(f)| \geqq P. \tag{8.13}$$

Then there are natural numbers $q_1, q_2, \ldots, q_k$ and integers $y_2, \ldots, y_k$ satisfying

$$q = q_1 \cdots q_k \leqq (MNP^{-1})^k N^\varepsilon, \tag{8.14}$$

$$|q_k q_{k-1} \cdots q_j \alpha_j - y_j| \leqq M^{-1}(MNP^{-1})^k N^{\varepsilon-j} \quad (j = 1, \ldots, k), \tag{8.15}$$

$$(y_j, q_j) = 1 \quad (j = 1, \ldots, k). \tag{8.16}$$

Moreover, for $j = 2, \ldots, k-1$,

$$q_j/(q_j, (q_{j+1} \cdots q_k)^k) \ll M(MNP^{-1})^j N^\varepsilon \tag{8.17}$$

and

$$q_1 \leqq MN^\varepsilon. \tag{8.18}$$

Proof. The hypotheses of Theorem 5.1 are satisfied with δ in place of ε, so let $q, v_1, \ldots, v_k$ be as in Theorem 5.1. Define $Q_k, \ldots, Q_2, Q_1$ inductively as follows. Let $(q, v_k) = Q_k$, and if Q_j has been defined, let

$$(Q_j, v_{j-1}) = Q_{j-1} \quad (j = k, k-1, \ldots, 2)$$

Thus

$$q = q_k Q_k \quad \text{and} \quad Q_j = q_{j-1} Q_{j-1} \quad (j = k, \ldots, 2)$$

where the q_j are integers. Note that

$$(q, v_k, \ldots, v_j) = (Q_k, v_{k-1}, \ldots, v_j) = \cdots = (Q_{j+1}, v_j) = Q_j$$

which shows that $Q_2 = (q_k, v_k, \ldots, v_2)$ and $Q_1 = 1$, $q_1 = Q_2$. Thus (8.18) follows from (5.11).

It is clear that (8.14) holds. For $j = 2, \ldots, k$ we have

$$q\alpha_j - v_j = Q_j(q_k q_{k-1} \cdots q_j \alpha_j - y_j),$$

where $y_j = v_j/Q_j$ satisfies (8.16). Thus

$$|q_k \cdots q_j \alpha_j - y_j| \leq |q\alpha_j - v_j|,$$

and (8.15) follows from (5.12).

It remains to derive the estimate (8.17). For any natural number $t \leq M$, let

$$d = d(t) = (q, tv_2, \ldots, tv_k) \leq tMN^\delta \leq M^2 N^\delta$$

from (5.11). We apply Lemma 4.4 to tf. After all, we have

$$q \leq (MNP^{-1})^k N^\delta \leq N^{1-\delta},$$

$$|qt\alpha_j - tv_j| = t|q\alpha_j - v_j|$$
$$\leq (MNP^{-1})^k N^{\delta-j} \leq N^{1-\delta-j}$$

from (8.12), (5.11), (5.12). Thus, if

$$T_t = \sum_{x=1}^{q} e_q(tv_k x^k + \cdots + tv_1 x),$$

we have

$$S_t(f) - q^{-1} T_t \int_0^N e(\beta_k ty^k + \cdots + \beta_1 ty) \, dy \ll q^{1-(1/k)+\delta} d(t)^{1/k}. \quad (8.19)$$

Here $\beta_j = \alpha_j - v_j/q$. Moreover, since $M \ll MNP^{-1}$, we find that

$$q^{1-(1/k)+\delta} \sum_{t=1}^{M} (d(t))^{1/k} < M^2 q^{1-(1/k)} N^\delta \leq M(MNP^{-1})^k N^\delta < PN^{-\delta} \quad (8.20)$$

from (8.12). (We need the inequality $k + 1 \leq K$.) Combining (8.13), (8.19) and (8.20) we obtain

$$\tfrac{1}{2} P \leq \sum_{t=1}^{M} q^{-1} |T_t| \left| \int_0^N e(\beta_{k,t} y^k + \cdots + \beta_{1,t} y) \, dy \right|$$

$$\leq N \sum_{t=1}^{M} q^{-1} |T_t|.$$

Thus

$$\sum_{t=1}^{M} |T_t| \geq \tfrac{1}{2} q P N^{-1}. \quad (8.21)$$

Let $1 \leq j < k$. We now apply Lemma 8.2 with $tv_k, \ldots, tv_1$ in place of

$u_k, \ldots, u_1$, taking $R = q_{j+1} \cdots q_k$ and $Q = Q_{j+1}$. Clearly $Q_{j+1} \mid (tv_{j+1}, \ldots, tv_k)$ and

$$D = D(t) = (Q_{j+1}, tv_j, \ldots, tv_2)$$
$$= (q, v_k, \ldots, v_{j+1}, tv_j, \ldots, tv_2) \leq q_1(q, t) \leq MN^\delta(q, t).$$

We deduce from Lemma 8.2 that

$$|T_t| \ll q^{1+\delta}((q, t)MN^\delta)^{1/j}Q_0^{-1/j} \tag{8.22}$$

where $Q_0 = Q_{j+1}/(Q_{j+1}, (q_{j+1} \cdots q_k)^k)$.

Combining (8.21), (8.22), we see that

$$qPN^{-1} \ll q^{1+\delta}(MN^\delta)^{1/j}Q_0^{-1/j} \sum_{t=1}^{M} (q, t)$$

$$\ll q^{1+\delta}(MN^\delta)^{1/j}Q_0^{-1/j} \sum_{d \mid q} d\phi\left(\left[\frac{M}{d}\right]\right)$$

$$\ll q^{1+\delta}(MN^\delta)^{1/j}Q_0^{-1/j}MN^\delta,$$

or

$$Q_0 \ll (MNP^{-1})^j MN^\varepsilon.$$

Since q_j is a divisor of Q_{j+1}, Q_0 is divisible by $q_j/(q_j, (q_{j+1} \cdots q_k)^k)$, and the inequality (8.17) follows.

8.3   An alternative lemma

Lemma 8.4. *Let $N > C_1(h, k, \varepsilon)$. Let Λ be a lattice in $\mathbb{R}^h$ with $\Lambda \cap K_0 = \{0\}$ and*

$$\Delta^{\max(K,h+\tau)+\varepsilon} \leq N. \tag{8.23}$$

Suppose that (8.3) is insoluble. Then either

(i) *there is a primitive point $\mathbf{p}$ in Π with $\Delta^{-1} \ll |\mathbf{p}| < 2N^\delta$, and a natural number q, such that*

$$q < |\mathbf{p}|^{-k} N^\eta \tag{8.24}$$

$$\|q\mathbf{pa}_j\| < |\mathbf{p}|^{-k+1} N^{\eta-j} \quad (j = 1, \ldots, k). \tag{8.25}$$

or

(ii) *there is an integer m, $2 \leq m \leq h$, a natural number q, and a point $\mathbf{p}$ in Π with*

$$\Delta^{-1} \ll |\mathbf{p}| \leq 4N^\delta, \tag{8.26}$$

$$q \leq \min(|\mathbf{p}|^{-km}, \Delta^k)N^\eta, \tag{8.27}$$

$$\|q\mathbf{a}_j\mathbf{p}\| \leq |\mathbf{p}|^{-km} N^{\eta-j} \quad (j = 1, \ldots, k). \tag{8.28}$$

Moreover, in case (ii), *there are linearly independent points* $\mathbf{p}_1, \ldots, \mathbf{p}_m$ *such that*

$$\max |\mathbf{p}_i| \leqq 2N^\delta, \tag{8.29}$$

$$\|q\mathbf{a}_j\mathbf{p}_i\| \leqq N^{n-j}\Delta^k \quad (j = 1, \ldots, k, \, i = 1, \ldots, m). \tag{8.30}$$

Obviously, this is a rather elaborate analogue of Lemma 7.7.

Proof. Let a, B and $\mathscr{B}$ be as in the proof of Lemma 7.5. There are two cases to consider. Suppose first that $B \geqq N^{1-2kh\delta}$. In this case, we select any $\mathbf{p}$ in $\mathscr{B}$. Then alternative (i) holds with the integer $q = q(\mathbf{p})$, which satisfied (7.19)–(7.25) in that proof, as we easily verify.

Now suppose that

$$B < N^{1-2kh\delta}. \tag{8.31}$$

This is the tricky case. By using Lemma 8.3 in place of Theorem 5.1, we strengthen (7.23)–(7.25) as follows. For each $\mathbf{p}$ in $\mathscr{B}$ there are integers $q_j = q_j(\mathbf{p})$, $y_j = y_j(\mathbf{p})$ $(j = 1, \ldots, k)$ satisfying

$$1 \leqq q(\mathbf{p}) = q_1 \cdots q_k < a^{-k}B^{-k}N^{k+n}; \tag{8.32}$$

$$|q_k \cdots q_j\mathbf{a}_j\mathbf{p} - y_j| \leqq \|q\mathbf{a}_j\mathbf{p}\| < a^{-k+1}B^{-k}N^{k-j+n} \quad (j = 1, \ldots, k), \tag{8.33}$$

$$q_j/(q_j, (q_{j+1} \cdots q_k)^k) < N^{-\eta}a^{-1}(a^{-1}B^{-1}N)^j \tag{8.34}$$

for $j = 2, \ldots, k-1$;

$$q_1 < N^{2\delta}a^{-1}, \tag{8.35}$$

and

$$(q_j, y_j) = 1 \quad (j = 1, \ldots, k). \tag{8.36}$$

Just as in the proof of Lemma 7.7, let b_j denote the maximum number of elements of $\mathscr{B}$ that lie in a subspace of dimension j, and let S_j denote a subspace such that $|\mathscr{B} \cap S_j| = b_j$. Thus $b_1 = 1$ or 2,

$$b_h = |\mathscr{B}| \geqq N^{2kh\delta}(\log N)^{-2}$$

from (7.22) and (8.31). Let m be the largest integer such that

$$b_m \geqq b_{m-1}N^{2k\delta}; \tag{8.37}$$

evidently such an integer exists, $2 \leqq m \leqq h$. Note that

$$b_m \geqq |\mathscr{B}| N^{-2k(h-1)\delta}. \tag{8.38}$$

We apply Lemma 7.6 to $\mathscr{A} = \mathscr{B} \cap S_m$. Clearly the dimension of the linear span of $\mathscr{A}$ is m, from (8.37). Here, and in subsequent applications of Lemma 7.6, we take

$$Z = (2a)^m,$$

as we may in view of (7.21). In the present application, we take

$$l(\mathbf{p}) = q_k(\mathbf{p}), \qquad w(\mathbf{p}) = y_k(\mathbf{p})$$

and, using (8.32), (8.33),

$$U = a^{-k}B^{-k}N^{k+\eta}, \qquad V = a^{-k+1}B^{-k}N^{\eta}$$

The crucial condition (7.30) is satisfied, since

$$ZU^mV\Delta N^{\delta} \ll (2a)^m(a^{-k}B^{-k}N^k)^m a^{-k+1}B^{-k}\Delta N^{\eta}$$
$$\ll (a^{-1}B^{-1}N)^{k(m+1)}\Delta N^{-k+\eta}$$
$$\ll \Delta^{k(h+1)+1}N^{-k+\eta} \ll 1$$

from (7.19), (7.20) and (8.23).

Thus there is a subset $\mathscr{C}_k$ of $\mathscr{B} \cap S_m$ having

$$|\mathscr{C}_k| \geq b_m N^{-\delta} \tag{8.39}$$

such that $q_k(\mathbf{p}) = q_k$ (say) for all $\mathbf{p}$ in $\mathscr{C}_k$. Since $(q_{k-1}(\mathbf{p}), q_k^k)$ is a divisor of q_k^k, there is a subset $\mathscr{D}_k$ of $\mathscr{C}_k$ having $|\mathscr{D}_k| \geq |\mathscr{C}_k| N^{-\delta}$ and

$$(q_{k-1}(\mathbf{p}), q_k^k) = r_k \qquad \text{(say)}$$

for all p in $\mathscr{D}_k$. Here q_k and r_k are, we emphasize, independent of $\mathbf{p}$.

By induction we may find a sequence of sets

$$\mathscr{C}_k \supset \mathscr{D}_k \supset \mathscr{C}_{k-1} \supset \mathscr{D}_{k-1} \supset \cdots \supset \mathscr{C}_2$$

with the following properties.

(I) $\qquad |\mathscr{D}_j| \geq |\mathscr{C}_j| N^{-\delta} \geq |\mathscr{D}_{j+1}| N^{-2\delta} \qquad (j = 2, \ldots, k-1)$;

(II) $q_j(\mathbf{p})$ takes a fixed value, say q_j, for all $\mathbf{p}$ in $\mathscr{C}_j$ $(j = 2, \ldots, k)$;

(III) $(q_{j-1}(p), (q_j \cdots q_k)^k)$ takes a fixed value, r_j say, for all p in $\mathscr{D}_j$ $(j = 2, \ldots, k)$.

We have already got $\mathscr{C}_k$ and $\mathscr{D}_k$. Let us suppose that $\mathscr{C}_k, \ldots, \mathscr{D}_j$ exist, where $3 \leq j \leq k$. We apply Lemma 7.6 to $\mathscr{A} = \mathscr{D}_j$. Clearly the linear span of $\mathscr{A}$ has dimension m, since it can be seen from (8.37), (8.39), and what is already available from (I), that

$$|\mathscr{D}_j| > b_{m-1}.$$

We take $\mathbf{e} = q_k \cdots q_j r_j \mathbf{a}_{j-1}$,

$$l(\mathbf{p}) = q_{j-1}(\mathbf{p})/r_j, \qquad w(\mathbf{p}) = y_{j-1}(\mathbf{p}),$$

so that $(l(\mathbf{p}), w(\mathbf{p})) = 1$ from (8.36). Using (8.33), (8.34) we may take

$$U = a^{-1}(a^{-1}B^{-1}N)^{j-1}N^{\eta}, \qquad V = a^{-k+1}B^{-k}N^{k-j+1+\eta}.$$

The condition (7.30) is satisfied, since

$$ZU^m V \Delta N^\delta \ll (2a)^m a^{-m} (a^{-1}B^{-1}N)^{(j-1)m} a^{-k+1} B^{-k} N^{k-j+1+\eta} \Delta$$
$$\ll \Delta^{(j-1)m+k+1} N^{-(j-1)+\eta} \ll 1.$$

Here we use the inequality (8.23) together with the observation that

$$((j-1)m + k + 1)/(j-1) \leq h + \tau$$

for $2 \leq m \leq h$, $3 \leq j \leq k$.

We deduce that there is a subset $\mathscr{C}_{j-1}$ of $\mathscr{D}_j$ such that $|\mathscr{C}_{j-1}| \geq |\mathscr{D}_j| N^{-\delta}$ and $q_{j-1}(\mathbf{p})/r_j$ takes a fixed value on $\mathscr{C}_{j-1}$—so does $q_{j-1}(\mathbf{p})$, of course! To construct $\mathscr{D}_{j-1}$ from $\mathscr{C}_{j-1}$ we proceed as we did in obtaining $\mathscr{D}_k$ from $\mathscr{C}_k$. (We don't actually need $\mathscr{D}_2$.)

One final use of Lemma 7.6 enables us to find a subset $\mathscr{E}$ of $\mathscr{C}_2$ having $|\mathscr{E}| \geq |\mathscr{C}_2| N^{-\delta}$ and $q_1(\mathbf{p}) = q_1$, say, for all $\mathbf{p}$ in $\mathscr{E}$. For this step we use (8.33), (8.35) and take

$$U = N^{2\delta} a^{-1}, \qquad V = a^{-k+1} B^{-k} N^{k-1+\eta}.$$

The criterion (7.30) is satisfied because

$$\Delta^{k+1} N^{-1+\eta} \ll \Delta^K N^{-1+\eta} \ll 1$$

from (8.23).

To sum up, $q(\mathbf{p}) = q$ for all $\mathbf{p}$ in $\mathscr{E}$; also

$$|\mathscr{E}| > \max (b_{m-1}, |\mathscr{B}| N^{-2kh\delta}) \tag{8.40}$$

by combining all the bounds in (8.38), (8.39) and (I).

There are certainly m linearly independent points $\mathbf{p}_1, \ldots, \mathbf{p}_m$ in $\mathscr{E}$, since $|\mathscr{E}| > b_{m-1}$. It is easy to check that $q, \mathbf{p}_1, \ldots, \mathbf{p}_m$ have the property (8.30), using (8.32), (8.33) and (7.20), and (8.29) follows from (7.21).

Let d be the least distance between two distinct members of $\mathscr{E}$. Because $\mathscr{E}$ lies in $S_m \cap (2aK_0)$, it is clear that

$$|\mathscr{E}| d^m \ll a^m.$$

(Consider the total measure in S_m of balls centred on the points of $\mathscr{E}$ of radius $d/2$.) Thus

$$d \ll a |\mathscr{E}|^{-1/m} \ll a |\mathscr{B}|^{-1/m} N^\eta$$
$$\ll a(NB^{-1})^{-1/m} N^\eta, \tag{8.41}$$

in view of (7.22) and (8.40).

Let $\mathbf{p}', \mathbf{p}''$ be two points of $\mathscr{E}$ with $|\mathbf{p}' - \mathbf{p}''| = d$, and let $\mathbf{p} = \mathbf{p}' - \mathbf{p}''$. Then (recalling (7.27))

$$\Delta^{-1} \ll |\mathbf{p}| \leq 4N^\delta, \qquad NB^{-1} \ll a^m |\mathbf{p}|^{-m} N^\eta$$

from (8.41). From (8.32), (8.33) and (8.41) we get

$$q < \min \left(|\mathbf{p}|^{-mk}, \Delta^k \right) N^\eta$$

$$\| q\mathbf{a}_j \mathbf{p} \| \leq \| q\mathbf{a}_j \mathbf{p}' \| + \| q\mathbf{a}_j \mathbf{p}'' \|$$

$$\ll |\mathbf{p}|^{-mk} N^{-j+\eta} \qquad (j = 1, \ldots, k).$$

Thus all the conditions required for alternative (ii) are satisfied. This completes the proof of Lemma 8.4.

8.4   Proof of Theorem 8.2

We proceed by induction on h. The case $h = 1$ is a rewording of Theorem 5.2. In the induction step let $h \geq 2$, let $N > C_1(h, k, \varepsilon)$, let $\mathbf{a}_1, \ldots, \mathbf{a}_k \in \mathbb{R}^h$ and suppose if possible that (8.3) is insoluble for some lattice Λ satisfying (8.2). We may appeal to Lemma 8.4, since $F(h) \geq \max(K, h + \tau)$.

Suppose first that alternative (i) holds. We apply Lemma 7.8 and Lemma 7.9 with $\mathbf{p}_1 = \mathbf{p}$, $T = $ linear span of $\mathbf{p}$. We have det $\Pi' = |\mathbf{p}|$ and, in (7.50),

$$c = 1, \tag{8.42}$$

because $\mathbf{p}$ is primitive. Also $\Lambda' = \Lambda \cap T^\perp$ has

$$d(\Lambda') \ll \Delta |\mathbf{p}|. \tag{8.43}$$

For later reference it is convenient to include powers of c in various expressions, even though $c = 1$.

For $j = 1, \ldots, k$, $cq^j a_j$ may be written in the form

$$cq^j \mathbf{a}_j = \mathbf{l}_j + \mathbf{s}_j + \mathbf{b}_j \tag{8.44}$$

where $\mathbf{l}_j \in \Lambda$, $\mathbf{s}_j \in T^\perp$ and

$$|\mathbf{b}_j| \ll |\mathbf{p}|^{-1} \max_j \| q^j \mathbf{a}_j \mathbf{p} \|$$

$$\ll |\mathbf{p}|^{-jk} N^{\eta-j} \tag{8.45}$$

in view of (8.24), (8.25) and $\| q^j \mathbf{a}_j \mathbf{p} \| \leq q^{j-1} \| q\mathbf{a}_j \mathbf{p} \|$.

We now apply Theorem 8.2 in the $(h-1)$-dimensional space $T^\perp$, replacing ε by δ, Λ by $2\Lambda'$, $\mathbf{a}_j$ by $2\mathbf{s}_j$, and N by $[d(2\Lambda')^{F(h-1)} N^\delta] \geq \max(C_1(h-1, k, \delta), d(2\Lambda')^{F(h-1)+\delta})$. Thus there is a natural number x,

$$x \leq d(2\Lambda')^{F(h-1)} N^\delta \ll |\mathbf{p}|^{F(h-1)} \Delta^{F(h-1)} N^\delta \tag{8.46}$$

such that

$$2x^k c^{k-1} \mathbf{s}_k + \cdots + 2x \mathbf{s}_1 \in 2\Lambda' + K_0.$$

This implies

$$x^k c^{k-1} \mathbf{s}_k + \cdots + x \mathbf{s}_1 \in \Lambda + \tfrac{1}{2} K_0. \tag{8.47}$$

Let $n = xcq$. Then

$$n \ll |\mathbf{p}|^{F(h-1)-k} \Delta^{F(h-1)} N^{\eta}$$

from (8.24), (8.42) and (8.46). Since $h \geqq 2$ we have $F(h-1) \geqq K$. Thus

$$n \ll \Delta^{F(h-1)} N^{\eta} \ll \Delta^{F(h)} N^{\eta} \tag{8.48}$$

and we may deduce that $n \leqq N$ in view of (8.2). (Monotonicity of F is almost evident.)

Now from (8.44), for some $\mathbf{l} \in \Lambda$,

$$n^k \mathbf{a}_k + \cdots + n \mathbf{a}_1 = \mathbf{l} + \sum_{j=1}^{k} x^j c^{j-1} \mathbf{s}_j + \sum_{j=1}^{k} x^j c^{j-1} \mathbf{b}_j. \tag{8.49}$$

Because of (8.47), to show that the left-hand side of (8.49) is in $K_0 + \Lambda$, we need only prove that

$$\sum_{j=1}^{k} x^j c^{j-1} |\mathbf{b}_j| \ll N^{-\delta}. \tag{8.50}$$

Moreover, for $1 \leqq j \leqq k$,

$$x^j c^{j-1} |\mathbf{b}_j| \ll |\mathbf{p}|^{jF(h-1)} \Delta^{jF(h-1)} |\mathbf{p}|^{-jk} N^{\eta-j}$$
$$\ll (|\mathbf{p}|^{F(h-1)-k} \Delta^{F(h-1)} N^{-1})^j N^{\eta}.$$

We can therefore deduce (8.50) in the same way as (8.48). Thus $n \leqq N$, $n^k \mathbf{a}_k + \cdots + n \mathbf{a}_1 \in K_0 + \Lambda$. This is a contradiction. Thus alternative (ii) must hold.

Let m be the natural number in alternative (ii), $2 \leqq m \leqq h$. By the definition of F, we either have

$$F(h) \geqq mk + 1, \tag{8.51}$$

or we have

$$F(h) \geqq F(h - m) + k. \tag{8.52}$$

Suppose first that (8.51) holds. Let q, $\mathbf{p}$ be as in (8.26)–(8.28), and write

$$\mathbf{p} = z \mathbf{p}_0$$

where $\mathbf{p}_0$ is primitive in Π and z is a natural number. We apply Lemmas 7.8 and 7.9 with $\mathbf{p}_1 = \mathbf{p}$, $T = $ linear span of $\mathbf{p}_0$. We have $\det \Pi' = |\mathbf{p}_0|$, and from (7.49) we get

$$c \ll z = |\mathbf{p}| |\mathbf{p}_0|^{-1}. \tag{8.53}$$

With $\Lambda' = \Lambda \cap T^{\perp}$, we have

$$d(\Lambda') \ll \Delta |\mathbf{p}_0|. \tag{8.54}$$

Again, we have (8.44), but this time

$$|\mathbf{b}_j| \ll |\mathbf{p}_0|^{-1} q^{j-1} \|q\mathbf{p}\mathbf{a}_j\|$$
$$\ll |\mathbf{p}_0|^{-1} |\mathbf{p}|^{-kmj} N^{\eta-j} \tag{8.55}$$

for $j = 1, \ldots, k$, from (7.51), (8.27) and (8.28).

Just as above there is a natural number x satisfying (8.47) with

$$x \leqq d(2\Lambda')^{F(h-1)} N^\delta \ll |\mathbf{p}_0|^{F(h-1)} \Delta^{F(h-1)} N^\delta \tag{8.56}$$

from (8.54). Again, to get a contradiction it suffices to prove that

$$xcq \ll N^{1-\delta} \tag{8.57}$$

and that (8.50) holds.

To get (8.57), we observe that

$$xcq \ll |\mathbf{p}_0|^{F(h-1)} \Delta^{F(h-1)} |\mathbf{p}| |\mathbf{p}_0|^{-1} |\mathbf{p}|^{-mk} N^\eta$$
$$\ll |\mathbf{p}_0|^{F(h-1)-1} |\mathbf{p}|^{-(mk-1)} \Delta^{F(h-1)} N^\eta$$
$$\ll |\mathbf{p}|^{F(h-1)-mk} \Delta^{F(h-1)} N^\eta. \tag{8.58}$$

Here we use in turn (8.56), (8.53), (8.27), and the observations that $F(h-1) \geqq 1$ and $|\mathbf{p}_0| \leqq |\mathbf{p}|$. Since

$$|\mathbf{p}|^{F(h-1)-mk} \ll \max(N^\eta, \Delta^{-F(h-1)+mk})$$

from (8.26), we deduce that

$$xcq \ll \max(\Delta^{F(h-1)}, \Delta^{mk}) N^\eta \tag{8.59}$$

and (8.57) follows in view of (8.51).

As for (8.50), we have from (8.56), (8.53), (8.55) that

$$x^j c^{j-1} |\mathbf{b}_j| \ll |\mathbf{p}_0|^{jF(h-1)} \Delta^{jF(h-1)} |\mathbf{p}|^{j-1} |\mathbf{p}_0|^{-j+1}$$
$$\times |\mathbf{p}_0|^{-1} |\mathbf{p}|^{-kmj} N^{\eta-j}$$
$$\ll |\mathbf{p}|^{jF(h-1)-jmk-1} \Delta^{jF(h-1)} N^{\eta-j}$$

as in the proof of (8.58). Now (8.50) follows in much the same way as (8.59). Thus (8.51) leads to a contradiction.

Now suppose that (8.52) holds. Let $q, \mathbf{p}_1, \ldots, \mathbf{p}_m$ be as in (8.27), (8.29), (8.30). We apply Lemmas 7.8 and 7.9 with $T = $ linear span of $\mathbf{p}_1, \ldots, \mathbf{p}_m$. Let us write $D = \det \Pi'$: we have

$$\Delta^{-1} \ll D \leqq N^\eta \tag{8.60}$$

from the proof of Lemma 7.6 and (8.29). Thus $\Lambda' = \Lambda \cap T^\perp$ has

$$d(\Lambda') \ll \Delta D, \tag{8.61}$$

unless $m = h$ when Λ' is not defined. In (7.49), we have

$$c \ll D^{-1} N^\eta. \tag{8.62}$$

As before, we have (8.44) with $\mathbf{l}_j \in \Lambda$, $\mathbf{s}_j \in T^{\perp}$, and

$$|\mathbf{b}_j| \ll D^{-1}N^{\eta} \max \|q^j \mathbf{p}_i \mathbf{a}_j\|$$
$$\ll D^{-1}\Delta^{kj}N^{-j+\eta} \tag{8.63}$$

from (8.27), (8.29), (8.30).

Suppose for a moment that $m < h$. We apply Theorem 8.2 in the $(h - m)$-dimensional space $T^{\perp}$, replacing ε by δ, Λ by $2\Lambda'$, $\mathbf{a}_j$ by $2c^{j-1}\mathbf{s}_j$, and N by $[d(2\Lambda')^{F(h-m)}N^{\delta}]$. Thus there is a natural number x,

$$x \leqq d(2\Lambda')^{F(h-m)}N^{\delta} \ll \Delta^{F(h-m)}D^{F(h-m)}N^{\delta} \tag{8.64}$$

such that (8.47) holds. To get a contradiction it suffices to prove (8.57) and (8.50).

To prove (8.57), we note that

$$xcq \ll D^{F(h-m)}\Delta^{F(h-m)}D^{-1}\Delta^k N^{\eta}$$
$$\ll \Delta^{F(h-m)+k}N^{\eta}$$

using in turn (8.64), (8.62), (8.27) and the upper bound for D in (8.60). Now (8.57) follows because of (8.52) and the initial hypothesis (8.2).

As for (8.50), we have

$$x^j c^{j-1} |\mathbf{b}_j| \ll D^{jF(h-m)}\Delta^{jF(h-m)}D^{-j+1}D^{-1}\Delta^{kj}N^{-j+\eta}$$

from (8.64), (8.62), (8.63). Just as above we may derive (8.50).

The argument is similar but a little simpler when $m = h$ (and each $\mathbf{s}_j = 0$). We replace x by 1 and derive (8.57), (8.50) using

$$D^{-1}\Delta^k \ll \Delta^{k+1} \ll \Delta^K \leqq N^{1-\eta}.$$

(This is where the lower bound for D in (8.60) is of service.)

Thus the assumption that (8.2) is insoluble leads to a contradiction. This completes the proof of the induction step, and Theorem 8.2 is proved.

8.5   More about exponential sums: monomials

We begin by stating the theorem which is our objective in the last section of the chapter.

Theorem 8.3. *Let* $k \geqq 3$, $1 \leqq h \leqq 2^{k-2}$. *Let* $N > C_3(h, k, \varepsilon)$. *Let* $\alpha_1, \ldots, \alpha_k$ *be real. Then there is a natural number* $n \leqq N$ *such that*

$$\|n^k \alpha_i\| < N^{\varepsilon - 1/(hK)} \qquad (i = 1, \ldots, h).$$

As pointed out in Chapter 1, this type of result is out of reach for general polynomials. To get Theorem 8.3, we exploit sharper versions of

Lemma 4.2 in the case $G(x) = ax^k$. These are all due to Hardy and Littlewood (1922).

Lemma 8.5. *Let*

$$S(a, q) = \sum_{x=1}^{q} e(ax^k/q)$$

where $q \geq 1$ and $(a, q) = 1$. Then we have

$$S(a, q) \ll q^{1/2+\varepsilon} \tag{8.65}$$

if q is square free,

$$S(a, q) \ll q^{1-(1/m)} \tag{8.66}$$

if $q = u^m$ and u is square free, $1 < m < k$.

Proof. For $q = q_1 \cdots q_m$, where $q_1, \ldots, q_m$ are relatively prime in pairs, we have

$$S(a, q) = S(a_1, q_1) \cdots S(a_r, q_r) \tag{8.67}$$

Here a runs over a reduced set of residues to the modulus q as the a_i run over reduced sets $(\bmod\, q_i)$. See Vaughan (1981), Lemma 2.10. The inequalities (8.65) and (8.66) now follow at once from Lemma 4.3 and Lemma 4.4 of Vaughan (1981).

Lemma 8.6. *Suppose that α is real and that there are integers $N > C_4(k, \varepsilon)$, q and a such that*

$$1 \leq q \leq N^{1-\delta}, \quad |q\alpha - a| \leq N^{1-k-\delta}, \quad (a, q) = 1, \tag{8.68}$$

Suppose further that

$$\left| \sum_{x=1}^{N} e(\alpha x^k) \right| \geq P \geq q^{1-(1/k)} N^{2\delta}. \tag{8.69}$$

Then there is a divisor r of q with

$$1 \leq r \leq (N/P)^2 N^\eta, \quad \|\alpha r^k\| \leq (N/P)^{2k} N^{-k+\eta}.$$

This should be compared with the case $M = 1$ of Theorem 5.1. The bound for $\|\alpha r^k\|$ is much smaller than the bound we can get for $\|\alpha q^k\|$:

$$\|\alpha q^k\| \leq q^{k-1} \|\alpha q\| \leq (N/P)^{k^2} N^{-k+k\delta}$$

Of course, nothing is asserted about $\|\alpha r\|$.

Proof of Lemma 8.6. We write q in the form

$$q = q_k \prod_{m=1}^{k-1} q_m^m,$$

where

(a) $q_1, \ldots, q_k$ are relatively prime in pairs;
(b) $q_1, \ldots, q_{k-1}$ are square free;
(c) for any prime p dividing q_k we have $p^k \mid q_k$.

Thus

$$q_k = t^k q', \qquad \text{where } q' \mid t^k.$$

From (8.67), Lemma 4.2 and Lemma 8.5,

$$S(a, q) \ll q^\delta q_k^{1-(1/k)} q_1^{1/2} \prod_{m=2}^{k-1} q_m^{m-1}. \tag{8.70}$$

Let $\beta = \alpha - a/q$. We apply Lemma 4.4 to $f(x) = \alpha x^k$ with $T = N$, $d = 1$. In view of (8.69) this yields

$$q^{-1} |S(a, q)| \left| \int_0^N e(\beta y^k) \, dy \right| > \tfrac{1}{2} P.$$

Using Lemma 4.5 and (8.70) we deduce that

$$P \ll q^\delta q_k^{-1/k} q_1^{-1/2} \prod_{m=2}^{k-1} q_m^{-1} N, \tag{8.71}$$

and

$$P \ll q^\delta q_k^{-1/k} q_1^{-1/2} \prod_{m=2}^{k-1} q_m^{-1} |\beta|^{-1/k}. \tag{8.72}$$

We write $r = t^2 \prod_{m=1}^{k-1} q_m$. Then $r \mid q$; also $q \mid r^k$, so

$$\|r^k \alpha\| = \|r^k \beta\|.$$

What is more,

$$r \leq q_k^{2/k} \prod_{m=1}^{k-1} q_m \leq q_k^{2/k} q_1 \prod_{m=2}^{k-1} q_m^2$$

$$\ll (N/P)^2 q^{2\delta}$$

from (8.71). Similarly,

$$|r^k \beta| = t^{2k} \prod_{m=1}^{k-1} q_m^k |\beta|$$

$$\leq q_k^2 q_1^k \prod_{m=2}^{k-1} q_m^{2k} |\beta|$$

$$\ll (N/P)^k P^{-k} q^{2k\delta}$$

from (8.71) and (8.72). This gives the desired estimate for $\|r^k \alpha\|$, and Lemma 8.6 is proved.

Now we prove a lattice theorem, from which Theorem 8.3 can be deduced at once.

Theorem 8.4. *Let $k \geq 3$, $1 \leq h \leq 2^{k-2}$. Let $N > C_3(h, k, \varepsilon)$. Let Λ be a lattice in $\mathbb{R}^h$ satisfying $\Lambda \cap K_0 = \{\mathbf{0}\}$ and*

$$\Delta^{K+\varepsilon} \leq N. \tag{8.73}$$

Then for any $\mathbf{a}$ in $\mathbb{R}^h$ there is a natural number n satisfying

$$1 \leq n \leq N, \qquad n^k \mathbf{a} \in \Lambda + K_0. \tag{8.74}$$

Proof. We may suppose that $k \geq 4$, otherwise the result is contained in Theorem 8.2. We proceed by induction on h; the case $h = 1$ is contained in Theorem 8.2 also. In the induction step, suppose the hypotheses are satisfied for some h, $2 \leq h \leq 2^{k-2}$, and suppose $\mathbf{a} \in \mathbb{R}^h$ exists such that (8.74) is insoluble. By Lemma 7.5 with $\mathbf{a}_k = \mathbf{a}$, $\mathbf{a}_{k-1} = \cdots = \mathbf{a}_1 = \mathbf{0}$, there are numbers a and B, and a subset $\mathcal{B}$ of Π^*, satisfying (7.19)–(7.22), and with the following property. For each $\mathbf{p}$ in $\mathcal{B}$ there are integers $q = q(\mathbf{p})$, $v = v(\mathbf{p})$ satisfying

$$1 \leq q < a^{-k} B^{-k} N^{k+\eta} \tag{8.75}$$

$$(q, v) = 1, \qquad |q \mathbf{a}_k \mathbf{p} - v| < a^{-k+1} B^{-k} N^{\eta} \tag{8.76}$$

(The condition $(q, v) = 1$ presents no difficulty as we may take $v_{k-1} = \cdots = v_1 = 0$.)

Suppose at first that $B \geq N^{1-2hk\delta}$. We easily obtain the inequalities of alternative (i) of Lemma 8.4. Then we proceed to derive a contradiction as in the proof of Theorem 8.2 (in the case of alternative (i)). We simply replace $F(h-1)$ by K in the argument. Thus we may assume from now on that

$$B < N^{1-2hk\delta}.$$

Since $h \leq K/2$, we certainly have (8.23). Thus we may proceed as in the proof of Lemma 8.4 to obtain a set $\mathcal{E}$ satisfying (8.40) such that $q(\mathbf{p}) = q$ for all $\mathbf{p}$ in $\mathcal{E}$, and moreover (8.26)–(8.30) are satisfied for certain points $\mathbf{p}, \mathbf{p}_1, \ldots, \mathbf{p}_m$ of Π^*. We also need the inequality

$$NB^{-1} \ll a^m |\mathbf{p}|^{-m} N^{\eta} \tag{8.77}$$

established at the end of Section 8.3; here $\mathbf{p} = \mathbf{p}' - \mathbf{p}''$ with $\mathbf{p}', \mathbf{p}''$ in $\mathcal{E}$.

If $m = h$, there is no difficulty in obtaining a contradiction by arguing as we did in this case in the proof of Theorem 8.2. So we suppose that $m < h$ from now on.

Let $\mathbf{p} \in \mathcal{E}$. Using (7.21), select an integer $y = y(\mathbf{p})$, $1 \leq y \leq N^{\delta} a^{-1}$ such that

$$\left| \sum_{n=1}^{N} e(y n^k \mathbf{a} \mathbf{p}) \right| \geq a B N^{-\delta}.$$

We would like to apply Lemma 8.6, taking $\alpha = y\mathbf{ap}$, $P = aBN^{-\delta}$. The inequalities (8.68) are valid, since

$$|qy\mathbf{ap} - yv| \leqq a^{-k}B^{-k}N^{\eta}$$
$$< \Delta^k N^{\eta-k} < N^{1-k-\delta},$$

$$q \leqq \Delta^k N^{\eta} \leqq N^{1-\delta}.$$

However, we may have $(q, yv) > 1$. This obstacle may be surmounted by replacing q by one of its divisors, for which we retain the symbol q. We wish to take the same divisor for each $\mathbf{p}$ in $\mathscr{E}$, and this may be accomplished with a reduction of $|\mathscr{E}|$ by a factor $< N^{\delta}$, without damaging any of the procedures used in the case $m = h$ or invalidating (8.77).

We observe that the integer r supplied by the proof of Lemma 8.6 takes a constant value on $\mathscr{E}$, since r depends only on the prime factorization of q and not on $\alpha = y\mathbf{ap}$. We have

$$r \leqq a^{-2}B^{-2}N^{2+\eta},$$

$$\|\mathbf{ap}'y(\mathbf{p}')r^k\| \leqq a^{-2k}B^{-2k}N^{k+\eta};$$

similarly for $\mathbf{p}''$. Thus if $u = y(\mathbf{p}')y(\mathbf{p}'')r$, we easily obtain the inequalities

$$u \leqq a^{-4}B^{-2}N^{2+\eta}, \tag{8.78}$$

$$\|\mathbf{ap}u^k\| \leqq a^{-4k+1}B^{-2k}N^{k+\eta}. \tag{8.79}$$

Let $\mathbf{p} = z\mathbf{p}_0$ where $\mathbf{p}_0$ is primitive in Π and z is a natural number. We apply Lemmas 7.8 and 7.9 with $\mathbf{p}_1 = \mathbf{p}$, $T = $ linear span of $\mathbf{p}_0$; the inequalities (8.53) and (8.54) apply. Thus we may write $cu^k\mathbf{a}$ in the form

$$cu^k\mathbf{a} = \mathbf{l} + \mathbf{s} + \mathbf{b}$$

where $\mathbf{l} \in \Lambda$, $\mathbf{s} \in T^{\perp}$ and, using (8.79),

$$|\mathbf{b}| \ll |\mathbf{p}_0|^{-1} \|u^k\mathbf{ap}\|$$
$$\ll |\mathbf{p}_0|^{-1} a^{-4k+1}B^{-2k}N^{k+\eta}. \tag{8.80}$$

We now apply Theorem 8.4 in the $(h-1)$-dimensional space $T^{\perp}$, obtaining a natural number x,

$$x \leqq d(2\Lambda')^K N^{\delta} \ll |\mathbf{p}_0|^K \Delta^K N^{\delta}, \tag{8.81}$$

such that

$$x^k c^{k-1}\mathbf{s} \in \Lambda + \tfrac{1}{2}K_0.$$

To get a contradiction it suffices to prove that

$$xcu \ll N^{1-\delta} \tag{8.82}$$

and that

$$x^k c^{k-1} |\mathbf{b}| \ll N^{-\delta} \tag{8.83}$$

(This should be obvious to the reader who has gone over the proof of Theorem 8.2.)

Now using (8.81), (8.53), (8.78), (8.77)

$$xcu \ll |\mathbf{p}_0|^K \Delta^K |\mathbf{p}_0|^{-1} |\mathbf{p}| a^{-4} B^{-2} N^{2+\eta}$$
$$\ll |\mathbf{p}_0|^{K-1} \Delta^K a^{-3} (B^{-1}N)^2 N^\eta$$
$$\ll |\mathbf{p}_0|^{K-1-2m} \Delta^K a^{2m-3} N^\eta \ll \Delta^K N^\eta \ll N^{1-\delta}.$$

(We have also used $2m + 1 < 2h \le K$, and

$$|\mathbf{p}_0| \le |\mathbf{p}| \le 4a \le 4N^\delta.)$$

Similarly, using (8.80) this time,

$$x^k c^{k-1} |\mathbf{b}| \ll |\mathbf{p}_0|^{Kk} \Delta^{Kk} |\mathbf{p}_0|^{-k+1} |\mathbf{p}|^{k-1} |\mathbf{p}_0|^{-1} a^{-4k+1} B^{-2k} N^{k+\eta}$$
$$\ll |\mathbf{p}_0|^{(K-1)k} \Delta^{Kk} a^{-3k} (B^{-1}N)^{2k} N^{\eta-k}$$
$$\ll (|\mathbf{p}_0|^{K-1-2m} \Delta^K a^{2m-3} N^{-1})^k N^\eta \ll N^{-\delta}.$$

We obtain a contradiction. This completes the induction step, and Theorem 8.4 is proved.

In conclusion we remark that Conjecture A is just an attempt to bring higher-dimensional results into line with one-dimensional work. Probably the inequality (1.15) could be weakened to

$$\psi_1 \cdots \psi_h \ge N^{-1+\varepsilon}.$$

It would be interesting to prove this even in an 'almost everywhere' form. That is, one might prove such a result is true except for polynomials $f_1, \ldots, f_h$ whose coefficients lie in a set of Lebesgue measure zero in $\mathbb{R}^{kh}$; one would assume $N > C(f_1, \ldots, f_h)$.

9
Quadratic forms

9.1 Self-adjoint linear forms

Let $Q(x_1, \ldots, x_s)$ be a quadratic form with real coefficients. In the present chapter we consider inequalities of the type

$$\|Q(n_1, \ldots, n_s)\| < N^{-\lambda}, \qquad 0 < \max(|n_1|, \ldots, |n_s|) \leqq N.$$

(Implied constants will depend at most on s.)

Naturally this leads to the study of exponential sums such as

$$S(Q) = \sum_{x_1=1}^{N} \cdots \sum_{x_s=1}^{N} e(Q(x_1, \ldots, x_s)). \tag{9.1}$$

These can be treated in the following manner (Davenport, (1956), (1958)). We may write

$$Q(x_1, \ldots, x_s) = \sum_{i=1}^{s} \sum_{j=1}^{s} \lambda_{ij} x_i x_j \tag{9.2}$$

with $\lambda_{ji} = \lambda_{ij}$. Let

$$L_i(x_1, \ldots, x_s) = \sum_{j=1}^{s} \lambda_{ij} x_j. \tag{9.3}$$

Then we have the identities

$$Q(\mathbf{x}) = \sum_{i=1}^{s} x_i L_i(\mathbf{x}),$$

$$Q(\mathbf{x} + \mathbf{z}) - Q(\mathbf{x}) = 2(x_1 L_1(\mathbf{z}) + \cdots + x_s L_s(\mathbf{z})) + Q(\mathbf{z}). \tag{9.4}$$

Squaring the modulus of the sum (9.1), we obtain

$$|S(Q)|^2 = \sum_{x_1=1}^{N} \cdots \sum_{x_s=1}^{N} \sum_{y_1=1}^{N} \cdots \sum_{y_s=1}^{N} e(Q(\mathbf{y}) - Q(\mathbf{x})).$$

In the sum over $\mathbf{y}$ we make the substitution

$$\mathbf{y} = \mathbf{x} + \mathbf{z}, \qquad \mathbf{z} = (z_1, \ldots, z_s).$$

Then

$$|S(Q)|^2 = \sum_{x_1,\ldots,x_s=1}^{N} \sum_{1 \le x_j+z_j \le N} e(2x_1 L_1(\mathbf{z}) + \cdots + 2x_s L_s(\mathbf{z}) + Q(\mathbf{z}))$$

from (9.4), and hence

$$|S(Q)|^2 \le \sum_{\substack{z_1,\ldots,z_s=-(N-1)}}^{N-1} \left| \sum_{\substack{1 \le x_1,\ldots,x_s \le N \\ 1 \le x_j+z_j \le N}} e(2(x_1 L_1 + \cdots + x_s L_s)) \right|$$

$$= \sum_{z_1,\ldots,z_s=-(N-1)}^{N-1} \prod_{j=1}^{s} \left| \sum_{x_j=\max(1,1-z_j)}^{\min(N-z_j,N)} e(2x_j L_j(z)) \right|$$

$$\le \sum_{z_1,\ldots,z_s=-(N-1)}^{N-1} \prod_{j=1}^{s} \min\left(N, \|2L_j(\mathbf{z})\|^{-1}\right) \qquad (9.5)$$

by (3.4).

Davenport's approach to the final sum in (9.5) is via the set of $2s$ linear forms

$$\left. \begin{aligned} \zeta_j(x_1,\ldots,x_{2s}) &= N(L_j(x_1,\ldots,x_s) - x_{s+j}) \\ \zeta_j(x_1,\ldots,x_{2s}) &= N^{-1} x_j \end{aligned} \right\} \qquad (j=1,\ldots,s)$$

on $\mathbb{R}^{2s}$.

Lemma 9.1. *Let $L_1,\ldots,L_s$ be linear forms on $\mathbb{R}^s$ given by (9.3) where $\lambda_{ij} = \lambda_{ji}(i,j=1,\ldots,s)$. Define linear forms $\zeta_1,\ldots,\zeta_{2s}$ on $\mathbb{R}^{2s}$ by*

$$\left. \begin{aligned} \zeta_j &= A(L_j(x_1,\ldots,x_s) - x_{s+j}) \\ \zeta_{s+j} &= A^{-1} x_j \end{aligned} \right\} \qquad (j=1,\ldots,s)$$

where $A \ge 1$. Then the lattice

$$\Pi = \{(\zeta_1(\mathbf{x}),\ldots,\zeta_{2s}(\mathbf{x}) : \mathbf{x} \in \mathbb{Z}^{2s}\}$$

has successive minima $\pi_1,\ldots,\pi_s$ with respect to the unit ball satisfying

$$1 \ll \pi_i \pi_{2s+1-i} \ll 1 \qquad (i=1,\ldots,2s). \qquad (9.6)$$

Proof. Define linear forms $\eta_1,\ldots,\eta_{2s}$ by

$$\left. \begin{aligned} \eta_j &= -A^{-1} y_{s+j} \\ \eta_{s+j} &= A(L_j' + y_j) \end{aligned} \right\} \qquad (j=1,\ldots,s),$$

where

$$L_j' = \sum_{i=1}^{s} \lambda_{ij} y_{s+i}.$$

Then we have

$$\zeta_1 \eta_1 + \cdots + \zeta_{2s} \eta_{2s} = x_1 y_1 + \cdots + x_{2s} y_{2s}$$

identically in $\mathbf{x}, \mathbf{y}$. After a little thought, one infers that the polar lattice of Π is

$$\Lambda = \{(\eta_1(\mathbf{y}), \ldots, \eta_{2s}(\mathbf{y})) : \mathbf{y} \in \mathbb{Z}^{2s}\}.$$

Now the forms $L'_1, \ldots, L'_s$ are derived from the forms $L_1, \ldots, L_s$ by merely replacing $x_1, \ldots, x_s$ by $y_{s+1}, \ldots, y_{2s}$, because of the condition $\lambda_{ij} = \lambda_{ji}$. It follows easily that Λ has the same successive minima as Π, and (9.6) is now a consequence of Mahler's inequalities (7.10).

Lemma 9.2. *With the hypotheses and notations of Lemma 9.1, let $0 < B < A/2$. Then the number of solutions of the inequalities*

$$\left.\begin{aligned}\|L_i(x_1, \ldots, x_s)\| &< BA^{-1} &&(i = 1, \ldots, s)\\ |x_j| &\leq AB &&(j = 1, \ldots, s)\end{aligned}\right\} \tag{9.7}$$

in integers $x_1, \ldots, x_s$, is

$$\ll 1 + (\pi_1 \cdots \pi_l)^{-1} B^l \tag{9.8}$$

for some l, $1 \leq l \leq 2s$.

Proof. The number of solutions of the inequalities (9.7) is at most equal to the number of lattice points $\mathbf{p}$ in Π with $|\mathbf{p}| \leq \sqrt{(2s)}B$, with Π as in the previous proof. Just as in the proof of Lemma 7.1, this number of $\mathbf{p}$ is 1 if $\sqrt{(2s)}B < \pi_1$ and is

$$\ll (\pi_1 \cdots \pi_l)^{-1}(B^l + \pi_l^l) \ll (\pi_1 \cdots \pi_l)^{-1} B^l \tag{9.9}$$

otherwise, with π_l the greatest successive minimum that is $\leq \sqrt{(2s)}B$. This proves Lemma 9.2.

Lemma 9.3. *Let $1 \leq N \leq A$. With the hypotheses and notations of (9.1), (9.2) and Lemma 9.1, we have*

$$\sum_{1 \leq m \leq (AN^{-1})^2/4} |S(mQ)|^2 \ll A^{2\delta}(1 + (AN^{-1})^l(\pi_1 \cdots \pi_l)^{-1})N^s, \tag{9.10}$$

for some l, $1 \leq l \leq 2s$.

Proof. Since $\pi_1 \cdots \pi_l \ll d(\Pi) = 1$, we may suppose that $N \geq 2$. By (9.5) with mQ in place of Q we have

$$\sum_{m \leq (AN^{-1})^2/4} |S(mQ)|^2 \leq \sum_{m \leq (AN^{-1})^2/4} \sum_{z_1, \ldots, z_s = -(N-1)}^{N-1} \times \prod_{j=1}^{s} \min(N, \|2mL_j(\mathbf{z})\|^{-1}). \tag{9.11}$$

For any particular integers $x_1, \ldots, x_s$ we collect those terms in the multiple sum on the right of (9.11) for which

$$x_j = 2mz_j \quad (j = 1, \ldots, s);$$

the number of such terms is $\ll A^\delta$. Hence the left-hand side of (9.11), which we denote by S, satisfies

$$S \ll A^\delta \sum_{\substack{x_1,\ldots,x_s \\ |x_j| \leq A^2 N^{-1/2}}} \prod_{i=1}^{s} \min\left(N, \|L_i(\mathbf{x})\|^{-1}\right). \qquad (9.12)$$

Let $k_1, \ldots, k_s$ be integers satisfying $0 \leq k_i < N$ and let $\mathscr{E}(k_1, \ldots, k_s)$ be the set of $\mathbf{x}$ with

$$\frac{k_i}{N} \leq \{L_i(\mathbf{x})\} < \frac{k_i + 1}{N} \qquad (i = 1, \ldots, s),$$

$$|x_j| \leq A^2 N^{-1}/2 \qquad (j = 1, \ldots, s).$$

Fix $\mathbf{x}'$ in $\mathscr{E}(k_1, \ldots, k_s)$. Then for $\mathbf{x}'' = \mathbf{x}' + \mathbf{x}$ in $\mathscr{E}(k_1, \ldots, k_s)$,

$$\|L_i(\mathbf{x})\| < N^{-1} \qquad (i = 1, \ldots, s),$$
$$|x_j| \leq A^2 N^{-1} \qquad (j = 1, \ldots, s).$$

Applying Lemma 9.2 with $B = AN^{-1}$ (we see that

$$\max_{k_1,\ldots,k_s} |\mathscr{E}(k_1, \ldots, k_s)| \ll 1 + (AN^{-1})^l (\pi_1 \cdots \pi_l)^{-1} \qquad (9.13)$$

for some l, $1 \leq l \leq 2s$.

Combining (9.12) and (9.13), and writing $k_i' = \min(k_i, N - 1 - k_i)$,

$$S \ll A^\delta \sum_{k_1,\ldots,k_s=0}^{N-1} |\mathscr{E}(k_1, \ldots, k_s)| \prod_{i=1}^{s} \min\left(N, \frac{N}{k_i'}\right)$$

$$\ll A^\delta (1 + (AN^{-1})^l (\pi_1 \cdots \pi_l)^{-1}) \sum_{k_1,\ldots,k_s=0}^{N-1} \prod_{i=1}^{s} \min\left(N, \frac{N}{k_i'}\right)$$

$$\ll A^\delta (1 + (AN^{-1})^l (\pi_1 \cdots \pi_l)^{-1}) (N \log N)^s,$$

and the lemma follows.

9.2 Danicic's theorem

We can now prove Danicic's theorem on fractional parts of quadratic forms, mentioned in Chapter I.

Theorem 9.1. *Let $N > C_1(s, \varepsilon)$. Let $Q(x_1, \ldots, x_s)$ be a real quadratic form. Then there are integers $n_1, \ldots, n_s$ with*

$$0 < \max(|n_1|, \ldots, |n_s|) \leq N, \quad \|Q(n_1, \ldots, n_s)\| < N^{-s/(s+1))+\varepsilon}. \quad (9.14)$$

Proof. Suppose if possible that (9.14) is insoluble. Let $M = [N^{s/(s+1)-\varepsilon}] + 1$. Applying Theorem 2.2 to the sequence of N^s numbers

$Q(x_1, \ldots, x_s)(1 \leqq x_1, \ldots, x_s \leqq N)$ we obtain

$$\sum_{m=1}^{M} |S(mQ)| \geqq N^s/6$$

in the notation of Section 9.1.

Let $A = 2M^{1/2}N$. By Cauchy's inequality and Lemma 9.3,

$$N^{2s}M^{-1} \ll \sum_{m=1}^{M} |S(mQ)|^2$$

$$\ll (1 + (AN^{-1})^l(\pi_1 \cdots \pi_l)^{-1})N^{s+3\delta}$$

$$\ll (1 + M^{l/2}(\pi_1 \cdots \pi_l)^{-1})N^{s+3\delta}.$$

Here $\pi_1, \ldots, \pi_l$ are successive minima of the lattice Π of determinant 1 in Lemma 9.1. Thus, applying Minkowski's inequalities (7.8), and (9.6),

$$N^{2s}M^{-1} \ll (1 + M^{l/2}\pi_{l+1} \cdots \pi_{2s})N^{s+3\delta}$$

$$\ll (1 + M^{l/2}\pi_1^{-(2s-l)})N^{s+3\delta} \qquad (9.15)$$

From our hypothesis that (9.14) is insoluble we also deduce that

$$\pi_1 > M^{-1/2}/(4s). \qquad (9.16)$$

For suppose the contrary; then there is a nonzero integer point $(x_1, \ldots, x_{2s})$ with

$$\max_{1 \leqq j \leqq s} |L_j(x_1, \ldots, x_s) - x_{s+j}| < A^{-1}(4sM^{1/2})^{-1}$$

$$< s^{-1}M^{-1}N^{-1}, \qquad (9.17)$$

$$\max_{1 \leqq j \leqq s} |x_j| < A(4sM^{1/2})^{-1} < N. \qquad (9.18)$$

We cannot have $(x_1, \ldots, x_s) = (0, \ldots, 0)$, otherwise $x_{s+j} = 0$ $(j = 1, \ldots, s)$ from (9.17). Thus there is a nonzero integer point $(x_1, \ldots, x_s)$ satisfying (9.18) and

$$\|Q(x_1, \ldots, x_s)\| \leqq \sum_{j=1}^{s} x_i \|L_i(\mathbf{x})\| < sN(sMN)^{-1} = M^{-1},$$

which is a contradiction. This establishes (9.16).

Combining (9.15) and (9.16), we get

$$N^{2s}M^{-1} \ll (1 + M^{l/2}M^{s-l/2})N^{s+3\delta}$$

or

$$M^{s+1} \gg N^{s-3\delta}.$$

This contradicts the definition of M, and Theorem 9.1 is proved.

9.3 Small solutions of quadratic congruences

In preparation for our proof of Theorem 9.3, which is sharper than Danicic's theorem for $s \geq 4$, we prove the existence of small solutions of a quadratic congruence with integer coefficients.

Lemma 9.4. *Let p be a prime, let $d > 1$, and let*

$$Q(x_1, \ldots, x_{2d+1}) = \sum_{i=1}^{2d+1} \sum_{j=1}^{2d+1} a_{ij} x_i x_j$$

be a quadratic form over $\mathbb{F}_p$, with $\det Q = \det (a_{ij}) \neq 0$. Then after a linear transformation of the variables we may write Q in the form

$$Q_0(\mathbf{y}) = y_1 y_2 + y_3 y_4 + \cdots + y_{2d-1} y_{2d} + a y_{2d+1}^2. \tag{9.19}$$

(We say Q is 'equivalent to Q_0'.) There is a subspace of $\mathbb{F}_p^{2d+1}$ having dimension d on which Q takes only the value 0.

Proof. Suppose first p is odd. By Chevalley's theorem (Borevich and Shafarevich (1966), p. 6) any quadratic form $G(z_1, \ldots, z_s)$ over $\mathbb{F}_p$ with $s \geq 3$ takes the value zero for some $\mathbf{z} \neq \mathbf{0}$ in $\mathbb{F}_p^s$. According to Theorem 7, p. 394 of Borevich and Shafarevich, this property of G, together with the further hypothesis that G has nonzero determinant, guarantees that G is equivalent to a form

$$w_1 w_2 + H(w_3, \ldots, w_s).$$

It is now fairly obvious how to proceed. Q is equivalent to a form

$$y_1 y_2 + H(t_3, \ldots, t_{2d+1}) \quad \text{(say)}.$$

It is clear that $\det H \neq 0$. If $\mathbf{t} = (t_3, \ldots, t_{2d+1})$ has only one coordinate we are done. If not, we apply the same procedure to H instead of Q, and continue until 'only one variable remains'. This proves the first assertion of Lemma 9.4 for odd p. The case $p = 2$ is rather obvious, as $x_1^2 \pm x_2^2 = (x_1 + x_2)(x_1 - x_2) = y_1 y_2$ (say).

As for the second assertion, we need only note that $Q(\mathbf{x}) = Q_0(\mathbf{y})$ vanishes whenever $y_1 = y_3 = \cdots = y_{2d-1} = y_{2d} = 0$. This completes the proof of Lemma 9.4.

Lemma 9.5. *(Minkowski's linear forms theorem). Suppose that β_{ij} $(1 \leq i, j \leq n)$ are real numbers with determinant ± 1. Suppose that $A_1, \ldots, A_n$ are positive with $A_1 \cdots A_n = 1$. Then there exists an integer point $x = (x_1, \ldots, x_n) \neq \mathbf{0}$ such that*

$$|\beta_{i1} x_1 + \cdots + \beta_{in} x_n| < A_i \quad (1 \leq i \leq n-1)$$

and

$$|\beta_{n1}x_1 + \cdots + \beta_{nn}x_n| \leqq A_n.$$

Proof. See Schmidt (1980*b*), p. 33.

Lemma 9.6. *Let h be an odd positive integer and let $Q(x_1, \ldots, x_h)$ be a quadratic form over $\mathbb{Z}$. Let m be a natural number. Let $K_1, \ldots, K_h$ be a positive real number with*

$$K_1 \cdots K_h \geqq m^{(h+1)/2}. \tag{9.20}$$

Then there are integers $x_1, \ldots, x_h$, not all zero, with

$$Q(x_1, \ldots, x_h) \equiv 0 \pmod{m} \tag{9.21}$$

and having

$$|x_i| \leqq K_i \qquad (i = 1, \ldots, h). \tag{9.22}$$

Proof. We first observe that the result is trivial if $h = 1$ or if $K_i \geqq m$ for some i; hence we suppose that $h \geqq 3$,

$$K_i < m \qquad (i = 1, \ldots, h).$$

We may assume that m is square-free. For any m may be written in the form

$$m = r^2 a$$

where a is square-free. The inequality (9.20) implies

$$(K_1/r) \cdots (K_h/r) \geqq a^{(h+1)/2}.$$

A solution $(y_1, \ldots, y_h)$ of $Q(\mathbf{y}) \equiv 0 \pmod{a}$ with $\mathbf{y} \neq \mathbf{0}$, $|y_i| \leqq K_i/r$, yields a solution $x_i = ry_i$ of (9.21) satisfying (9.22), $\mathbf{x} \neq \mathbf{0}$.

For a square-free m, we consider a prime p dividing m. Let $d = (h-1)/2$. With a little thought, it can be seen that the second assertion of Lemma 9.4 is true even without the hypothesis that the quadratic form has non-zero determinant. Re-wording this assertion, we can find integer vectors $\mathbf{r}_1^{(p)}, \ldots, \mathbf{r}_d^{(p)}$ which are linearly independent modulo p (that is, $\sum_{i=1}^d x_i \mathbf{r}_i^{(p)} \equiv \mathbf{0} \pmod{p}$ implies all $x_i \equiv 0 \pmod{p}$) and with the following property:

$$Q(a_1 \mathbf{r}_1^{(p)} + \cdots + a_d \mathbf{r}_d^{(p)}) \equiv 0 \pmod{p}$$

whenever $a_1, \ldots, a_d$ are integers.

By the Chinese remainder theorem, there are integer vectors $\mathbf{r}_1, \ldots, \mathbf{r}_d$ having

$$\mathbf{r}_i \equiv \mathbf{r}_i^{(p)} \pmod{p} \tag{9.23}$$

for each prime p dividing m. Write $\mathbf{r}_i = (r_{i1}, \ldots, r_{ih})$.

Appealing to Minkowski's linear forms theorem, there are integers $a_1, \ldots, a_d, z_1, \ldots, z_h$, not all zero, with

$$|a_i| < m \qquad (i = 1, \ldots, d), \qquad (9.24)$$

$$\left| \sum_{k=1}^{d} a_k r_{kj} + m z_j \right| \leq K_j \qquad (j = 1, \ldots, h) \qquad (9.25)$$

(Here we recall the inequality (9.20), in the form

$$\frac{K_1}{m} \cdots \frac{K_h}{m} m^d \geq 1.\big)$$

Put $\mathbf{x} = a_1 \mathbf{r}_1 + \cdots + a_d \mathbf{r}_d + m\mathbf{z}$, where $\mathbf{z} = (z_1, \ldots, z_h)$. Then clearly (9.21) holds, and (9.22) follows from (9.25). Since $K_i < m$ we easily see that $(a_1, \ldots, a_d) \neq 0$, say $a_1 \neq 0$. Since m is square-free, there is a prime factor p of m with $a_1 \not\equiv 0 \pmod{p}$. This immediately yields $\mathbf{x} \not\equiv \mathbf{0} \pmod{p}$ by (9.23) and the linear independence property discussed above. Thus $\mathbf{x} \neq \mathbf{0}$, and Lemma 9.6 is proved.

The special case $K_1 = \cdots = K_h = m^{(h+1)/2h}$ of Lemma 9.6 was given by Schinzel *et al.* (1980). Baker and Harman (1982a) noted that it was as easy to prove the above lemma, and more efficient to have freedom in the choice of K_i in the application to real quadratic forms modulo 1.

Note that for an *even* number h of variables, this method has no more to offer than 'set one variable equal to zero', so that (9.20) would be replaced by a condition (e.g)

$$K_1 \cdots K_{h-1} \geq m^{(h-1)/2}$$

(say, $K_1 = \cdots = K_h = m^{h/(2h-2)}$).

Lemma 9.6 is best possible for $h = 3$, even for diagonal Q. One may see this in the following way (Baker (1983)).

Let X be a large positive number. We select three distinct primes p_1, p_2, p_3 in the interval $(X, (1-\varepsilon)^{-1/2}X)$, and write $m = p_1 p_2 p_3$. Let b_j be any quadratic non-residue $\pmod{p_j}$. Consider the set of congruences

$$x_1^2 - b_1 x_2^2 \equiv 0 \pmod{p_1}, \ x_2^2 - b_2 x_3^2 \equiv 0 \pmod{p_2}, \\ x_3^2 - b_3 x_1^2 \equiv 0 \pmod{p_3}. \Bigg\} \qquad (9.26)$$

Any *simultaneous* solution $\mathbf{x} = (x_1, x_2, x_3)$ of these congruences has $p_1 \mid (x_1, x_2)$, $p_2 \mid (x_2, x_3)$, $p_3 \mid (x_3, x_1)$. Thus all of x_1, x_2, x_3 are divisible by two of the p_i. Either $\mathbf{x} = \mathbf{0}$, or

$$\max_j |x_j| > X^2 > (1-\varepsilon) m^{2/3}. \qquad (9.27)$$

By the Chinese remainder theorem, there is a diagonal form $Q(\mathbf{x}) = a_1 x_1^2 + a_2 x_2^2 + a_3 x_3^2$ for which the congruence (9.21) is equivalent to the set of congruences (9.26). This yields the lower bound (9.27) for the size of nonzero solutions of (9.21).

Lemma 9.6 is also best possible for $s = 3$ and *prime modulus p* (Heath-Brown, (to appear)). Let $p > C_2(\varepsilon)$. Let a be a quadratic non-residue (mod p) and let $b = [p^{1/3}]$. We take

$$Q = (x_1 - bx_2)^2 - a(x_2 - bx_3)^2.$$

Then if $p \mid Q$ we must have $x_1 \equiv bx_2 \pmod{p}$ and $x_2 \equiv bx_3 \pmod{p}$. Now if $x_1 \neq bx_2$ we have $|x_1 - bx_2| \geq p$, whence

$$(1 + b) \max(|x_1|, |x_2|) \geq p.$$

Similarly if $x_2 \neq bx_3$ then

$$(1 + b) \max(|x_2|, |x_3|) \geq p.$$

It follows that

$$\max_{1 \leq i \leq 3} |x_i| \geq (1 + b)^{-1}p \geq (1 - \varepsilon)p^{2/3}$$

unless $x_1 = bx_2$ and $x_2 = bx_3$. In the latter case a non-zero solution must have $x_3 \neq 0$, whence

$$\max_{1 \leq i \leq 3} |x_i| \geq |x_1| = b^2 |x_3| \geq b^2 \geq p^{2/3}(1 - \varepsilon).$$

This gives the lower bound $p^{2/3}(1 - \varepsilon)$ for nonzero solutions of $Q(\mathbf{x}) \equiv 0 \pmod{p}$.

Lemma 9.6 is probably far from the truth for $h \geq 4$; one might conjecture that a permissible lower bound for $K_1 \cdots K_h$ is

$$K_1 \cdots K_h \geq m^{(h/2)+\varepsilon},$$

at least for $m \geq C_3(\varepsilon)$. In this direction we shall prove the following theorem of Heath-Brown (to appear).

Theorem 9.2. *Let $Q(x_1, x_2, x_3, x_4)$ be a quadratic form over $\mathbb{Z}$ in four variables. For any prime p the congruence*

$$Q(\mathbf{x}) \equiv 0 \pmod{p} \tag{9.28}$$

has a solution with

$$0 < |\mathbf{x}| \leq C_4 p^{1/2} \log p. \tag{9.29}$$

The constant C_4 is numerical.

A notable feature of the proof is the use of the Poisson summation formula, and we now prove a simple form of this.

Lemma 9.7.   *Let f be a continuous function in $L^1(\mathbb{R}^h)$ and suppose that the series*

$$g(\mathbf{x}) = \sum_{\mathbf{n} \in \mathbb{Z}^h} f(\mathbf{x} + \mathbf{n}) \tag{9.30}$$

converges absolutely for $\mathbf{x} \in \mathbb{R}^h$ and defines a continuous function g. Suppose further that

$$\sum_{\mathbf{m} \in \mathbb{Z}^h} |\hat{f}(\mathbf{m})| < \infty. \tag{9.31}$$

Then

$$\sum_{\mathbf{n} \in \mathbb{Z}^h} f(\mathbf{n}) = \sum_{\mathbf{m} \in \mathbb{Z}^h} \hat{f}(\mathbf{m}). \tag{9.32}$$

Proof.   Evidently g has period one in each variable. Moreover,

$$\hat{g}(\mathbf{m}) = \hat{f}(\mathbf{m})$$

for $\mathbf{m} \in \mathbb{Z}^h$, just as in the proof of Lemma 2.6. In view of (9.31) we can apply the Fourier formula

$$g(\mathbf{x}) = \sum_{\mathbf{m} \in \mathbb{Z}^h} \hat{g}(\mathbf{m}) e(\mathbf{m}\mathbf{x}) = \sum_{\mathbf{m} \in \mathbb{Z}^h} \hat{f}(\mathbf{m}) e(\mathbf{m}\mathbf{x})$$

(see the start of Section 7.3). In particular, for $\mathbf{x} = \mathbf{0}$, we get (9.32). This proves Lemma 9.7.

In our applications the condition on f in (9.30), and the bound (9.31), are easily verified and we shall not dwell on this.

We shall also need a standard result about the *Gauss sum*

$$G = \sum_{m=1}^{p-1} \left(\frac{m}{p}\right) e_p(m)$$

where p is an odd prime and $\left(\dfrac{m}{p}\right)$ is the Legendre symbol. Evidently

$$G = \sum_{R} e_p(R) - \sum_{N} e_p(N),$$

where R runs over quadratic residues (mod q) and N over non-residues. Since

$$\sum_{R} e_p(R) + \sum_{N} e_p(N) + 1 = \sum_{m=1}^{p} e_p(m) = 0, \tag{9.33}$$

we see that

$$G = 1 + 2\sum_{R} e_p(R) = \sum_{x=0}^{p-1} e_p(x^2). \tag{9.34}$$

After all, x^2 assumes the value 0 once and assumes each value R twice.

Davenport (1980; p. 12) proves that

$$G^2 = \pm p \tag{9.35}$$

and later evaluates G exactly, although we shall not need this. We do need the formula

$$\sum_{u=1}^{p} e_p(au^2 + bu) = G\left(\frac{a}{p}\right)e_p(-\overline{4ab}^2) \quad (p \nmid 4a). \tag{9.36}$$

Here, and in the rest of this section, $\bar{x}$ denotes inverse $(\bmod\, p)$, for $p \nmid x$. To get (9.36), we put

$$\sum_{u=1}^{p} e_p(au^2 + bu) = \sum_{u=1}^{p} e_p(a(u + \overline{2ab})^2 - \overline{4ab}^2)$$

$$= e_p(-\overline{4ab}^2)\sum_{v=1}^{p} e_p(av^2).$$

The last sum over v obviously matches that in (9.34) over x, if $\left(\dfrac{a}{p}\right) = 1$, whereas if $\left(\dfrac{a}{p}\right) = -1$ it is $1 + 2\sum_N e_p(N)$; we get (9.36) in either case on appealing to (9.33).

As a final preparation for the proof of Theorem 9.2 we give a sharper result for a subclass of quaternary forms.

Lemma 9.8. *Let* $Q = Q(x_1, x_2, x_3, x_4)$ *be a quadratic form over* $\mathbb{Z}$ *and* p *an odd prime. If* $p \mid \det Q$ *or* $\left(\dfrac{\det Q}{p}\right) = 1$ *then* $Q(\mathbf{x}) \equiv 0 \pmod{p}$ *for some* $\mathbf{x} \in \mathbb{Z}^4$ *with*

$$0 < \max_i |x_i| \leq p^{1/2}.$$

Proof. This resembles that of Lemma 9.6. We begin by showing that there are two linear forms $L_1(\mathbf{x})$, $L_2(\mathbf{x})$ such that $p \mid Q(\mathbf{x})$ whenever $L_1(\mathbf{x}) \equiv L_2(\mathbf{x}) \equiv 0 \pmod{p}$. To do this we work in the field $\mathbb{F}_p$, and look for a form $Q'(x_1', \ldots, x_4')$, equivalent to Q, such that $Q' = 0$ when $x_1' = x_2' = 0$. If Q has rank 2 or less this is immediate since Q is equivalent to a form $Q'(x_1', x_2')$. If Q has rank 3, then it can be transformed into $x_1'x_3' + ax_2'^2$ as in the proof of Lemma 9.4, and this suffices for our assertion. Finally, if Q is nonsingular, then as in the proof of Lemma 9.4 Q is equivalent to $Q' = x_1'x_2' + Q_0(x_3', x_4')$. Moreover, $\det Q_0 = -\det Q$, so that $-\det Q_0$ is a square in $\mathbb{F}_p$. Thus Q_0 factorizes as $Q_0 = y_1'y_2'$, whence $Q' = 0$ for $x_1' = y_1' = 0$. The existence of L_1, L_2 now follows in all cases.

With L_1, L_2 as above we now solve the inequalities

$$|L_i(\mathbf{x}) + py_i| < 1 \qquad (i = 1, 2)$$
$$|x_j| \leq p^{1/2} \qquad (j = 1, 2, 3, 4)$$

with $(\mathbf{x}, \mathbf{y}) \neq \mathbf{0}$, by appealing to Lemma 9.5. It is easy to see that $\mathbf{x} \neq \mathbf{0}$, and Lemma 9.8 follows.

Throughout the remainder of this section we assume that $p \geqq 3$ and $Q = Q(x_1, x_2, x_3, x_4)$, with $\left(\dfrac{\det Q}{p}\right) = -1$. We also think of the quadratic form as a matrix, also denoted by Q with entries in $\mathbb{F}_p$. With this convention Q^{-1} will be another quadratic form, with coefficients defined $(\bmod\, p)$. It is convenient to write $\mathbf{x}$ for a column vector in $\mathbb{R}^4$, giving

$$Q(\mathbf{x}) = \mathbf{x}^T Q \mathbf{x}.$$

Lemma 9.9. *Suppose that $f \in L^1(\mathbb{R}^4)$ and that f and $f(p\mathbf{x})$ satisfy the conditions of Lemma 9.7. Then*

$$\sum_{\mathbf{x} \in \mathbb{Z}^4, p | Q(\mathbf{x})} f(\mathbf{x}) = p^{-5} \sum_{\mathbf{y} \in \mathbb{Z}^4} S_p(\mathbf{y}) \hat{f}\left(\frac{1}{p}\mathbf{y}\right), \tag{9.37}$$

where

$$S_p(\mathbf{y}) = \sum_{s=1}^{p} \sum_{\mathbf{t}(\bmod\, p)} e_p(sQ(\mathbf{t}) + \mathbf{y}\mathbf{t}). \tag{9.38}$$

Proof. The left-hand side of (9.37) is

$$\frac{1}{p}\sum_{s=1}^{p} \sum_{\mathbf{x} \in \mathbb{Z}^4} e_p(sQ(\mathbf{x}))f(\mathbf{x}) = \frac{1}{p}\sum_{s=1}^{p} \sum_{\mathbf{t}(\bmod\, p)} e_p(sQ(\mathbf{t})) \sum_{\mathbf{u} \in \mathbb{Z}^4} f(\mathbf{t}+p\mathbf{u})$$

on writing lattice points $\mathbf{x}$ in the form $\mathbf{t}+p\mathbf{u}$. We apply the Poisson summation formula (9.32) to $f(\mathbf{t}+p\mathbf{x}) = F(\mathbf{x})$ rather than f, this function also obeys the conditions of Lemma 9.7.
    We get

$$\sum_{\mathbf{u} \in \mathbb{Z}^4} F(\mathbf{u}) = \sum_{\mathbf{y} \in \mathbb{Z}^4} \hat{F}(\mathbf{y}).$$

Since

$$\hat{F}(\mathbf{y}) = p^{-4} e_p(\mathbf{y}\mathbf{t}) \hat{f}\left(\frac{1}{p}\mathbf{y}\right),$$

Lemma 9.9 follows.

We note that the left-hand side of (9.37) counts solutions of the congruence we are interested in, with a 'smooth weight' $f(\mathbf{x})$ attached to the solutions.

Lemma 9.10. *We have*

$$S_p(\mathbf{y}) = p^2 + p^4 Y(\mathbf{y}) - p^3 Z(\mathbf{y})$$

where

$$Y(\mathbf{y}) = \begin{cases} 1 & \text{if } p \mid \mathbf{y} \\ 0 & \text{if } p \nmid \mathbf{y}, \end{cases} \qquad Z(\mathbf{y}) = \begin{cases} 1 & \text{if } p \mid Q^{-1}(\mathbf{y}) \\ 0 & \text{if } p \nmid Q^{-1}(\mathbf{y}). \end{cases}$$

Proof. We begin by diagonalizing Q. Choose R, invertible (mod p) such that $Q = R^T D R$, with $D = \text{Diag}\,(d_1, d_2, d_3, d_4)$. We substitute $R\mathbf{t} = \mathbf{u}$ in (9.38), so that

$$Q(\mathbf{t}) = \mathbf{t}^T Q \mathbf{t} = \mathbf{u}^T D \mathbf{u} = D(\mathbf{u}).$$

We also have

$$\mathbf{yt} = \mathbf{y}^T \mathbf{t} = \mathbf{y}^T R^{-1} \mathbf{u} = \mathbf{v}^T \mathbf{u} = \mathbf{vu},$$

with

$$\mathbf{v} = (R^{-1})^T \mathbf{y}. \tag{9.39}$$

Thus

$$S_p(\mathbf{y}) = \sum_{s=1}^{p} \sum_{\mathbf{u}(\text{mod } p)} e_p(sD(\mathbf{u}) + \mathbf{vu})$$

$$= p^4 Y(\mathbf{v}) + \sum_{s=1}^{p-1} \prod_{i=1}^{4} e_p(sd_i u_i^2 + v_i u_i). \tag{9.40}$$

Here the term $Y(\mathbf{v})$ is the contribution from $s = p$. From (9.39) we have $Y(\mathbf{v}) = Y(\mathbf{y})$.

By (9.36) we may write, for $1 \leq s \leq p - 1$,

$$\prod_{i=1}^{4} \sum_{u_i=1}^{p} e_p(sd_i u_i^2 + v_i u_i) = G^4 \prod_{i=1}^{4} \left(\frac{sd_i}{p}\right) e_p\left(-\overline{4s} \sum_{i=1}^{p} \bar{d}_i v_i^2\right). \tag{9.41}$$

Moreover, $G^4 = p^2$ from (9.35) and

$$\prod_{i=1}^{4} \left(\frac{sd_i}{p}\right) = \left(\frac{\det D}{p}\right) = \left(\frac{\det Q}{p}\right) = -1. \tag{9.42}$$

We deduce from (9.40)–(9.42) that

$$S_p(\mathbf{y}) = p^4 Y(\mathbf{y}) - p^2 \sum_{s=1}^{p-1} e_p(-\overline{4s}D^{-1}(\mathbf{v})).$$

Finally we observe that

$$\sum_{s=1}^{p-1} e_p(-\overline{4s}k) = \sum_{t=1}^{p-1} e_p(tk) = \begin{cases} p - 1 & \text{if } p \mid k \\ -1 & \text{if } p \nmid k, \end{cases}$$

and that

$$D^{-1}(\mathbf{v}) = \mathbf{v}^T D^{-1} \mathbf{v} = \mathbf{y}^T Q^{-1} \mathbf{y}.$$

Lemma 9.10 now follows.

Lemma 9.9 and 9.10 together yield, for suitable f,

$$\sum_{\mathbf{x}\in\mathbb{Z}^4, p\mid Q(\mathbf{x})} f(\mathbf{x}) = p^{-3}\sum_{\mathbf{y}\in\mathbb{Z}^4}\hat{f}\left(\frac{1}{p}\mathbf{y}\right) + p^{-1}\sum_{\mathbf{y}\in\mathbb{Z}^4}\hat{f}(\mathbf{y})$$
$$- p^{-2}\sum_{\mathbf{y}\in\mathbb{Z}^4, p\mid Q^{-1}(\mathbf{y})}\hat{f}\left(\frac{1}{p}\mathbf{y}\right).$$

We may apply the Poisson summation formula again to the first two sums on the right to produce

Lemma 9.11. *Let f satisfy the hypotheses of Lemma 9.9, then*

$$\sum_{\mathbf{x}\in\mathbb{Z}^4, p\mid Q(\mathbf{x})} f(\mathbf{x}) = p^{-1}\sum_{\mathbf{x}\in\mathbb{Z}^4} f(\mathbf{x}) + p\sum_{\mathbf{x}\in\mathbb{Z}^4} f(p\mathbf{x})$$
$$- p^{-2}\sum_{\mathbf{y}\in\mathbb{Z}^4, p\mid Q^{-1}(\mathbf{y})}\hat{f}\left(\frac{1}{p}\mathbf{y}\right).$$

Our choice of f will be based on the function h of the next lemma.

Lemma 9.12. *Define* $g(x) = \max(0, 1 - |x|)$. *Let* $h(x) = (g * g * g)(x)$. *Then*

(i) $h(x) = 0$ *for* $|x| > 3$,
(ii) $0 \leq h(x) \leq 1$ *for all* x,
(iii) $h(x) \geq 2^{-7}$ *for* $|x| \leq \frac{1}{4}$,
(iv) $\hat{h}(y) = (\sin \pi y)^6/(\pi y)^6$.

Proof. In the statement of the lemma we used the *convolution* operation defined in general by

$$(F * G)(x) = \int_{-\infty}^{\infty} F(x - t)G(t)\, dt$$

($F, G \in L^1(\mathbb{R})$). Background material about this can be found in Chapter XVI of Zygmund (1968). For the present context we need only the basic facts that $F * G \in L^1(\mathbb{R})$ for $F, G \in L^1(\mathbb{R})$, and

$$(F * G)\hat{} = \hat{F} \cdot \hat{G}. \tag{9.43}$$

The operation is therefore commutative and associative.

To prove (i)–(iv), we note that

$$h(x) = \int_{-\infty}^{\infty}\int_{-\infty}^{\infty} g(u)g(v - u)g(x - v)\, du\, dv. \tag{9.44}$$

Thus if $h(x) \neq 0$, there must exist u, v such that $|u| \leq 1$, $|v - u| \leq 1$, and $|x - v| \leq 1$. This requires $|x| \leq 3$, giving part (i). The lower bound $h \geq 0$ is obvious from (9.44). Moreover,

$$h(x) \le \int_{-\infty}^{\infty} \int_{-\infty}^{\infty} g(u)g(v-u) \, du \, dv = \int_{-\infty}^{\infty} \int_{-\infty}^{\infty} g(u)g(w) \, du \, dw = 1^2 = 1,$$

which establishes part (ii). For part (iii) we note that if $|u|, |v|, |x| \le \frac{1}{4}$, then $g(u), g(v-u), g(x-v) \ge \frac{1}{2}$, while the corresponding area of integration in (9.44) is $(\frac{1}{4})^2$.

Now g itself can be written $g = \chi * \chi$ where χ is the indicator function of $[-\frac{1}{2}, \frac{1}{2}]$. Also $\hat{\chi}(y) = (\sin \pi y)/\pi y$ by a simple calculation; now

$$\hat{g}(y) = \left(\frac{\sin \pi y}{\pi y}\right)^2, \qquad \hat{h}(y) = \left(\frac{\sin \pi y}{\pi y}\right)^6. \tag{9.45}$$

Note that a form of convolution occurs in Sections 2.1 and 7.3, but the functions there are periodic.

Proof of Theorem 9.2. Implied constants in this proof are numerical. We begin by applying Lemma 9.11 with the function

$$f(\mathbf{x}) = f_D(\mathbf{x}) = \prod_{i=1}^{4} h\left(\frac{x_i}{D}\right).$$

From Lemma 9.12, parts (i) and (ii), we have, for $D \ge 1$,

$$\sum_{\mathbf{x}} f_D(\mathbf{x}) \le |\{\mathbf{x} \in \mathbb{Z}^4 : |x_i| \le 3D\}| \ll D^4,$$

$$\sum_{\mathbf{x}} f_D(p\mathbf{x}) \le |\{\mathbf{x} \in \mathbb{Z}^4 : |x_i| \le 3D/p\}| = 1 \qquad (D < p/3).$$

By Lemma 9.12, part (iii), we have $f_D(x) \gg 1$ for $|x_i| \le D/4$, and by part (iv) we have $\hat{f}_D(\mathbf{y}) \ge 0$. We deduce

Lemma 9.13. *If* $\left(\dfrac{\det Q}{p}\right) = -1$ *and* $D < p/3$, *then*

$$|\{\mathbf{x} : |x_i| \le D/4, \, p \,|\, Q(\mathbf{x})\}| \ll p^{-1}D^4 + p.$$

Since $Q(\mathbf{x}) \equiv 0 \pmod{p}$ has $O(p^3)$ solutions $\pmod p$, the lemma is clearly true for $D > p/3$ too.

We can improve Lemma 9.13 for small values of D. Suppose that $D \le p^{1/2}/4$ and put $P = p^{1/2}(2D)^{-1}$. Consider primes q in the range $P < q \le 2P$. If $p \,|\, Q(\mathbf{x})$ with $|x_i| \le D$, then $p \,|\, Q(q\mathbf{x})$ and $|qx_i| \le p^{1/2}$. Hence

$$(\pi(2P) - \pi(P)) \,|\, \{\mathbf{x} \ne \mathbf{0} : |x_i| \le D, \, p \,|\, Q(\mathbf{x})\}$$

$$\le \sum_{\substack{0 < \max|y_i| \le p^{1/2} \\ p|Q(y)}} \left(\sum_{\substack{y=q\mathbf{x} \\ P < q \le 2P, \, \mathbf{x} \in \mathbb{Z}^4}} 1 \right).$$

The inner sum here is easily seen to be at most $\log p^{1/2}/(\log P)$. (If, say, $y_1 \neq 0$ and the inner sum is N, then y_1 has N distinct prime factors greater than P, so $y_1 > P^N$.) Moreover, since $P \geq 2$, we have

$$\pi(2P) - \pi(P) \gg P/(\log P).$$

Hence

$$\{\mathbf{x} \neq \mathbf{0}: |x_i| \leq D, \, p \mid Q(\mathbf{x})\}$$
$$\ll P^{-1}(\log p) \, |\{\mathbf{y}: |y_i| \leq p^{1/2}, \, p \mid Q(\mathbf{y})\}|$$
$$\ll P^{-1}p(\log p)$$

by Lemma 9.13. On using Lemma 9.13 itself for $D \geq p^{1/2}/4$, we now have

Lemma 9.14. *If* $\left(\dfrac{\det Q}{p}\right) = -1$, *then*

$$|\{\mathbf{x} \in \mathbb{Z}^4: |x_i| \leq D, \, p \mid Q(\mathbf{x})\} \ll D^4 p^{-1} + D p^{1/2}(\log p).$$

We apply this not to Q but to Q^{-1}, noting that

$$\left(\frac{\det Q^{-1}}{p}\right) = \left(\frac{\det Q}{p}\right) = -1.$$

We take $f = f_B$, with $p^{1/2} < B < p$, in Lemma 9.11, whence

$$\hat{f}_B\left(\frac{1}{p}\mathbf{y}\right) = B^4 \prod_{i=1}^{4} \left(\frac{\sin \pi y_i B/p}{\pi y_i B/p}\right)^6$$
$$\ll B^4 \prod_{i=1}^{4} \min\left(1, \left(\frac{p}{B\,|y_i|}\right)^6\right)$$
$$\ll B^4 \min\left(1, \left(\frac{p/B}{\max |y_i|}\right)^6\right).$$

We proceed to bound

$$\sum_{\mathbf{y} \in \mathbb{Z}^4, p \mid Q^{-1}(\mathbf{y})} \hat{f}\left(\frac{1}{p}\mathbf{y}\right).$$

The term $\mathbf{y} = \mathbf{0}$ contributes $O(B^4)$. We group the remaining terms into ranges

$$\tfrac{1}{2}D < \max |y_i| \leq D,$$

where D is a power of 2. In such a range, the number of terms is

$$\ll D^4 p^{-1} + D p^{1/2}(\log p)$$

according to Lemma 9.14, while each is of magnitude

$$\ll B^4 \min\left(1, (p/BD)^6\right).$$

The total for $D \leqq p/B$ is thus

$$\ll B^4 \sum_D (D^4 p^{-1} + Dp^{1/2} \log p)$$

$$\ll B^4 \left(\left(\frac{p}{B}\right)^4 p^{-1} + \left(\frac{p}{B}\right) p^{1/2} \log p \right)$$

$$\ll p^3 + p^{3/2} B^3 \log p,$$

while for $D \geqq p/B$ it is

$$\ll B^{-2} p^6 \sum_D (D^{-2} p^{-1} + D^{-5} p^{1/2} (\log p))$$

$$\ll B^{-2} p^6 \left(\left(\frac{p}{B}\right)^{-2} p^{-1} + \left(\frac{p}{B}\right)^{-5} p^{1/2} (\log p) \right)$$

$$\ll p^3 + p^{3/2} B^3 (\log p).$$

Hence

$$p^{-2} \sum_{\mathbf{y} \in \mathbb{Z}^4, p | Q^{-1}(\mathbf{y})} \hat{f}_B \left(\frac{1}{p} \mathbf{y}\right) \ll p^{-2} B^4 + p + p^{-1/2} B^3 (\log p)$$

$$\ll p^{-1/2} B^3 (\log p) \qquad (9.46)$$

since $p^{1/2} < B < p$.

On the other hand $p \sum_{\mathbf{x}} f_B(p\mathbf{x}) \geqq 0$, and, by part (iii) of Lemma 9.12,

$$p^{-1} \sum_{\mathbf{x} \in \mathbb{Z}^4} f_B(\mathbf{x}) \geqq 2^{-7} p^{-1} |\{\mathbf{x} \in \mathbb{Z}^4 : |x_i| \leqq B/4\}| \gg p^{-1} B^4. \qquad (9.47)$$

If the implied constants in (9.46) and (9.47) are C_5 and C_6 respectively, then

$$\sum_{\mathbf{x} \in \mathbb{Z}^4, p | Q(\mathbf{x})} f_B(\mathbf{x}) \geqq C_6 p^{-1} B^4 - C_5 p^{-1/2} B^3 (\log p) \qquad (9.48)$$

$$> (1/2) C_6 p^{-1} B^4$$

provided that $B \geqq 2C_5 C_6^{-1} p^{1/2} (\log p)$. Assuming, as we may, that p is large, the term $\mathbf{x} = \mathbf{0}$ contributes only $1 = o(p^{-1} B^4)$. It follows from Lemma 9.12, part (i), that $p | Q(\mathbf{x})$ with some $\mathbf{x} \neq \mathbf{0}$ for which $|x_i| \leqq 6C_5 C_6^{-1} p^{1/2} \log p$. This completes the proof of Theorem 9.2.

Note that it would not have been sufficient to use Lemma 9.13 in place of Lemma 9.14. The subtracted term in (9.48) would then have to incorporate a multiple of $p^{-1} B^4$, and no non-trivial result would be obtained.

9.4 Fractional parts of quadratic forms again

The following result ties with Theorem 9.1 for $s = 3$ and is stronger for $s \geqq 4$.

Theorem 9.3 (Baker and Harman (1982a)). *Let* $s \geq 3$ *and let* $Q(x_1, \ldots, x_s)$ *be a real quadratic form. Let* $N \geq C_3(s, \varepsilon)$. *Then there are integers* $n_1, \ldots, n_s$ *with*

$$0 < \max(|n_1|, \ldots, |n_s|) \leq N, \ \|Q(n_1, \ldots, n_s)\| < N^{-c(s)+\varepsilon} \quad (9.49)$$

Here

$$c(s) = \begin{cases} 2s/(s+5) & \text{for odd } s \\ 2s(s-1)/(s^2 + 4s - 4) & \text{for even } s. \end{cases}$$

The curious disparity between odd and even s which was forced on us in Lemma 9.6 naturally pops up again here, since Lemma 9.6 is used in the proof.

Proof of Theorem 9.3. As usual, we suppose that there are no integers $n_1, \ldots, n_s$ satisfying (9.49). Ultimately we shall obtain a contradiction. Arguing as in the first paragraph of the proof of Theorem 9.1, we may select a natural number m,

$$1 \leq m \leq L = [N^{c(s)-\varepsilon}] + 1, \quad (9.50)$$

such that

$$|S(m)| = \left| \sum_{n_1=1}^{N} \cdots \sum_{n_s=1}^{N} e(mQ(n_1, \ldots, n_s)) \right| > N^s/(6L). \quad (9.51)$$

We apply Lemma 9.3 with $A = 2N$, to estimate $S(m)$ from above. Let $L_1, \ldots, L_s$ be the linear forms in Lemma 9.3, *with the quadratic form Q replaced by mQ*. Then with $\pi_1, \ldots, \pi_l$ as in that lemma,

$$|S(m)|^2 \ll N^{s+2\delta}(1 + (\pi_1 \cdots \pi_l)^{-1}). \quad (9.52)$$

Combining (9.51) and (9.52), and noting that $N^{2s}L^{-2}$ is of larger order of magnitude than $N^{s+2\delta}$, we deduce that

$$(\pi_1 \cdots \pi_l)^{-1} \gg N^{s-2\delta}L^{-2}. \quad (9.53)$$

In fact, $\pi_s \ll 1$, $\pi_{s+1} \gg 1$ from (9.6). Considering separately the cases $l \leq s$ and $l > s$, we easily deduce from (9.53) that

$$(\pi_1 \cdots \pi_s)^{-1} \gg (\pi_1 \cdots \pi_l)^{-1} \gg N^{s-2\delta}L^{-2}. \quad (9.54)$$

We now consider the cases of odd and even s separately.

Case I. Odd s. By the definition of successive minima, we can find s linearly independent integer vectors $\mathbf{r}'_\mu$ in $2s$-dimensional space with

$$|mL_j(\mathbf{r}_\mu) - r_{j+s,\mu}| \leq (2N)^{-1}\pi_\mu \quad (9.55)$$

$$|r_{j,\mu}| \leq 2N\pi_\mu \quad (9.56)$$

for $j = 1, \ldots, s,$ $\mu = 1, \ldots, s.$ Here $\mathbf{r}'_\mu = (r_{1\mu}, \ldots, r_{2s\mu})$ and $\mathbf{r}_\mu = (r_{1\mu}, \ldots, r_{s\mu}).$

Let us write

$$K_\mu = m^{(s+1)/(2s)} L^{2/s} \pi_\mu^{-1} N^{-1+\delta}, \tag{9.57}$$

then

$$K_1 \cdots K_s \geq m^{(s+1)/2} L^2 (\pi_1 \cdots \pi_s)^{-1} N^{-s+3\delta} \tag{9.58}$$
$$\geq m^{(s+1)/2}$$

in view of (9.53). We also write

$$\theta_{\mu v} = m \sum_{j=1}^{s} r_{j\mu} L_j(\mathbf{r}_v) \qquad (\mu, v = 1, \ldots, s),$$

so that

$$\|\theta_{\mu v}\| \leq s \pi_\mu \pi_v \tag{9.59}$$

from (9.55) and (9.56). Let $b_{\mu v}$ be integers with

$$\|\theta_{\mu v}\| = |\theta_{\mu v} - b_{\mu v}| \qquad (\mu, v = 1, \ldots, s).$$

By Lemma 9.6 and (9.58) there are integers $x_1, \ldots, x_s$ not all zero, with

$$|x_\mu| \leq K_\mu \qquad (\mu = 1, \ldots, s) \tag{9.60}$$

and

$$\sum_{\mu=1}^{s} \sum_{v=1}^{s} b_{\mu v} x_\mu x_v \equiv 0 \qquad (\text{mod } m). \tag{9.61}$$

Put

$$\mathbf{n} = (n_1, \ldots, n_s) = \sum_{\mu=1}^{s} x_\mu \mathbf{r}_\mu.$$

The point of considering $\theta_{\mu v}$ and $b_{\mu v}$ is seen in the following expression for $Q(\mathbf{n})$:

$$Q(\mathbf{n}) = \sum_{\mu=1}^{s} \sum_{v=1}^{s} \left(\sum_{i=1}^{s} L_i(\mathbf{r}_\mu) r_{iv} \right) x_\mu x_v$$

$$= m^{-1} \sum_{\mu=1}^{s} \sum_{v=1}^{s} \theta_{\mu v} x_\mu x_v$$

$$= m^{-1} \sum_{\mu=1}^{s} \sum_{v=1}^{s} b_{\mu v} x_\mu x_v + m^{-1} \sum_{\mu=1}^{s} \sum_{v=1}^{s} (\theta_{\mu v} - b_{\mu v}) x_\mu x_v.$$

The first part of the last expression is an integer, by (9.61). Thus

$$\|Q(\mathbf{n})\| \leq m^{-1} \sum_{\mu=1}^{s} \sum_{v=1}^{s} |\theta_{\mu v} - b_{\mu v}| \, |x_\mu x_v|$$

$$\leq \frac{s}{m} \sum_{\mu=1}^{s} \sum_{v=1}^{s} \pi_\mu \pi_v K_\mu K_v$$

by (9.59), (9.60) and the definition of $b_{\mu\nu}$. Now (9.57) yields

$$\pi_\mu \pi_\nu K_\mu K_\nu = m^{(s+1)/s} L^{4/s} N^{-2+2\delta},$$

so

$$\|Q(\mathbf{n})\| \leqq s^3 m^{1/s} L^{4/s} N^{-2+2\delta}$$
$$< L^{5/s} N^{-2+3\delta} < L^{-1}$$

by definition of L.

Moreover, we have

$$|n_i| = \left| \sum_{\mu=1}^{s} r_{i\mu} x_\mu \right| \leqq \sum_{\mu=1}^{s} \pi_\mu 2N K_\mu$$
$$= 2sm^{(s+1)/(2s)} L^{2/s} N^{\delta}$$
$$< L^{(s+5)/(2s)} N^{2\delta} < N$$

from (9.56), (9.57), (9.60). By hypothesis, then, we must have

$$(n_1, \ldots, n_s) = \mathbf{0},$$

so that $\sum_{\mu=1}^{s} x_\mu \mathbf{r}_\mu = \mathbf{0}$ and consequently

$$\sum_{\mu=1}^{s} x_\mu L_j(\mathbf{r}_\mu) = 0 \qquad (j = 1, \ldots, s). \tag{9.62}$$

Combining (9.62) with (9.55) we obtain

$$\left| \sum_{\mu=1}^{s} x_\mu r_{j+s,\mu} \right| \leqq N^{-1} \sum_{\mu=1}^{s} \pi_\mu |x_\mu|$$
$$\leqq N^{-1} \sum_{\mu=1}^{s} \pi_\mu K_\mu < 1$$

as we already saw above. Hence

$$\sum_{\mu=1}^{s} x_\mu r_{j\mu} = 0$$

is true not only for $j = 1, \ldots, s$ but for $j = s+1, \ldots, 2s$ also. This contradicts the linear independence of $\mathbf{r}_1', \ldots, \mathbf{r}_s'$.

Thus the theorem is proved in Case I.

Case II. Even s. From (9.54) and $\pi_1 \leqq \cdots \leqq \pi_s$, we obtain

$$(\pi_1 \cdots \pi_{s-1})^{-1} \geqq (\pi_1 \cdots \pi_s)^{-(s-1)/s} \gg N^{s-1-2\delta} L^{-2(s-1)/s}. \tag{9.63}$$

Let $\mathbf{r}_\mu'$, $\mathbf{r}_\mu$, $\theta_{\mu\nu}$, $b_{\mu\nu}$ be as in Case I. Let

$$H_\mu = L^{2/s} m^{s/(2s-2)} \pi_\mu^{-1} N^{-1+2\delta}.$$

Then it follows from (9.63) that

$$H_1 \cdots H_{s-1} \geqq m^{s/2}.$$

By Lemma 9.6 there are integers $x_1, \ldots, x_{s-1}$, not all zero, such that

$$\sum_{\mu-1}^{s-1} \sum_{v=1}^{s-1} b_{\mu v} x_\mu x_v \equiv 0 \qquad (\text{mod } m)$$

and

$$|x_\mu| \leqq H_\mu \qquad (\mu = 1, \ldots, s-1).$$

Let

$$\mathbf{n} = (n_1, \ldots, n_s) = \sum_{\mu=1}^{s-1} x_\mu \mathbf{r}_\mu.$$

Then, much as in Case I,

$$|n_i| \leqq 2s \sum_{\mu=1}^{s} \pi_\mu N H_\mu \leqq 2 L^{(2/s)+s/(2s-2)} N^{2\delta} < N$$

from the definition of L. Moreover,

$$\|Q(\mathbf{n})\| \leqq m^{-1} \sum_{\mu=1}^{s-1} \sum_{v=1}^{s-1} \|\theta_{\mu v}\| \, |x_\mu x_v|$$

$$\leqq \frac{s^3}{m} \left(\max_\mu \pi_\mu H_\mu \right)^2 < L^{(4/s)+1/(s-1)} N^{-2+5\delta}$$

$$< L^{-1}.$$

The argument used in Case I can be repeated to obtain

$$\sum_{\mu=1}^{s-1} x_\mu \mathbf{r}'_\mu = \mathbf{0},$$

which is a contradiction. This proves the theorem in Case II.

In the case of a diagonal quadratic form

$$Q(x) = \alpha_1 x_1^2 + \cdots + \alpha_s x_s^2$$

with real $\alpha_1, \ldots, \alpha_s$, the exponent $c(s)$ of Theorem 9.3 can be increased somewhat. The reason is that the sum $S(mQ)$ factors as a product of s sums in an obvious way. We then have the option of studying rational approximation to an individual α_i, and can exploit the fact that (for instance)

$$Q(x_1, 0, \ldots, 0)$$

is small if x_1 and $\|\alpha_1 x_1\|$ are both small. (This technique will be illustrated in the next chapter.) See Baker (1983) for the sharpest known results for a diagonal quadratic form. Schmidt (1977b) gives a much stronger inequality, with exponent tending to minus infinity, under the extra hypothesis that $\alpha_1, \ldots, \alpha_s$ are 'not very well approximable'.

10
Simultaneous approximation for quadratic forms and additive forms

10.1 Introduction

The chapter is devoted to proving two theorems.

Theorem 10.1. *Let $h \geq 1$. Let $Q_1(x_1, \ldots, x_s), \ldots, Q_h(x_1, \ldots, x_s)$ be real quadratic forms, where $s > C_1(h, \varepsilon)$. Then for any integer $N > C_2(h, \varepsilon)$ there exist integers $n_1, \ldots, n_s$ with*

$$0 < \max(|n_1|, \ldots, |n_s|) \leq N, \tag{10.1}$$

$$\|Q_i(n_1, \ldots, n_s)\| < N^{-(2/h)+\varepsilon} \qquad (i = 1, \ldots, h). \tag{10.2}$$

The first theorem of this 'simultaneous' type for quadratic forms was given by Danicic (1967) with $h = 2$. In place of $-1 + \varepsilon$ he had the exponent $-\frac{1}{3} + V(s) + \varepsilon$, where $V(s)$ is an explicitly given function tending to zero as $s \to \infty$. We could cast Theorem 10.1 in an analogous 'explicit' form, but it would be a rather clumsy result.

It was pointed out in Chapter 1 that the constant $2/h$ in (10.2) is 'best possible' (and similarly in the next theorem). See the example after (1.30).

Theorem 10.2. *Let $h \geq 1$. Let*

$$F_i(x) = \alpha_{i1}x_1^k + \cdots + \alpha_{is}x_s^k \qquad (i = 1, \ldots, s)$$

be real diagonal forms of degree k, where $s > C_3(h, k, \varepsilon)$. Then for any integer $N > C_4(h, k, \varepsilon)$ there exist non-negative integers $n_1, \ldots, n_s$ with (10.1) and

$$\|F_i(n_1, \ldots, n_s)\| < N^{-(k/h)+\varepsilon} \qquad (i = 1, \ldots, h). \tag{10.3}$$

Theorem 10.2 is a special case of Theorem 10.1 when $k = 2$. In practice it is easier to prove Theorem 10.2 first, and then deduce Theorem 10.1 by a fairly straightforward 'almost-diagonalization' argument (see Section 10.3).

The proof of Theorem 10.2 depends on the following lemma. In case $k = 2$, we are already familiar with the result (Lemma 9.6). For $k \geq 3$, the lemma is much more difficult and we defer its proof to Chapter 11.

Lemma 10.1. *Let m be a natural number. Given integers $a_1, \ldots, a_s$ where $s \geqq C_5(k, \varepsilon)$, there exist non-negative integers $z_1, \ldots, z_s$ with*

$$a_1 z_1^k + \cdots + a_s z_s^k \equiv 0 \qquad (\mathrm{mod}\, m), \tag{10.4}$$

$$0 < \max (z_1, \ldots, z_s) \leqq m^{(1/k)+\varepsilon}. \tag{10.5}$$

10.2 Proof of Theorem 10.2

We are going to use the lattice method developed in Chapters 7 and 8. For this reason we write down a more general version of Theorem 10.2.

Proposition 10.1. *Let $N > C_4(h, k, \varepsilon)$. Let Λ be a lattice in $\mathbb{R}^h$ having $\Lambda \cap K_0 = \{\mathbf{0}\}$, and*

$$d(\Lambda)^{(1/k)+\varepsilon} \leqq N. \tag{10.6}$$

Let $\boldsymbol{\alpha}_1, \ldots, \boldsymbol{\alpha}_s$ be points of $\mathbb{R}^h$ and

$$\mathbf{F}(x_1, \ldots, x_s) = x_1^k \boldsymbol{\alpha}_1 + \cdots + x_s^k \boldsymbol{\alpha}_s, \tag{10.7}$$

where $s \geqq C_6(h, k, \varepsilon)$. Then there are non-negative integers $n_1, \ldots, n_s$ with (10.1) having

$$\mathbf{F}(n_1, \ldots, n_s) \in \Lambda + K_0. \tag{10.8}$$

This proposition implies Theorem 10.2 (and the reader may not need to be shown how). Let $s \geqq C_5(h, k, \varepsilon/k^2)$. Let

$$\Lambda = N^{(k-\varepsilon)/h} \mathbb{Z}^h,$$

then $\Lambda \cap K_0 = \{\mathbf{0}\}$, and (10.6) holds with ε/k^2 instead of ε. Now let

$$\boldsymbol{\alpha}_j = N^{(k-\varepsilon)/h}(\alpha_{1j}, \ldots, \alpha_{hj}) \qquad (j = 1, \ldots, s),$$

so that

$$\mathbf{F}(x_1, \ldots, x_s) = N^{(k-\varepsilon)/h}(F_1(x_1, \ldots, x_s), \ldots, F_h(x_1, \ldots, x_s)).$$

Let $n_1, \ldots, n_s$ be as in (10.8); then plainly

$$\|F_i(n_1, \ldots, n_s)\| < N^{-(k-\varepsilon)/h} \qquad (i = 1, \ldots, h).$$

This shows that Theorem 10.2 is true with $C_3(h, k, \varepsilon) = C_6(h, k, \varepsilon/k^2)$.

Proof of the Proposition. We proceed by induction on h. Suppose first that $h = 1$. Then, using the abbreviation $d(\Lambda) = \Delta$ as in previous chapters, we have

$$\Lambda = \Delta \mathbb{Z}.$$

Suppose if possible that a real form

$$F(\mathbf{x}) = \Delta \alpha_1 x_1^k + \cdots + \Delta \alpha_s x_s^k,$$

with $s \geq C_6(1, k, \varepsilon)$, satisfies $F(\mathbf{n}) \notin \Lambda + K_0$ whenever $\mathbf{n}$ is a vector satisfying (10.1) with $n_1 \geq 0, \ldots, n_s \geq 0$. Then for such an $\mathbf{n}$,

$$\|\alpha_1 n_1^k + \cdots + \alpha_s n_s^k\| \geq \Delta^{-1} \geq N^{-k/(1+\varepsilon)}. \tag{10.9}$$

We shall obtain a contradiction by an argument similar to that in Section 9.4. Let $L = [N^{k/(1+\varepsilon)}] + 1$, then by the argument leading to (9.51) we have

$$\left| \sum_{n_1=1}^{N} \cdots \sum_{n_s=1}^{N} e(m(\alpha_1 n_1^k + \cdots + \alpha_s n_s^k)) \right| \geq N^s/(6L) \tag{10.10}$$

for some natural number m, $1 \leq m \leq L$.

We may write the multiple sum on the left-hand side of (10.10) in the form $T_1 \cdots T_s$, where

$$T_i = \sum_{n=1}^{N} e(m\alpha_i n^k).$$

Thus

$$|T_1| \cdots |T_s| \geq N^s/(6L). \tag{10.11}$$

We may suppose without loss of generality that

$$|T_1| \geq \cdots \geq |T_s|.$$

The left-hand side of (10.11) is then at most

$$N^{c-1} |T_c|^{s-c+1}.$$

Here $c = [C_6(1, k, \varepsilon)/2]$. It follows that for $j = 1, \ldots, c$ we have

$$|T_j| \geq |T_c| \geq N(6L)^{-1/(s+1-c)} = P,$$

say. We have

$$P \geq N^{1-(1/K)+\delta},$$

since

$$NP^{-1} = (6L)^{1/(s+1-c)} < N^{2k/C_6} < N^\delta \tag{10.12}$$

by a suitable choice of C_6. By Theorem 5.1 with $M = 1$, there exist natural numbers q_j having

$$q_j < (NP^{-1})^k N^\delta < N^\eta, \tag{10.13}$$

$$\|q_j m\alpha_j\| < (NP^{-1})^k N^{\delta-k} < N^{\eta-k} \tag{10.14}$$

for $j = 1, \ldots, c$.

We write a_j for the nearest integer to $q_j^k m\alpha_j$ and select non-negative integers $x_1, \ldots, x_s$ such that

$$a_1 x_1^k + \cdots + a_c x_c^k \equiv 0 \pmod{m} \tag{10.15}$$

$$0 < \max x_i \leq m^{(1/k)+\delta}. \tag{10.16}$$

This is possible by Lemma 10.1 provided that

$$c = [C_6/2] \geqq C_5(k, \delta).$$

Now we use the identity

$$\sum_{j=1}^{c} x_j^k q_j^k \alpha_j = m^{-1} \sum_{j=1}^{c} a_j x_j^k + m^{-1} \sum_{j=1}^{c} x_j^k (q_j^k m \alpha_j - a_j).$$

The first expression on the right-hand side is an integer, by (10.15). Thus, appealing to (10.14), (10.16), we have

$$\left\| \sum_{j=1}^{c} x_j^k q_j^k \alpha_j \right\| \leqq m^{-1} \sum_{j=1}^{k} x_j^k \left\| q_j^k m \alpha_j \right\|$$

$$< m^{-1} m^{1+k\delta} N^{\eta-k} < N^{-k/(1+\varepsilon)}, \qquad (10.17)$$

since $m \leqq 2N^{k-\varepsilon}$. Moreover, $x_1 q_1, \ldots, x_c q_c$ are non-negative integers with

$$0 < \max x_j q_j \leqq m^{(1/k)+\delta} N^{\eta}$$

$$\leqq N^{\eta+k((1/k)+\eta)/(1+\varepsilon)} \leqq N.$$

Here we use the inequalities (10.13) and (10.16). Thus (10.17) contradicts our hypothesis (10.9). The case $h = 1$ of the Proposition is proved.

Now suppose that $h > 1$ and the proposition has been proved for $h - 1$ diagonal forms. Let

$$t = C_5(k, \delta) \qquad \text{and} \qquad u = C_6(h - 1, k, \delta).$$

We take

$$C_6(h, k, \varepsilon) = tu + (k^2/\delta). \qquad (10.18)$$

Let N, Λ and $\mathbf{F}$ be as in the statement of the proposition and suppose if possible that $\mathbf{F}(n_1, \ldots, n_s)$ lies outside $\Lambda + K_0$ whenever $n_1, \ldots, n_s$ are non-negative integers satisfying (10.1). We apply Lemma 7.4 to obtain the lower bound

$$\sum_{\substack{\mathbf{p} \in \Pi \\ 0 < |\mathbf{p}| < N^{\delta}}} \left| \sum_{n_1=1}^{N} \cdots \sum_{n_s=1}^{N} e(\mathbf{p}\mathbf{F}(n_1, \ldots, n_s)) \right| \gg N^s$$

for a sum in the dual lattice Π.

The number of terms in the sum over $\mathbf{p}$ is $\ll \Delta N^{h\delta}$, by Lemma 7.1. Let $\mathbf{p}'$ be an element of Π for which $0 < |\mathbf{p}'| < N^{\delta}$ and

$$\left| \sum_{n_1=1}^{N} \cdots \sum_{n_s=1}^{N} e(\mathbf{p}'\mathbf{F}(n_1, \ldots, n_s)) \right| = |T_1| \cdots |T_s| \geqq N^{s-2h\delta} \Delta^{-1}.$$

Here

$$T_j = \sum_{n=1}^{N} e(\mathbf{p}' n^k \alpha_j) \qquad (j = 1, \ldots, s).$$

Just as in the case $h = 1$, we derive the inequalities

$$|T_j| \geq |T_{tu}| \geq N(\Delta N^{2h\delta})^{-1/(s+1-tu)} = P \qquad (10.19)$$

for $j = 1, \ldots, tu$. We have

$$NP^{-1} = (\Delta N^{2h\delta})^{1/(s+1-tu)} \leq (N^k)^{1/(s+1-tu)} < N^\delta,$$

on appealing to (10.18).

By the case $M = 1$ of Theorem 5.1, for $j = 1, \ldots, tu$ there are integers q_j having

$$1 \leq q_j \leq N^\delta(NP^{-1})^k < N^{2\delta} \qquad (10.20)$$

$$\|q_j \mathbf{p}' \boldsymbol{\alpha}_j\| \leq N^{\delta-k}(NP^{-1})^k < N^{2\delta-k}. \qquad (10.21)$$

Hence

$$|q_j^k \mathbf{p}' \boldsymbol{\alpha}_j - b_j| \leq q_j^{k-1} \|q_j \mathbf{p}' \mathbf{a}_j\| < N^{\eta-k} \qquad (10.22)$$

for some integers b_j $(j = 1, \ldots, tu)$.

There is a natural number m such that $\mathbf{p} = m^{-1}\mathbf{p}'$ is a primitive point of Π. Note that

$$m < |\mathbf{p}|^{-1} N^\delta \ll \Delta N^\delta < N^k \qquad (10.23)$$

(recall (7.27) for the lower bound $|\mathbf{p}| \gg \Delta^{-1}$). We are now going to give small solutions of congruences in u variables (modulo m) using Lemma 10.1.

It is convenient to renumber the points $\boldsymbol{\alpha}_j$ $(j = 1, \ldots, tu)$ as

$$\boldsymbol{\alpha}_{iv} \qquad (i = 1, \ldots, t; v = 1, \ldots, u)$$

and to define q_{iv}, b_{iv} similarly. For each $v = 1, \ldots, u$ there are non-negative integers $x_{1v}, \ldots, x_{tv}$ such that

$$\sum_{i=1}^{t} x_{iv}^k b_{iv} \equiv 0 \pmod{m} \qquad (10.24)$$

and (recalling that $t = C_5(k, \delta)$)

$$0 < \max(x_{1v}, \ldots, x_{tv}) \leq m^{(1/k)+\delta}. \qquad (10.25)$$

The inequalities (10.20), (10.22) may be rewritten in the form

$$1 \leq q_{iv} \leq N^{2\delta} \qquad (i = 1, \ldots, t; v = 1, \ldots, u), \qquad (10.26)$$

$$|q_{iv}^k m \mathbf{p} \boldsymbol{\alpha}_{iv} - b_{iv}| < N^{\eta-k} \qquad (i = 1, \ldots, t; v = 1, \ldots, u). \qquad (10.27)$$

Now we use the identity

$$\sum_{i=1}^{t} x_{iv}^k q_{iv}^k \boldsymbol{\alpha}_{iv} \mathbf{p} = m^{-1} \sum_{i=1}^{t} x_{iv}^k b_{iv} + m^{-1} \sum_{i=1}^{t} x_{iv}^k (q_{iv}^k m \boldsymbol{\alpha}_{iv} \mathbf{p} - b_{iv})$$

to establish that

$$\left\| \sum_{i=1}^{t} x_{iv}^{k} q_{iv}^{k} \alpha_{iv} \mathbf{p} \right\| \leq m^{-1} \sum_{i=1}^{t} x_{iv}^{k} \left| q_{iv}^{k} m \mathbf{p} \alpha_{iv} - b_{iv} \right|$$
$$\leq t m^{-1} m^{1+k\delta} N^{\eta-k} < N^{\eta-k}. \tag{10.28}$$

Here we used in turn (10.24), (10.25), (10.27) and (10.23).

We still have the option of multiplying each x_{iv}^{k} $(i = 1, \ldots, t)$ by a variable y_{v}^{k}, for each value of $v = 1, \ldots, u$, and then considering the sum

$$\sum_{v=1}^{u} y_{v}^{k} \sum_{i=1}^{t} x_{iv}^{k} q_{iv}^{k} \alpha_{iv} = \mathbf{F}(y_1 x_{11} q_{11}, \ldots, y_u x_{tu} q_{tu}, 0, \ldots, 0)$$

modulo Λ. This technique of 'dividing the variables into blocks' in a problem where a very large number of variables is available, can be traced back at least to Linnik (1943) (see Gelfond and Linnik (1966), Chapter 2 for an account in English). More recently it was taken up by Birch (1970) and Schmidt (1979*a*, *b*). We shall come across it again in several subsequent chapters.

By the case $t = 1$, $\mathbf{p}_1 = \mathbf{p}$, $T =$ linear span of $\mathbf{p}$, of Lemma 7.9, we may write

$$\sum_{i=1}^{t} x_{iv}^{k} q_{iv}^{k} \alpha_{iv} = \mathbf{l}_v + \mathbf{s}_v + \mathbf{b}_v, \tag{10.29}$$

where $\mathbf{l}_v \in \Lambda$, $\mathbf{s}_v \in T^{\perp}$ and

$$|\mathbf{b}_v| \ll d(\Pi')^{-1} \left\| \mathbf{p} \sum_{i=1}^{t} x_{iv}^{k} q_{iv}^{k} \alpha_{iv} \right\|$$
$$\ll |\mathbf{p}|^{-1} N^{\eta-k} \tag{10.30}$$

from (10.28).

We apply the induction hypothesis to the form

$$\mathbf{F}^*(y_1, \ldots, y_u) = 2y_1^{k} \mathbf{s}_1 + \cdots + 2y_u^{k} \mathbf{s}_u$$

having coefficients in the $(h - 1)$-dimensional space $T^{\perp}$. In doing so we replace ε by δ, Λ by $2\Lambda'$ and N by

$$N^* = (|\mathbf{p}| \, \Delta)^{1/k} N^{2k\delta} \gg N^{2k\delta}.$$

Note that $(2\Lambda') \cap K_0 = \{\mathbf{0}\}$ and, recalling (7.45),

$$d(2\Lambda')^{(1/k)+\delta} < (|\mathbf{p}| \, \Delta)^{(1/k)+\delta} N^{\delta} < N^*.$$

Since $u = C_6(h - 1, k, \delta)$, there are non-negative integers $y_1, \ldots, y_u$ such that

$$\mathbf{F}^*(y_1, \ldots, y_u) \in 2\Lambda' + (K_0 \cap T^{\perp}) \tag{10.31}$$

and

$$0 < \max (y_1, \ldots, y_u) \leq N^*. \tag{10.32}$$

Let us combine this with (10.29). Writing $n_{iv} = q_{iv} x_{iv} y_v$, we obtain

$$\sum_{i=1}^{t} \sum_{v=1}^{u} n_{iv}^k \mathbf{a}_{iv} = \sum_{v=1}^{u} y_v^k \mathbf{l}_v + \sum_{v=1}^{u} y_v^k \mathbf{s}_v + \sum_{v=1}^{u} y_v^k \mathbf{b}_v. \qquad (10.33)$$

Of the three summands on the right-hand side of (10.33), the first is a point of Λ, and the second is in

$$\Lambda + \tfrac{1}{2} K_0$$

in view of (10.31). The third is in $\frac{1}{2} K_0$, since from (10.30), (10.32), (10.6), we have

$$\left| \sum_{v=1}^{u} y_v^k \mathbf{b}_v \right| \leqq (N^*)^k |\mathbf{p}|^{-1} N^{\eta - k}$$

$$\leqq |\mathbf{p}| \, \Delta \, |\mathbf{p}|^{-1} N^{\eta - k} < \tfrac{1}{2}.$$

We conclude that the right-hand side of (10.33) is in $\Lambda + K_0$. But its value is

$$F(n_{11}, \ldots, n_{tu}, 0, \ldots, 0).$$

Moreover, taking into account (10.26), (10.25), (10.32) and (10.23) we have

$$0 < \max (n_{11}, \ldots, n_{tu}) \leqq N^{2\delta} m^{(1/k) + \delta} (|\mathbf{p}| \, \Delta)^{1/k} N^{2k\delta}$$

$$\leqq N^{\eta} (m \, |\mathbf{p}|)^{1/k} \Delta^{1/k}$$

$$\leqq N^{\eta} \Delta^{1/k} \leqq N.$$

This contradicts the hypothesis about the insolubility of (10.8) and completes the induction step. The proposition is proved.

10.3 Application to a set of quadratic forms

The type of argument in the present section has quite a long history; compare, for example, Birch and Davenport (1958*b*). Their source of inspiration was a paper of Brauer (1945) on the solubility of homogeneous equations in *p*-adic fields. The present application can be thought of as a dry run for a much more sophisticated 'almost diagonalization' of Schmidt, which occurs in Chapter 14.

Lemma 10.2. *Let r, w be natural numbers and $Z \geqq 1$. Let θ_{ij} be real $(i = 1, \ldots, r; j = 1, \ldots, w)$. Then there are integers $z_1, \ldots, z_w$ with*

$$\|\theta_{i1} z_1 + \cdots + \theta_{iw} z_w\| < Z^{-w/r} \qquad (i = 1, \ldots, r),$$

$$0 < \max (|z_1|, \ldots, |z_w|) \leqq Z.$$

Proof. This follows from Lemma 9.5 applied to the $r + w$ linear forms
$$\theta_{i1}z_1 + \cdots + \theta_{iw}z_w + z_{w+i} \quad (i = 1, \ldots, r) \qquad \text{and} \qquad z_j \quad (j = 1, \ldots, w).$$
Note that $\mathbf{z} = (z_1, \ldots, z_{w+r}) \neq 0$, $\mathbf{z} \in \mathbb{Z}^{r+w}$, together with

$$|\theta_{i1}z_1 + \cdots + \theta_{iw}z_w + z_{w+i}| < Z^{-w/r} \qquad (i = 1, \ldots, r)$$

implies $(z_1, \ldots, z_w) \neq \mathbf{0}$.

Proof of Theorem 10.1. We write $b = [C_3(h, 2, \delta)] + 1$, and take $N \geq b^{1/\delta} + C_4(h, 2, \delta)^2$. Let

$$w = [5h(b-1)/\delta] + 1. \tag{10.34}$$

Suppose that

$$s \geq C_1(h, \varepsilon) = bw,$$

and let $Q_1(x_1, \ldots, x_s), \ldots, Q_h(x_1, \ldots, x_s)$ be arbitrary quadratic forms with real coefficients.

Let $\mathscr{S}_1, \ldots, \mathscr{S}_b$ be disjoint subsets of $\{1, \ldots, s\}$ each having w members. Let $\mathbf{z}_1 = (z_{11}, \ldots, z_{1s}), \ldots, \mathbf{z}_b = (z_{b1}, \ldots, z_{bs})$ be non-zero integer s-tuples, to be chosen below, having

$$z_{ij} = 0 \qquad \text{if } j \notin \mathscr{S}_i \tag{10.35}$$

for $i = 1, \ldots, b$. Let us write

$$(n_1, \ldots, n_s) = y_1\mathbf{z}_1 + \cdots + y_b\mathbf{z}_b, \tag{10.36}$$

where $\mathbf{y} = (y_1, \ldots, y_b)$ is a non-zero integer b-tuple, to be chosen below. We have

$$Q_l(X_1, \ldots, X_s) = \sum_{1 \leq i, j \leq s} \lambda_{ij}^{(l)} X_i X_j \qquad (\lambda_{ij}^{(l)} = \lambda_{ji}^{(l)}).$$

We write $B_l(\mathbf{X}, \mathbf{Y})$ for the bilinear form

$$B_l(\mathbf{X}, \mathbf{Y}) = \sum_{1 \leq i, j \leq s} \lambda_{ij}^{(l)} X_i Y_j.$$

Note that

$$B_l(\mathbf{X}, \mathbf{Y}) = B_l(\mathbf{Y}, \mathbf{X}). \tag{10.37}$$

From (10.36) we easily deduce that

$$Q_l(n_1, \ldots, n_s) = \sum_{p=1}^{b} \sum_{q=1}^{b} B_l(\mathbf{z}_p, \mathbf{z}_q) y_p y_q. \tag{10.38}$$

Now we are ready to choose $\mathbf{z}_1, \ldots, \mathbf{z}_b$. We do this in such a way that the quadratic form in $\mathbf{y}$ appearing on the right-hand side of (10.38) is 'almost diagonal', at least considered modulo 1. Namely, let z_{1j} be 1 or 0 according as $j \in \mathscr{S}_1$ or not. If $\mathbf{z}_1, \ldots, \mathbf{z}_{p-1}$ have been chosen, where $1 < p \leqq b$, Lemma 10.2 allows us to choose $\mathbf{z}_p$ so that (10.35) holds, and in addition

$$\|B_i(\mathbf{z}_j, \mathbf{z}_p)\| < N^{-\delta w/(h(p-1))} \qquad (j = 1, \ldots, p-1; i = 1, \ldots, h) \quad (10.39)$$

and

$$0 < \max_i |z_{pi}| \leqq N^\delta. \tag{10.40}$$

Having chosen $\mathbf{z}_1, \ldots, \mathbf{z}_b$, we select $\mathbf{y}$ so that

$$\left\| \sum_{p=1}^{b} B_i(\mathbf{z}_p, \mathbf{z}_p) y_p^2 \right\| < N^{-(1-\delta)((2/h)-\delta)} \tag{10.41}$$

for $i = 1, \ldots, h$, and

$$0 < \max (|y_1|, \ldots, |y_b|) \leqq N^{1-\delta}. \tag{10.42}$$

This can be done, by Theorem 10.2, since $b > C_3(h, 2, \delta)$, and

$$N^{1-\delta} > N^{1/2} > C_4(h, 2, \delta).$$

Combining (10.38)–(10.42), and taking account of (10.37), we see that, for $i = 1, \ldots, h$,

$$\|Q_i(n_1, \ldots, n_s)\| \leqq 2 \sum_{1 \leqq p < q \leqq b} \|B_i(\mathbf{z}_p, \mathbf{z}_q)\| \, |y_p| \, |y_q|$$

$$+ \left\| \sum_{p=1}^{b} B_i(\mathbf{z}_p, \mathbf{z}_p) y_p^2 \right\|$$

$$\leqq b^2 N^{2-2\delta} N^{-\delta w/h(b-1)} + N^{-(1-\delta)(2/h-\delta)}$$

$$\leqq N^2 N^{-5} + N^{-(2/h)+3\delta} \leqq N^{-(2/h)+\varepsilon}.$$

Moreover, it follows from (10.35), (10.40) and (10.42) that $n_1, \ldots, n_s$ satisfy (10.1). This completes the proof of Theorem 10.1.

11
Nonnegative solutions of additive equations

11.1 A direct application of the circle method

We are interested in small solutions of the equation

$$c_1 x_1^k + \cdots + c_r x_r^k - (c_{r+1} x_{r+1}^k + \cdots + c_s x_s^k) = 0 \qquad (11.1)$$

in *non-negative* integers $x_1, \ldots, x_s$, where $c_1, \ldots, c_s$ are given natural numbers. Our goal is the following result of Schmidt (1979a):

Theorem 11.1. *Let* $\min(r, s-r) > C_1(k, \varepsilon)$. *There is a solution* $x_1, \ldots, x_s$ *of* (11.1) *with* $x_i \geq 0$ $(i = 1, \ldots, s)$,

$$0 < \max_i x_i \leq m^{(1/k)+\varepsilon}. \qquad (11.2)$$

Here $m = \max(c_1, \ldots, c_s)$.

A weaker formulation (with x_i of both signs) was given earlier by Birch (1970). There is no hope of improving the constant $1/k$ in (11.2). For if a, b are coprime integers and $s > r$, every non-trivial non-negative solution of

$$a(x_1^k + \cdots + x_r^k) - b(x_{r+1}^k + \cdots + x_s^k) = 0$$

has

$$x_1^k + \cdots + x_r^k \geq b \qquad \text{and} \qquad x_{r+1}^k + \cdots + x_s^k \geq a,$$

whence

$$\max(x_1, \ldots, x_r, x_{r+1}, \ldots, x_s) \geq C_2(s, k) m^{1/k}.$$

Lemma 10.1 is an easy corollary of Theorem 11.1. In that lemma, $a_1, \ldots, a_s$ can be assumed to satisfy $1 \leq a_i \leq m$. If $x_1, \ldots, x_{2s}$ is a non-negative solution of

$$a_1 x_1^k + \cdots + a_s x_s^k - m(x_{s+1}^k + \cdots + x_{2s}^k) = 0$$

with (11.2), then evidently $(x_1, \ldots, x_s) \neq \mathbf{0}$. Thus $\mathbf{z} = (x_1, \ldots, x_s)$ does what is required in Lemma 10.1.

It is a well-known consequence of the Hardy–Littlewood method that if $s \geq C_3(k)$, $1 \leq r < s$, then solutions $x_1, \ldots, x_s$ of (11.1) do exist. As to how many variables are required merely to *solve* (11.1), see Vaughan (1977) for recent results and references. A more helpful starting point from the point of view of Theorem 11.1 is a result of Pitman (1971b) that for $s \geq C_4(k)$, (11.1) is soluble with $x_i \geq 0$ ($i = 1, \ldots, s$),

$$0 < \max_i c_i x_i^k < C_5(k)(c_1 \cdots c_s)^{C_6(k)}. \tag{11.3}$$

The question of how small one may take $C_6(k)$ with (say) $C_4(k) = 2^k + 1$ is an interesting one, and has applications to solving Diophantine inequalities of the form

$$|\lambda_1 x_1^k + \cdots + \lambda_s x_s^k| < 1,$$

with x_i satisfying

$$0 < |\lambda_1 x_1^k| + \cdots + |\lambda_s x_s^k| < C_7(k) |\lambda_1 \cdots \lambda_s|^{C_8(k)}.$$

Here $\lambda_1, \ldots, \lambda_s$ are given real numbers with $|\lambda_i| \geq 1$ for all i, not all of the same sign. See Birch and Davenport (1958a), Pitman and Ridout (1967) and Pitman (1971a, b). Theorem 11.1 also has an application of this general type (Schlickewei (1979)). We shall be considering similar inequalities for equations of odd degree in Chapter 14, en route to our main results there.

We shall need solutions of (11.1) subject to some bound (11.3) in order to start an inductive proof of Theorem 11.1. We aim for a straightforward proof rather than for good bounds for $C_4(k)$ or $C_6(k)$.

Proposition 11.1.   *Let* $s \geq 36Kk$. *Let* $c_1, \ldots, c_s$ *be natural numbers and let* $1 \leq r < s$. *The equation* (11.1) *has a solution in non-negative integers* $x_1, \ldots, x_s$ *satisfying*

$$0 < \max_i c_i x_i^k \leq C_9(k)(c_1 \cdots c_s)^{16k}. \tag{11.4}$$

Proof.   Suppose first that each of $c_1, \ldots, c_s$ is k-free (that is, not divisible by p^k for any prime p). We use the Hardy–Littlewood circle method. We may suppose that $s = 36Kk$. Let

$$A = [C_9(k)^{1/k}(c_1 \cdots c_s)^{16}],$$

and

$$S_i(\alpha) = \sum_{1 \leq x \leq Ac_i^{-1/k}} e(\alpha c_i x^k).$$

Then the integral

$$Z = \int_0^1 S_1(\alpha) \cdots S_r(\alpha) \overline{S_{r+1}(\alpha)} \cdots \overline{S_s(\alpha)} \, d\alpha \tag{11.5}$$

counts solutions of (11.1) subject to

$$1 \leq x_i \leq Ac_i^{-1/k} \qquad (i = 1, \ldots, s).$$

This is a consequence of the identity

$$\int_0^1 e(nx) \, dx = \begin{cases} 0 & \text{for } n \in \mathbb{Z}, \ n \neq 0 \\ 1 & \text{for } n = 0. \end{cases}$$

We are going to show that $Z > 0$.

It is convenient to write

$$\omega = \tfrac{1}{8}.$$

When u, q are integers, $1 \leq u \leq q \leq A^\omega$ and $(u, q) = 1$, we define

$$\mathcal{M}(q, u) = \{\alpha: |\alpha - u/q| \leq A^{\omega - k}\}.$$

The $\mathcal{M}(q, u)$ are called *major arcs*. It is convenient to work on the interval $I = [A^{\omega - k}, 1 + A^{\omega - k}]$ rather than $[0, 1]$ as in (11.5). The set

$$\mathcal{M} = \bigcup_{1 \leq u \leq q \leq A^\omega} \mathcal{M}(q, u)$$

is contained in I; note that the $\mathcal{M}(q, u)$ are pairwise disjoint. The complement of $\mathcal{M}$ in I forms the *minor arcs*.

We have

$$|S_1(\alpha) \cdots \overline{S_s(\alpha)}| < A^{s-k-1} \tag{11.6}$$

throughout the minor arcs. For suppose that $\alpha \in I$ and

$$|S_1(\alpha) \cdots \overline{S_s(\alpha)}| \geq A^{s-k-1}.$$

Then plainly

$$|S_i(\alpha)| \geq A^{1-(k+1)/s} > A^{1-(1/K)+(1/s)}$$

for some i. By the case $M = 1$ of Theorem 5.1, there exist coprime integers q and u with

$$q \leq (A^{(k+1)/s} c_i^{-1/k})^k A^{\omega/6} \leq A^{\omega/2},$$

$$|\alpha c_i q - u| \leq (A^{(k+1)/s} c_i^{-1/k})^k A^{\omega/6} (Ac_i^{-1/k})^{-k} < A^{\omega - k}.$$

Since $c_i q \leq A^\omega$, α is seen to lie on a major arc. This proves our assertion.

It follows from (11.6) that

$$Z = \left(\int_\mathcal{M} + \int_{I \setminus \mathcal{M}} \right) S_1(\alpha) \cdots \overline{S_s(\alpha)} \, d\alpha$$

$$= \sum_{q \leq A^\omega} \sum_{\substack{u=1 \\ (u,q)=1}}^{q} \int_{\mathcal{M}(q,u)} S_1(\alpha) \cdots \overline{S_s(\alpha)} \, d\alpha + O(A^{s-k-1}). \tag{11.7}$$

(Implied constants in the present section depend at most on k.)

We need a result along the lines of Lemma 4.4. With later work in mind, we use a cruder but more widely applicable estimate. We write

$$I(B, q, z) = \{\zeta: 0 < q\zeta + z \leqq B\}.$$

Then for any integers $B \geqq 1$, $q \geqq 1$ and u, and $\alpha = (u/q) + \beta$,

$$\sum_{x=1}^{B} e(c_i \alpha x^k) = \sum_{z=1}^{q} \sum_{y \in I(B,q,z)} e(c_i \alpha (qy + z)^k)$$

$$= \sum_{z=1}^{q} e\left(\frac{c_i u}{q} z^k\right) \sum_{y \in I(B,q,z)} e(c_i \beta (qy + z)^k). \qquad (11.8)$$

We bound the error in replacing $\sum_y$ by the integral over the same interval in the following way. The function

$$g(\zeta) = e(c_i \beta (q\zeta + z)^k)$$

has

$$|g'(\zeta)| \leqq 2\pi q \cdot c_i |\beta| B^{k-1}, \qquad |g(\zeta)| \leqq 1 \qquad (\zeta \in I(B, q, z)).$$

As $I(B, q, z)$ has length B/q, we have

$$\left| \sum_{y \in I(B,q,z)} e(g(y)) - \int_{I(B,q,z)} e(g(\zeta)) \, d\zeta \right|$$

$$\leqq (B/q)(2\pi q c_i |\beta| B^{k-1}) + 3$$

$$= 2\pi c_i B^k |\beta| + 3. \qquad (11.9)$$

Let us write

$$S_{q,b} = \sum_{z=1}^{q} e\left(\frac{bz^k}{q}\right).$$

Using (11.9) in the equation (11.8), we find that

$$\sum_{x=1}^{B} e(c_i \alpha x^k) = \sum_{z=1}^{q} e\left(\frac{c_i u z^k}{q}\right) \int_{I(B,q,z)} e(c_i \beta (q\zeta + z)^k) \, d\zeta$$

$$+ O(q(c_i B^k |\beta| + 1))$$

$$= c_i^{-1/k} q^{-1} S_{q,c_i u} \int_0^{Bc_i^{1/k}} e(\beta t^k) \, dt + O(qc_i |\beta| B^k + q)$$

$$\qquad (11.10)$$

on substituting $c_i^{1/k}(q\zeta + z) = t$. We prefer an alternative function of β in place of the integral. It is shown on p. 16 of Vaughan (1981) that

$$\int_0^{n^{1/k}} e(\beta t^k) \, dt = \sum_{y=1}^{n} \frac{1}{k} y^{(1/k)-1} e(\beta y) + O(1 + n |\beta|)$$

for natural numbers $n \geqq 1$ and real β. We apply this to (11.10) with

$B = [Ac_i^{-1/k}]$, $n = A^k$. Writing

$$v(\beta) = k^{-1} \sum_{y=1}^{A^k} y^{(1/k)-1} e(\beta y),$$

we obtain

$$S_i(\alpha) = c_i^{-1/k} q^{-1} S_{q,c_i u} \int_0^{[Ac_i^{-1/k}]c_i^{1/k}} e(\beta t^k)\, dt + O(q\, |\beta|\, A^k + q)$$

$$= c_i^{-1/k} q^{-1} S_{q,c_i u} \int_0^A e(\beta t^k)\, dt + O(q\, |\beta|\, A^k + q)$$

$$= c_i^{-1/k} q^{-1} S_{q,c_i u} v(\beta) + O(q\, |\beta|\, A^k + q). \tag{11.11}$$

We conclude that

$$S_i(\alpha) = c_i^{-1/k} q^{-1} S_{q,c_i u} v(\beta) + O(A^{2\omega})$$

whenever $\alpha = (u/q) + \beta$ lies on a major arc. Indeed, since $S_i(\alpha) = O(A)$ we find that

$$S_1(\alpha) \cdots \overline{S_s(\alpha)} = (c_1 \cdots c_s)^{-1/k} q^{-s} S_{q,c_1 u} \cdots \overline{S_{q,c_s u}}\, v(\beta)^r \overline{v(\beta)}^{s-r}$$
$$+ O(A^{s-1+2\omega})$$

for $\alpha \in \mathcal{M}(u, q)$ and $\beta = \alpha - (u/q)$. We integrate over $\mathcal{M}(u, q)$ and sum over the distinct major arcs. Recalling (11.7), we get

$$Z = (c_1 \cdots c_s)^{-1/k} \mathfrak{S}(A) J(A) + O(A^{s-1-k+5\omega}), \tag{11.12}$$

where

$$\mathfrak{S}(A) = \sum_{q \leq A^\omega} \sum_{\substack{u=1 \\ (u,q)=1}}^{q} q^{-s} S_{c_1 u, q} \cdots \overline{S_{c_s u, q}},$$

$$J(A) = \int_{-A^{\omega-k}}^{A^{\omega-k}} v(\beta)^r \overline{v(\beta)}^{s-r}\, d\beta.$$

The natural next step is to complete the series $\mathfrak{S}(A)$ to infinity. Lemma 4.2, with $d = (c_i u, q) \leq c_i$, yields

$$\sum_{q > A^\omega} \sum_{\substack{u=1 \\ (u,q)=1}}^{q} q^{-s} S_{c_1 u, q} \cdots \overline{S_{c_s u, q}}$$

$$\ll \sum_{q > A^\omega} \sum_{u=1}^{q} q^{-s/(2k)} (c_1 \cdots c_s)^{1/k}$$

$$\ll (c_1 \cdots c_s)^{1/k} \sum_{q > A^\omega} q^{-3} \ll A^\omega \cdot A^{-2\omega}. \tag{11.13}$$

Thus

$$Z = (c_1 \cdots c_s)^{-1/k} (\mathfrak{S} + O(A^{-\omega})) J(A) + O(A^{s-1-k+5\omega}). \tag{11.14}$$

Here $\mathfrak{S}$ is the *singular series* associated with equation (11.1), defined by

$$\mathfrak{S} = \sum_{q=1}^{\infty} S(q), \qquad S(q) = \sum_{\substack{u=1 \\ (u,q)=1}}^{q} q^{-s} S_{c_1 u, q} \cdots \overline{S_{c_s u, q}},$$

and $S(q_1 q_2) = S(q_1)S(q_2)$ for coprime q_1, q_2, by a variant of Lemma 2.11 of Vaughan (1981). By Theorem 286 of Hardy and Wright (1979), $\mathfrak{S}$ can be expressed as a product over the primes:

$$\mathfrak{S} = \prod_p T(p), \qquad T(p) = \sum_{h=0}^{\infty} S(p^h). \tag{11.15}$$

We have

$$T(p) - 1 \ll \sum_{h=1}^{\infty} p^{-3h} \ll p^{-3}$$

by the argument leading to (11.13). Thus we easily find that

$$\prod_{p > C_{10}(k)} |T(p)| > \tfrac{1}{2}. \tag{11.16}$$

Just as in Lemma 2.12 of Vaughan (1981), we have

$$T(p) = \lim_{l \to \infty} p^{l(1-s)} M(p^l), \tag{11.17}$$

where $M(p^l)$ denotes the number of solutions of the congruence

$$c_1 x_1^k + \cdots + c_r x_r^k - (c_{r+1} x_{r+1}^k + \cdots + c_s x_s^k) \equiv 0 \pmod{p^l}$$

with $1 \leqq x_j \leqq p^l$. In the proof of his Theorem 9.1, Vaughan establishes that

$$M(p^l) > C_{11}(p, c_1, \ldots, c_s) p^{l(s-1)} \tag{11.18}$$

for sufficiently large l. (He requires a condition $s \geqq 4k^2 - k + 1$, which holds here.) On examining the proof of (11.18) we find that $C_{11}(p, c_1, \ldots, c_s)$ is independent of $c_1, \ldots, c_s$ when $c_1, \ldots, c_s$ are k-free, as we suppose here. We may therefore deduce from (11.15)–(11.18) that

$$\mathfrak{S} \gg 1. \tag{11.19}$$

Next, we would like to replace $J(A)$ in (11.12) by the *singular integral*

$$J = \int_{-1/2}^{1/2} v(\beta)^r \overline{v(\beta)}^{s-r} \, d\beta.$$

From Lemma 2.8 of Vaughan (1981)

$$v(\beta) \ll |\beta|^{-1/k}.$$

Thus

$$J - J(A) = \int_{A^{\omega-k}<|\beta|\leq\frac{1}{2}} v(\beta)\overline{v(\beta)}^{s-r}\, d\beta$$

$$\ll \int_{A^{\omega-k}}^{\infty} |\beta|^{-s/k}\, d\beta$$

$$\ll A^{(k-\omega)((s/k)-1)} \leq A^{s-k-\omega s/(2k)}. \tag{11.20}$$

Moreover, it can be seen that

$$J = k^{-s} \sum_{m_1,\ldots,m_s} (m_1 \cdots m_s)^{(1/k)-1}, \tag{11.21}$$

where the summation is restricted to integers $m_1, \ldots, m_s$ such that

$$1 \leq m_1, \ldots, m_s \leq A^k,$$

$$m_1 + \cdots + m_r = m_{r+1} + \cdots + m_s.$$

If we take m_{r+1} in $(A^k/2, (2/3)A^k)$, take all the rest of $m_2, \ldots, m_s$ to lie between 1 and $A^k/(3s)$, and define

$$m_1 = -(m_2 + \cdots + m_r) + m_{r+1} + \cdots + m_s,$$

then $m_1, \ldots, m_s$ will be counted in (11.21). Thus

$$J \gg A^{k(s-1)}A^{-s(k-1)} = A^{s-k},$$

which in tandem with (11.20) yields

$$J(A) \gg A^{s-k}. \tag{11.22}$$

Since $(c_1 \cdots c_s)^{1/k} \leq A^\omega$, we now find from (11.14), (11.19), (11.22) that

$$Z \gg (c_1 \cdots c_s)^{-1/k}A^{s-k}.$$

In particular, there are non-negative solutions of (11.1) satisfying (11.4).

Now let $c_1, \ldots, c_s$ be arbitrary natural numbers. Write $c_i = a_i q_i^k$ where a_i is k-free. There is a solution of

$$a_1 y_1^k + \cdots + a_r y_r^k - (a_{r+1}y_{r+1}^k + \cdots + a_s y_s^k) = 0$$

with

$$0 < \max_i a_i y_i^k \leq C_9(k)(a_1 \cdots a_s)^{16k}.$$

Let $q = q_1 \cdots q_s$ and $x_i = (q/q_i)y_i$; then

$$c_1 x_1^k + \cdots - c_s x_s^k = q^k(a_1 y_1^k + \cdots - a_s y_s^k) = 0,$$

while $\mathbf{x} \neq 0$ and, for each i,

$$c_i x_i^k \leq C_9(k) q^k (a_1 \cdots a_s)^{16k}$$
$$\leq C_9(k)(q_1^k a_1 \cdots q_s^k a_s)^{16k} = C_9(k)(c_1 \cdots c_s)^{16k}.$$

This proves Proposition 11.1.

11.2 An inductive argument

Proposition 11.2. *Let* $\lambda \geq 1/k$, $0 < \varepsilon < 1$ *and* $s \geq C_{12}(k, \lambda, \varepsilon)$. *Let* $a_1, \ldots, a_s$, $b_1, \ldots, b_s$ *be natural numbers with* $a = \max(a_1, \ldots, a_s)$, $b = \max(b_1, \ldots, b_s)$, $m = \max(a, b)$. *Then the equation*

$$a_1 x_1^k + \cdots + a_s x_s^k = b_1 y_1^k + \cdots + b_s y_s^k \qquad (11.23)$$

has a solution in non-negative integers $x_1, \ldots, x_s, y_1, \ldots, y_s$ *with*

$$0 < \max(x_1, \ldots, x_s, y_1, \ldots, y_s) \leq m^{\lambda + \varepsilon}.$$

The case $\lambda = 1/k$ implies Theorem 11.1. In fact, as a moment's thought shows, the proposition is equivalent to Theorem 11.1.

It will suffice to prove the proposition when m is large, say $m \geq C_{13}(k, \lambda, \varepsilon)$. For if $m < C_{13}$ and if s is large, then the a_i will assume the same value a at least m times, and the b_i will assume the same value b at least m times. Since a occurs at least b times and b occurs at least a times, one can construct a solution of the equation consisting of zeros and ones only.

We now come to a cunning device of Schmidt (1979a). This is the

Inductive assertion. *If* $\lambda > 1/k$, *and if Proposition* 11.2 *is true for* λ *then it is true for some* $\lambda' < \lambda$.

Suppose we have proved the inductive assertion. Let E be the set of λ in $[1/k, \infty)$ for which Proposition 11.2 is true. E is not empty; it contains $\lambda = 600kK$, as an easy consequence of Proposition 11.1. Also, E is closed, because of the ε in the statement of Proposition 11.2. Thus E has a least member which, by the inductive assertion, must be $1/k$. We conclude that Proposition 11.2 is true for all $\lambda \geq 1/k$.

In what follows, λ will be a fixed number $> 1/k$ for which Proposition 11.2 is true. Pick μ so small that

$$\frac{1}{k} + 4000kK\mu < \lambda \qquad \text{and} \qquad \mu < \tfrac{1}{20}, \qquad (11.24)$$

and put

$$\lambda' = \max(\lambda(1 - (\mu/2)) + \mu/(2k), (1/k) + 4000kK\mu),$$

so that indeed $\lambda' < \lambda$. We proceed to prove Proposition 11.2 for λ'.

Reduction.   *In proving Proposition 11.2 for λ' we may suppose that*

$$\tfrac{1}{2}a \leqq a_i \leqq a \qquad and \qquad \tfrac{1}{2}b \leqq b_i \leqq b \qquad (i = 1, \ldots, s).$$

To see this, let us suppose Proposition 11.2 has been established for λ', for this special type of equation. Now consider a general equation (11.23). Let

$$\gamma = \min\left(\frac{\varepsilon}{8\lambda}, \frac{\varepsilon}{4}\right).$$

We divide the interval $0 \leqq x \leqq 1$ into at most $(1/\gamma) + 1$ subintervals J of length not exceeding γ. If s is large, one of these intervals J will be such that many of the coefficients $a_1, \ldots, a_s$ are of the type $a_i = m^{\alpha_i}$ with $\alpha_i \in J$. We may therefore suppose without loss of generality that $a_j/a_i \leqq m^\gamma$ $(1 \leqq i, j \leqq s)$. Similarly we may suppose that $b_i/b_j \leqq m^\gamma$ $(1 \leqq i, j \leqq s)$. Put $a_0 = m^\gamma \max(a_1, \ldots, a_s)$ and $b_0 = m^\gamma \max(b_1, \ldots, b_s)$. Let p_i, q_i respectively be the largest integers with

$$a_i p_i^k \leqq a_0 \qquad and \qquad b_i q_i^k \leqq b_0 \qquad (i = 1, \ldots, s).$$

Now $a_0/a_i \geqq m^\gamma$, and if m is large (which we may suppose) then $p_i \geqq 2^{-1/k}(a_0/a_i)^{1/k}$, so that $a_i p_i^k \geqq \tfrac{1}{2}a_0$. Similarly, $b_i q_i^k \geqq \tfrac{1}{2}b_0$.

With $a_i' = a_i p_i^k$, $b_i' = b_i q_i^k$ and $x_i = p_i x_i'$, $y_i = q_i y_i'$ $(i = 1, \ldots, s)$, (11.23) becomes

$$a_1' x_1'^k + \cdot + a_s' x_s'^k = b_1' y_1'^k + \cdots + b_s' y_s'^k. \tag{11.25}$$

Proposition 11.2 holds for the particular equation (11.25), so we have a nontrivial non-negative solution with

$$\max(x_1', \ldots, x_s', y_1', \ldots, y_s') \leqq (\max(a_0, b_0))^{\lambda' + (\varepsilon/4)}$$
$$\leqq (m^{1+\gamma})^{\lambda' + (\varepsilon/4)} \leqq m^{\lambda' + \varepsilon/2}.$$

But clearly $a_i \geqq a_0 m^{-2\gamma}$, so that $p_i \leqq p_i^k \leqq m^{2\gamma} \leqq m^{\varepsilon/2}$, whence $x_i \leqq m^{\lambda' + \varepsilon}$ $(i = 1, \ldots, s)$ and similarly $y_i \leqq m^{\lambda' + \varepsilon}$, as desired.

From now on we concentrate on equations satisfying the condition of the reduction.

11.3   Two cases

In Section 11.3, h will be the integer

$$h = C_{12}(k, \lambda, \varepsilon)$$

occurring in Proposition 11.2. We suppose s much larger than h. Write

$$v = \mu/(2k). \tag{11.26}$$

We distinguish two cases, A and B.

A. *There is a subset of h elements among* $a_1, \ldots, a_s$, *say* $a_1, \ldots, a_h$, *and there is a subset of h elements among* $b_1, \ldots, b_s$, *say* $b_1, \ldots, b_h$, *and there are natural numbers*

$$p_1, \ldots, p_h, q_1, \ldots, q_h \leqq m^\nu \qquad (11.27)$$

such that the greatest common divisor

$$d = (a_1 p_1, \ldots, a_h p_h, b_1 q_1, \ldots, b_h q_h) \geqq m^\mu.$$

In this case put

$$x_i = p_i x_i', \quad y_i = q_i y_i' \quad (i = 1, \ldots, h)$$

and

$$x_{h+1} = y_{h+1} = \cdots = x_s = y_s = 0.$$

After division by d, (11.23) becomes

$$a_1' x_1'^k + \cdots + a_h' x_h'^k = b_1' y_1'^k + \cdots + b_h' y_h'^k \qquad (11.28)$$

where $a_i' = a_i p_i^k / d$ and $b_i' = b_i q_i^k / d$ $(i = 1, \ldots, h)$.

Because of the truth of Proposition 11.2 for λ, and by our choice of h, (11.28) has a non-trivial non-negative solution with

$$\max (x_1', \ldots, x_h', y_1', \ldots, y_h') \leqq (m^{1+k\nu-\mu})^{\lambda+\varepsilon},$$

so that

$$\max (x_1, \ldots, x_s, y_1, \ldots, y_s) \leqq m^{(1+k\nu-\mu)(\lambda+\varepsilon)+\nu} \leqq m^{\lambda'+\varepsilon}.$$

So we have a solution of (11.20) as required for the inductive assertion, in Case A.

In proving the inductive assertion, then, we can confine ourselves to case

B. *For any h elements, say* $a_1, \ldots, a_h$, *among* $a_1, \ldots, a_s$, *and for any h elements, say* $b_1, \ldots, b_h$, *among* $b_1, \ldots, b_s$, *and given* (11.27), *we have greatest common divisor*

$$(a_1 p_1, \ldots, a_h p_h, b_1 q_1, \ldots, b_h q_h) < m^\mu. \qquad (11.29)$$

Condition B depends on h, m, μ and ν, and if ν is given by (11.26), it is a condition $B(k, h, m, \mu)$.

Proposition 11.3. *Let* $h \geqq 1$, $k \geqq 2$, $0 < \mu < 1/20$. *Let* $0 < a$, $b \leqq m$ *and let* $a_1, \ldots, a_s$, $b_1, \ldots, b_s$ *be integers with*

$$\tfrac{1}{2} a \leqq a_i \leqq a, \qquad \tfrac{1}{2} b \leqq b_i \leqq b \qquad (i = 1, \ldots, s) \qquad (11.30)$$

with property $B(k, h, m, \mu)$. *Then if* $s \geqq C_{14}(k, h, \mu)$, *the equation*

$$a_1 x_1^k + \cdots + a_s x_s^k - (b_1 y_1^k + \cdots + b_s y_s^k) = z \qquad (11.31)$$

has a solution in non-negative integers $x_1, \ldots, x_s, y_1, \ldots, y_s, z$ *with*

$$0 < \max(x_1, \ldots, x_s, y_1, \ldots, y_s) \leqq m^{(1/k)+20\mu}, \qquad z \leqq m^{6\mu}.$$

This proposition implies the inductive assertion, as we now proceed to show. Let μ, ν be as in (11.24), (11.26). We may suppose that m is large, (11.30) holds, and that we are in the case $B = B(k, h, m, \mu)$ with $h = C_{12}(k, \lambda, \varepsilon)$. Suppose that

$$s = nu$$

where

$$n = 36kK, \qquad u = C_{14}(k, h, \mu).$$

After a change of notation, (11.23) becomes

$$\sum_{i=1}^{n} (a_{i1}x_{i1}^k + \cdots + a_{iu}x_{iu}^k - b_{i1}y_{i1}^k - \cdots - b_{iu}y_{iu}^k) = 0. \qquad (11.32)$$

For each i, $1 \leqq i \leqq n$, the coefficients $a_{i1}, \ldots, a_{iu}, b_{i1}, \ldots, b_{iu}$ satisfy the conditions of Proposition 11.3. Hence there are non-negative $x'_{i1}, \ldots, x'_{iu}, y'_{i1}, \ldots, y'_{iu}$, not all zero, having

$$a_{i1}x_{i1}'^k + \cdots + a_{iu}x_{iu}'^k - b_{i1}y_{i1}'^k - \cdots - b_{iu}y_{iu}'^k = z_i \qquad (11.33)$$

with

$$\max(x'_{i1}, \ldots, x'_{iu}, y'_{i1}, \ldots, y'_{iu}) \leqq m^{(1/k)+20\mu} \qquad \text{and} \qquad 0 \leqq z_i \leqq m^{6\mu}.$$

Let's modify this last condition to

$$0 \leqq z_i \leqq m^{6\mu} \qquad (i = 1, \ldots, n-1), \qquad -m^{6\mu} \leqq z_n \leqq 0.$$

This is not asking for too much, in view of the symmetry between the $+$ and $-$ terms in (11.33). If some $z_i = 0$, we get a small solution of (11.32) straight away. If $z_1, \ldots, z_n$ are each non-zero, then Proposition 11.1 gives non-negative $w_1, \ldots, w_n$ not all zero, with

$$z_1 w_1^k + \cdots + z_n w_n^k = 0,$$

having

$$\max(w_1, \ldots, w_n) \leqq C_9(k)(m^{6\mu})^{16n} \leqq m^{3600\mu kK},$$

since m is large.

Putting $x_{ij} = w_i x'_{ij}$, $y_{ij} = w_i y'_{ij}$ $(1 \leqq i \leqq n, 1 \leqq j \leqq u)$ we obtain a non-trivial solution of (11.32) with

$$\max(x_{ij}, y_{ij}) \leqq m^{(1/k)+20\mu+3600\mu kK} \leqq m^{\lambda'}.$$

Thus Proposition 11.2 is proved for λ'.

All we have to do now is prove Proposition 11.3. This we shall do in the next three sections. The sneaky little variable z in (11.31) has been introduced in order to avoid difficulties with the singular series, which is going to be 'truncated very low down' in accordance with our choice of major arcs.

11.4 The circle method again

Arguing as in the last section, it will suffice to prove Proposition 11.3 for $m \geq C_{15}(k, h, \mu)$. Put

$$s = h + [8kK/v], \qquad A = [(ab)^{1/k}m^{10\mu}], \qquad H = [m^{6\mu}]. \quad (11.34)$$

Write Z for the number of solutions of (11.31) in integers $x_1, \ldots, x_s$, $y_1, \ldots, y_s$, z subject to

$$1 \leq x_i \leq Aa_i^{-1/k}, \qquad 1 \leq y_i \leq Ab_i^{-1/k}, \qquad 1 \leq z \leq H. \quad (11.35)$$

Note that $Aa_i^{-1/k} > (m^{10\mu})/2$, $Ab_i^{-1/k} > (m^{10\mu})/2$, so that lengths of intervals (11.35) are large as a function of k, h and μ. Also, (11.35), (11.30) combine to give

$$x_i \leq 2b^{1/k}m^{10\mu} \leq m^{(1/k)+20\mu}$$

and similarly for y_i. Thus we have only to show that Z is positive.

As in Section 11.1, we have

$$Z = \int_0^1 S_1(\alpha) \cdots S_s(\alpha)\bar{T}_1(\alpha) \cdots \bar{T}_s(\alpha)U(\alpha)\,d\alpha$$

$$= \int_0^1 f(\alpha)\,d\alpha,$$

say, where

$$S_i(\alpha) = \sum_{1 \leq x \leq Aa_i^{-1/k}} e(\alpha a_i x^k), \quad (11.36)$$

$$T_i(\alpha) = \sum_{1 \leq y \leq Ab_i^{-1/k}} e(\alpha b_i y^k), \quad (11.37)$$

and

$$U(\alpha) = \sum_{z=1}^H e(-\alpha z). \quad (11.38)$$

We define the *major arcs* to be the intervals of the type

$$\mathcal{M}(q, u) = \{\alpha: |\alpha - u/q| \leq m^{2\mu}A^{-k}\},$$

where $1 \leq u \leq q \leq m^{\mu}$ and $(q, u) = 1$. Again, $\mathcal{M}$ will denote the union of the major arcs (which are pairwise disjoint); this time,

$$\mathcal{M} \subset I = [m^{2\mu}A^{-k}, 1 + m^{2\mu}A^{-k}].$$

Implied constants in the rest of the chapter will depend at most on k, h and μ.

Lemma 11.1. *Suppose that* $\alpha \in I$ *and*

$$|f(\alpha)| \geq H a^{-s/k} b^{-s/k} A^{2s-k} m^{-\mu}.$$

Then α *lies in a major arc.*

Proof. From the hypothesis on $|f(\alpha)|$,

$$|S_1(\alpha)| \cdots |S_s(\alpha)| \, |T_1(\alpha)| \cdots |T_s(\alpha)| \geq a^{-s/k} b^{-s/k} A^{2s-k} m^{-\mu}.$$

$$(11.39)$$

If, say, $|S_1(\alpha)| \geq \cdots \geq |S_s(\alpha)|$, then the left-hand side of (11.39) is

$$\leq |S_h(\alpha)|^{s-h+1} (2a^{-1/k}A)^{h-1}(2b^{-1/k}A)^s,$$

so that (11.39) yields

$$|S_i(\alpha)| \geq |S_h(\alpha)| \geq a^{-1/k} A (A^k m^{2\mu})^{-1/(s-h+1)} = P$$

for $i = 1, \ldots, h$. Observe that $A \leq m^2$,

$$Aa_i^{-1/k} P^{-1} \leq A^{(k+1)/(s-h+1)} \leq A^{(k+1)v/(8kK)}$$
$$\leq m^{v/(2K)} \leq m^{\mu/K} \leq (Aa_i^{-1})^{1/(2K)}$$

from (11.34). Hence we may deduce from Theorem 5.1 that there are natural numbers $p_1, \ldots, p_h$ with

$$p_i \leq (Aa_i^{-1/k}P^{-1})^k m^{v/2} \leq m^v,$$
$$\|\alpha a_i p_i\| \leq (Aa_i^{-1/k}P^{-1})^k m^{v/2} (Aa_i^{-1/k})^{-k}$$
$$\leq m^v A^{-k} a$$

for $i = 1, \ldots, h$. Similarly, after a possible reordering of $b_1, \ldots, b_s$, there are natural numbers $q_1, \ldots, q_h$ having

$$q_j \leq m^v, \qquad \|\alpha b_j q_j\| \leq m^v A^{-k} b \qquad (j = 1, \ldots, h).$$

There are integers $u_1, \ldots, u_h, v_1, \ldots, v_h$ with

$$\|\alpha a_i p_i\| = |\alpha a_i p_i - u_i| \qquad (i = 1, \ldots, h)$$

and

$$\|\alpha b_j q_j\| = |\alpha b_j q_j - v_j| \qquad (j = 1, \ldots, h).$$

It follows that

$$|u_i b_j q_j - v_j a_i p_i| \leq b_j q_j \|\alpha a_i p_i\| + a_i p_i \|\alpha b_j q_j\|$$
$$\leq b m^v m^v A^{-k} a + a m^v m^v A^{-k} b$$
$$\leq 2m^{2v-9\mu k} < 1,$$

since m is large. Thus the $2h$ non-zero vectors $(a_i p_i, u_i)$ $(i = 1, \ldots, h)$ and $(b_j q_j, v_j)$ $(j = 1, \ldots, h)$ are proportional to each other. They are integer multiples of some vector (q, u) where $q > 0$ and q, u are coprime. Since q is a common divisor of $a_1 p_1, \ldots, a_h p_h, b_1 q_1, \ldots, b_h q_h$, condition (11.29) of case B yields $q < m^\mu$. If, say, the vector $(a_i p_i, u_i)$ is l_i times (q, u), then $l_i \geq \frac{1}{2} a q^{-1}$, whence

$$
\begin{aligned}
|\alpha q - u| &= l_i^{-1} |\alpha a_i p_i - u_i| \\
&\leq 2a^{-1} q \, \|\alpha a_i p_i\| \\
&\leq 2a^{-1} q m^\nu A^{-k} a < m^{2\mu} A^{-k}.
\end{aligned}
$$

Thus α lies in $\mathcal{M}(q, u)$, and Lemma 11.1 follows.

11.5 The major arcs

In view of Lemma 11.1, we have

$$
\begin{aligned}
Z &= \int_{\mathcal{M}} f(\alpha) \, d\alpha + O(Ha^{-s/k} b^{-s/k} A^{2s-k} m^{-\mu}) \\
&= \sum_{q \leq m^\mu} \sum_{\substack{u=1 \\ (u,q)=1}}^{q} \int_{\mathcal{M}(q,u)} f(\alpha) \, d\alpha + O(Ha^{-s/k} b^{-s/k} A^{2s-k} m^{-\mu}). \quad (11.40)
\end{aligned}
$$

As in Section 11.1, write

$$
v(\beta) = k^{-1} \sum_{y=1}^{A^k} y^{(1/k)-1} e(\beta y).
$$

Let $\alpha = (u/q) + \beta$ be a point of a major arc. The approximation (11.11) is easily seen to apply to our present situation; it just needs to be rewritten as

$$
\begin{aligned}
S_i(\alpha) &= a_i^{-1/k} q^{-1} S_{q, a_i u} v(\beta) + O(q \, |\beta| \, A^k + q) \\
&= a_i^{-1/k} q^{-1} S_{q, a_i u} v(\beta) + O(m^{3\mu}),
\end{aligned}
$$

and similarly

$$
T_i(\alpha) = b_i^{-1/k} q^{-1} S_{q, b_i u} v(\beta) + O(m^{3\mu}).
$$

Since

$$
S_i(\alpha) = O(A a^{-1/k}), \qquad T_i(\alpha) = O(A b^{-1/k})
$$

and

$$
m^{3\mu} < \min (A a^{-1/k}, A b^{-1/k}),
$$

this leads to

$$S_1(\alpha) \cdots S_s(\alpha) \bar{T}_1(\alpha) \cdots \bar{T}_s(\alpha) e(-\alpha z)$$

$$= (a_1 \cdots a_s b_1 \cdots b_s)^{-1/k} q^{-2s} S_{q,a_1 u} \cdots \overline{S_{q,b_s u}}\, e(-\alpha z)\, |v(\beta)|^{2s}$$

$$+ O\!\left(A^{2s} a^{-s/k} b^{-s/k} \max\left(\frac{m^{3\mu}}{A a^{-1/k}}, \frac{m^{3\mu}}{A b^{-1/k}} \right) \right). \tag{11.41}$$

The error term here is easily seen to be

$$O(A^{2s-1} a^{-s/k} b^{-s/k} m^{3\mu} (ab)^{1/k}).$$

We sum (11.41) over $z = 1, \ldots, H$ and integrate over $\mathscr{M}(q, u)$:

$$\int_{\mathscr{M}(q,u)} f(\alpha)\, \mathrm{d}\alpha = \sum_{z=1}^{H} (a_1 \cdots a_s b_1 \cdots b_s)^{-1/k} q^{-2s} S_{q,a_1 u} \cdots \overline{S_{q,b_s u}}\, e\!\left(-\frac{uz}{q} \right)$$

$$\times \int_{-m^{2\mu} A^{-k}}^{m^{2\mu} A^{-k}} |v(\beta)|^{2s}\, e(-\beta z)\, \mathrm{d}\beta$$

$$+ O(H A^{2s-k-1} a^{-s/k} b^{-s/k} m^{5\mu} (ab)^{1/k}). \tag{11.42}$$

The last error term is

$$O(H A^{2s-k} a^{-s/k} b^{-s/k} m^{-3\mu}).$$

In the integral in (11.42) we replace $e(-\beta z)$ by 1. The error introduced into (11.42) is

$$\ll H a^{-s/k} b^{-s/k} A^{2s} \int_{-m^{2\mu} A^{-k}}^{m^{2\mu} A^{-k}} H\, |\beta|\, \mathrm{d}\beta$$

$$\ll H^2 a^{-s/k} b^{-s/k} A^{2s-2k} m^{4\mu}$$

$$\leqq H a^{-s/k} b^{-s/k} A^{2s-k} m^{-3\mu}.$$

We now sum, over q and u, the modified formula

$$\int_{\mathscr{M}(q,u)} f(\alpha) = \sum_{z=1}^{H} (a_1 \cdots a_s b_1 \cdots b_s)^{-1/k} q^{-2s} S_{q,a_1 u} \cdots \overline{S_{q,b_s u}}\, e\!\left(-\frac{uz}{q} \right)$$

$$\times \int_{-m^{2\mu} A^{-k}}^{m^{2\mu} A^{-k}} |v(\beta)|^{2s}\, \mathrm{d}\beta + O(H a^{-s/k} b^{-s/k} A^{2s-k} m^{-3\mu}).$$

Taking (11.40) into account, we find that

$$Z = (a_1 \cdots a_s b_1 \cdots b_s)^{-1/k} \mathfrak{S}(A, H) J_0(A) + O(H a^{-s/k} b^{-s/k} A^{2s-k} m^{-\mu}), \tag{11.43}$$

where

$$\mathfrak{S}(A, H) = \sum_{z=1}^{H} \sum_{q \leqq m^\mu} \sum_{\substack{u=1 \\ (u,q)=1}}^{q} q^{-2s} S_{q,a_1 u} \cdots \overline{S_{q,b_s u}}\, e\!\left(-\frac{uz}{q} \right), \tag{11.44}$$

$$J_0(A) = \int_{-m^{2\mu} A^{-k}}^{m^{2\mu} A^{-k}} |v(\beta)|^{2s}\, \mathrm{d}\beta.$$

11.6 The singular integral and the singular series

The singular integral $J_0(A)$ bears a close resemblance to $J(A)$ in Section 11.1. Repeating the argument leading to (11.22) with s replaced by $2s$ and A^ω replaced by $m^{2\mu}$, we find that

$$J_0(A) \gg A^{2s-k}. \tag{11.45}$$

As for the singular series $\mathfrak{S}(A, H)$ given by (11.44), we observe at once that the summands with $q = 1$ give the contribution H.
When $q > 1$,

$$\left| \sum_{z=1}^{H} e\left(-\frac{uz}{q}\right) \right| < q$$

for $(u, q) = 1$. Thus the summands with fixed $q > 1$ contribute $O(q^2)$. Since Σq^2 over $q \leqq m^\mu$ is $O(m^{3\mu})$, we obtain

$$\mathfrak{S}(A, H) = H + O(m^{3\mu}) \gg H. \tag{11.46}$$

Combining (11.46), (11.45) and (11.30), we have

$$(a_1 \cdots a_s b_1 \cdots b_s)^{-1/k} \mathfrak{S}(A, H) J_0(A) \gg a^{-s/k} b^{-s/k} H A^{2s-k}.$$

We therefore deduce from (11.43) that

$$Z \gg a^{-s/k} b^{-s/k} H A^{2s-k}.$$

Thus Z is positive, and Proposition 11.3 is proved.

Finally, we remark that the use of the inductive assertion is in a sense 'non-constructive'. We can get around this quite simply. Let $0 < \theta < 1$. Suppose that

$$\theta + (1/k) \leqq \lambda \leqq 600kK$$

and that Proposition 11.2 is true for λ. Use of the (proof of the) inductive assertion enables us to pass from λ to λ', where

$$\lambda' = \max\left(\lambda\left(1 - \frac{\mu}{2}\right) + \frac{\mu}{2k}, \frac{1}{k} + 4000kK\mu\right),$$

with

$$\mu = \frac{\lambda - (1/k)}{18\,000kK}.$$

Now

$$\lambda' \leqq \max\left(\lambda - \frac{\mu}{2}\left(\lambda - \frac{1}{k}\right), \frac{1}{k} + \frac{1}{2}\left(\lambda - \frac{1}{k}\right)\right)$$

$$\leqq \lambda - \theta^2/(36\,000kK).$$

In at most $10^8 k^2 K^2 \theta^{-2}$ steps we can descend from $\lambda = 600kK$ to a value of λ between $\dfrac{1}{k} + \theta$ and $\dfrac{1}{k} + 2\theta$, for which Proposition 11.2 holds. At each stage the details of the proof permit the effective computation of $C_{12}(k, \lambda, \varepsilon)$. This shows that a value for $C_1(k, 2\theta + \varepsilon)$ is effectively computable. Since $\varepsilon > 0$ and $\theta > 0$ are arbitrary, $C_1(k, \varepsilon)$ is effectively computable. Similar remarks apply to the constants in Chapters 12 and 14, where we shall be using the same style of inductive assertion.

12
Small solutions of additive congruences

12.1 An inductive assertion

Chapter 12 is devoted to the proof of a sharp theorem on non-negative solutions of simultaneous diagonal congruences.

Theorem 12.1. *Let*

$$F_j(\mathbf{x}) = a_{j1}x_1^k + \cdots + a_{js}x_s^k \qquad (j = 1, \ldots, h) \tag{12.1}$$

be forms with integer coefficients, where $s \geq C_1(h, k, \varepsilon)$. *Let* m *be a natural number. Then there are non-negative integers* $n_1, \ldots, n_s$ *such that*

$$F_j(n_1, \ldots, n_s) \equiv 0 \pmod{m} \qquad (j = 1, \ldots, h), \tag{12.2}$$

$$0 < \max(n_1, \ldots, n_s) \leq m^{(1/k)+\varepsilon}. \tag{12.3}$$

We shall make a good deal of use of the methods and results of Chapter 11. However, instead of the circle method we use a 'discrete' argument based on the identity

$$\sum_{u=1}^{m} e_m(ua) = 0$$

which holds whenever a is an integer, $a \not\equiv 0 \pmod{m}$. (Compare Section 9.3.)

We shall prove Theorem 12.1 by induction on h. The case $h = 1$ is, of course, Lemma 10.1. Until the end of the chapter, h denotes a fixed natural number, $h \geq 2$, and we suppose that the theorem has been proved for $h - 1$ additive forms.

Next, we state a proposition containing a parameter λ. The case $\lambda = 1/k$ is the induction step required to complete the proof of Theorem 12.1. ·

Proposition 12.1. *Let* $1/k \leq \lambda \leq 1$ *and* $s \geq C_2(h, k, \lambda, \varepsilon)$ *Let* $F_j(x_1, \ldots, x_s)$ *be as in* (12.1) *with integer coefficients. Then for each natural number* m, *there are non-negative integers* $n_1, \ldots, n_s$ *with* (12.2) *and*

$$0 < \max(n_1, \ldots, n_s) \leq m^{\lambda+\varepsilon}.$$

It is enough to prove Proposition 12.1 when $m \geqq C_3(h, k, \lambda)$. For suppose that $m < C_3$. Writing

$$(F_1(\mathbf{x}), \ldots, F_h(\mathbf{x})) = \mathbf{a}_1 x_1^k + \cdots + \mathbf{a}_s x_s^k,$$

we refer to the $\mathbf{a}_i = (a_{1i}, \ldots, a_{hi})$ as the *coefficient vectors* of $(F_1, \ldots, F_h)$. There are m^h possible values for these vectors (mod m). If $s \geqq C_3^{h+1}$, one coefficient vector must be repeated (mod m) at least m times; say $\mathbf{a}_i \equiv \mathbf{b}$ for $i \in \mathcal{S}$, where $|\mathcal{S}| = m$. Define $n_i = 1$ if $i \in \mathcal{S}$, $n_i = 0$ otherwise; then

$$(F_1(\mathbf{n}), \ldots, F_h(\mathbf{n})) \equiv |\mathcal{S}| \, \mathbf{b} \equiv \mathbf{0} \pmod{m},$$

in an obvious notation.

For the rest of the chapter, then, let $m \geqq C_3(h, k, \lambda)$.

Inductive assertion. Let $1/k < \lambda \leqq 1$. *If Proposiltion 12.1 is true for* λ, *then it is true for some number* $\lambda' < \lambda$.

It is trivial that Proposition 12.1 is true for $\lambda = 1$, because we can take $(n_1, \ldots, n_s) = (m, \ldots, m)$. By the argument used in Section 11.2, Proposition 12.1 follows from the inductive assertion.

Reduction. Let $\lambda \geqq 1/k$. *In proving Proposition 12.1 for* λ', *we may assume that* m *is square-free.*

(The proof of this reduction was supplied by Glyn Harman in conversation.)

To see this, suppose that Proposition 12.1 has been established for λ', for congruences with square-free moduli in at least C_4 variables, $C_4 = C_4(h, k, \lambda', \varepsilon)$. We proceed in the following way to show that the Proposition is also valid for λ' and j-free moduli, provided that $s \geqq C_4^{j-1}$. Naturally this involves an induction; we have the case $j = 2$ to begin with.

In the step from j to $j + 1$ (where $j \geqq 2$) write the $(j + 1)$-free integer m as

$$m = m_1 m_2$$

where m_1 is j-free and m_2 is square-free; this is not difficult. A set of diagonal forms in C_4^j variables may be written

$$F_j(x_{11}, \ldots, x_{Cr}) = \sum_{i=1}^{C} (a_{i1}(j) x_{i1}^k + \cdots + a_{ir}(j) x_{ir}^k)$$

where $r = C_4^{j-1}$, $C = C_4$. We choose non-negative $x_{i1}', \ldots, x_{ir}'$ so that the integer $a_i(j)$ defined by

$$a_i(j) = a_{i1}(j) x_{i1}'^k + \cdots + a_{ir}(j) x_{ir}'^k$$

is divisible by m_1 for $j = 1, \ldots, h$, with

$$0 < \max_{t \leq r} x_{it}' \leq m_1^{\lambda' + \varepsilon}.$$

We do this for $i = 1, \ldots, C$. Then we choose non-negative y_i $(i \leq C)$ so that

$$\sum_{i=1}^{C} \frac{a_i(j)}{m_1} y_i^k \equiv 0 \pmod{m_2} \qquad (j = 1, \ldots, h),$$

$$0 < \max_{i \leq C} y_i \leq m_2^{\lambda' + \varepsilon}.$$

Evidently $x_{it} = y_i x_{it}'$ gives

$$F_j(x_{11}, \ldots, x_{Cr}) \equiv 0 \pmod{m} \qquad (j = 1, \ldots, k)$$

$$0 < \max_{i,t} x_{it} \leq m^{\lambda' + \varepsilon}.$$

This completes the proof of the inductive step.

In particular, Proposition 12.1 works for λ' if $s \geq C_4^{k-1}$ and m is k-free. Now it is an easy step to an arbitrary natural number m. We may write

$$m = g^k v,$$

where v is k-free. A solution of

$$F_j(n_1', \ldots, n_s') \equiv 0 \pmod{v} \qquad (j = 1, \ldots, h)$$

with non-negative n_i',

$$0 < \max (n_1', \ldots, n_s') \leq v^{\lambda' + \varepsilon},$$

promptly yields a solution $n_i = g n_i'$ of

$$F_j(n_1, \ldots, n_s) \equiv 0 \pmod{m} \qquad (j = 1, \ldots, h).$$

After all, $F_j(\mathbf{n}) = g^k F_j(\mathbf{n}')$. Moreover,

$$0 < \max n_i \leq g v^{\lambda' + \varepsilon} \leq (g^k v)^{\lambda' + \varepsilon} = m^{\lambda' + \varepsilon}.$$

Thus we can take $C_2(h, k, \lambda', \varepsilon) = C_4^{k-1}$. This establishes the reduction.

We now give some hypotheses and notations which will hold for the rest of the chapter. We suppose that Proposition 12.1 is true for λ, where $1/k < \lambda \leq 1$. We write

$$c = 36kK,$$

$$\mu = (\lambda - (1/k))/\{(600k^2K)^h(2h + 7)\}, \tag{12.4}$$

$$\rho = (\lambda - (1/k))\mu/(2h). \tag{12.5}$$

We also define

$$d = C_1(h - 1, k, \rho) \tag{12.6}$$

(don't forget our hypothesis about $h - 1$ forms!);

$$w = C_2(h, k, \lambda, \rho). \qquad (12.7)$$

Let

$$\lambda' = \lambda - \rho. \qquad (12.8)$$

Note that

$$\lambda' > \frac{1}{k} + (600k^2 K)^h (2h + 6)\mu. \qquad (12.9)$$

Let m be a square-free natural number with

$$m \geqq C_3(h, k, \lambda). \qquad (12.10)$$

We shall prove Proposition 12.1 for λ' with this modulus m.

12.2 Division into two cases

Let us first of all write down a condition on h forms

$$F_j(x) = a_{j1}x_1^k + \cdots + a_{js}x_{js}^k \qquad (j = 1, \ldots, h)$$

which is, roughly speaking, going to play the role of the condition $B(k, h, m, \mu)$ in Chapter 11.

A vector $\mathbf{v} = (v_1, \ldots, v_h) \in \mathbb{Z}^h$ is said to be *prime to* the natural number q if the greatest common divisor $(v_1, \ldots, v_h, q)$ is 1.

Condition $D(h, k, t, m, \mu)$. *For any t coefficient vectors, say* $\mathbf{a}_1, \ldots, \mathbf{a}_t$, *for any divisor q of m, and any $\mathbf{v}$ in $\mathbb{Z}^h$ prime to q such that*

$$\mathbf{v}\mathbf{a}_i \equiv 0 \pmod{q} \qquad (i = 1, \ldots, t)$$

we have

$$q \leqq m^\mu. \qquad (12.11)$$

The following result corresponds to Proposition 11.3. We defer the proof until Section 12.3.

Proposition 12.2. *Let t be a natural number. Let the forms $F_j (j = 1, \ldots, h)$ have property $D(h, k, t, m, \mu)$. Then if $s \geqq C_5(h, k, \lambda, t)$, the set of congruences*

$$F_j(x_1, \ldots, x_s) \equiv z_j \pmod{m} \qquad (j = 1, \ldots, h) \qquad (12.12)$$

has a solution in nonnegative integers $x_1, \ldots, x_s, z_1, \ldots, z_h$ with

$$0 < \max(x_1, \ldots, x_s) \leqq m^{(1/k)+\mu}, \qquad \max(z_1, \ldots, z_h) \leqq m^{(2h+5)\mu}.$$

$$(12.13)$$

Note that dependence of C_5 on λ rather than μ is due to the definition (12.4).

We also need a result on 'bounded solutions' corresponding to Proposition 11.1. We shall deduce this from Proposition 11.1 itself by a straightforward use of the idea of 'blocks of variables'.

Lemma 12.1. *Let r be a natural number. There is a matrix*

$$G_r = \begin{pmatrix} g(1, 1) \cdots g(1, c^r) \\ \vdots \qquad \vdots \\ g(r, 1) \cdots g(r, c^r) \end{pmatrix}$$

whose entries are ± 1, having the following property. Given non-negative integers $b(j, i)(j = 1, \ldots, r; i = 1, \ldots, c^r)$ bounded above by b, there is a $\mathbf{y} = (y(1), \ldots, y(c^r))$ with non-negative integer coordinates, satisfying

$$\sum_{i=1}^{c^r} g(j, i)b(j, i)y(i)^k = 0 \qquad (j = 1, \ldots, r) \tag{12.14}$$

and

$$0 < \max_i y(i) \le C_6(k, r) \exp\left((600k^2K)^r \log b\right). \tag{12.15}$$

Proof. By induction on r. Obviously the case $r = 1$ follows from Proposition 11.1, with

$$G_1 = (-1 \quad 1 \quad \cdots \quad 1).$$

(Note that the case where one of the $b(1, i)$ vanishes is trivial.) Let $r \ge 2$. In the induction step let G_r be formed from G_{r-1} as shown below:

$$G_r = \begin{pmatrix} & G_{r-1} & & G_{r-1} & \cdots & G_{r-1} \\ -1 & -1 & \cdots & -1 \mid 1 & \cdots & 1 \mid \cdots \mid 1 & \cdots & 1 \end{pmatrix}.$$

The vertical lines here divide G_r into c blocks. The variables $y(i)$ $(i = 1, \ldots, c^r)$ are also divided into c blocks $\mathscr{B}_1 = \{1, \ldots, c^{r-1}\}$, $\mathscr{B}_2, \ldots, \mathscr{B}_c$ in an obvious way.

Take non-negative integers $b(j, i)(j = 1, \ldots, r; i = 1, \ldots, c^r)$. For any block $\mathscr{B}_v$, we can choose non-negative $y'(i)(i \in \mathscr{B}_v)$ such that

$$\sum_{i \in \mathscr{B}_v} \pm b(j, i)y'(i)^k = 0 \qquad (j = 1, \ldots, r - 1). \tag{12.16}$$

Here the $\pm$ signs are simply the $g(j, i')$ from G_{r-1}, with $i' \equiv i \pmod{c^{r-1}}$. The size of $y'(i)$ may be made to satisfy

$$0 < \max_{i \in \mathscr{B}_v} y'(i) \le C_6(k, r - 1) \exp\left((600k^2K)^{r-1} \log b\right). \tag{12.17}$$

For $1 \leqq v \leqq c$, let a_v be the non-negative integer given by

$$a_v = \sum_{i \in \mathcal{B}_v} b(r, i) y'(i)^k.$$

Using Proposition 11.1, we solve

$$-a_1 z_1^k + a_2 z_2^k + \cdots + a_c z_c^k = 0 \tag{12.18}$$

in non-negative integers z_v. Here we can take

$$0 < \max z_v \leqq C_7(k) \left(\max_v a_v \right)^{576kK}$$

$$\leqq C_7(k) (c^{r-1} b \cdot C_6(k, r-1)^k \exp{(k(600k^2 K)^{r-1} \log b)})^{576kK} \tag{12.19}$$

in view of the obvious bound for a_v.

Define $y(i) = z_v y'(i)$ $(i \in \mathcal{B}_v)$. It is obvious from (12.16) that (12.14) holds for $j = 1, \ldots, r - 1$. From (12.18), we deduce that (12.14) also holds for $j = r$. As for the bounds (12.15), these are easily deduced from (12.17) and (12.19). This completes the proof of Lemma 12.1.

For the remainder of Section 12.2, let

$$t = dw = C_1(h - 1, k, \rho) C_2(h, k, \lambda, \rho).$$

We now show that Proposition 12.1 is true for λ' whenever $(F_1, \ldots, F_h)$ satisfies the condition $D(h, k, t, m, \mu)$. Suppose that

$$s = c^h C_5(h, k, \lambda, t) = c^h u,$$

say. After a change of notation, (12.2) becomes

$$\sum_{r=1}^{c^h} (a_{r1}(j) x_{r1}^k + \cdots + a_{ru}(j) x_{ru}^k) \equiv 0 \pmod{m} \qquad (j = 1, \ldots, h).$$

$$\tag{12.20}$$

For each r, $1 \leqq r \leqq c^h$, consider the set of h forms

$$g(j, r)(a_{r1}(j) x_{r1}^k + \cdots + a_{ru}(j) x_{ru}^k) \qquad (j = 1, \ldots, h)$$

in u variables. These forms satisfy condition $D(h, k, t, m, \mu)$. The introduction of the $\pm$ signs $g(1, r), \ldots, g(h, r)$ (defined in Lemma 12.1) is not going to affect this, because $(\pm v_1, \ldots, \pm v_h)$ is prime to q whenever $(v_1, \ldots, v_h)$ is prime to q.

By Proposition 12.2, we have

$$g(j, r)(a_{r1}(j) x_{r1}'^k + \cdots + a_{ru}(j) x_{ru}'^k) \equiv z_{rj} \pmod{m}$$

for $j = 1, \ldots, h$, with non-negative integers $x_{r1}', \ldots, x_{ru}', z_{r1}, \ldots, z_{rh}$

satisfying

$$0 < \max (x'_{r1}, \ldots, x'_{ru}) \leqq m^{(1/k)+\mu}, \qquad \max (z_{r1}, \ldots, z_{rh}) \leqq m^{(2h+5)\mu}. \tag{12.21}$$

We can re-write these congruences as

$$a_{r1}(j)x'^{k}_{r1} + \cdots + a_{ru}(j)x'^{k}_{ru} \equiv g(j, r)z_{rj} \pmod{m}$$

$$(j = 1, \ldots, h; r = 1, \ldots, c^h). \tag{12.22}$$

By Lemma 12.1, there are non-negative integers $y(r)$ $(r = 1, \ldots, c^h)$ such that

$$\sum_{r=1}^{c^h} g(j, r)z_{rj}y(r)^k = 0 \qquad (j = 1, \ldots, h) \tag{12.23}$$

and

$$0 < \max (y(1), \ldots, y(c^h)) \leqq C_6(k, h)m^E, \tag{12.24}$$

where $E = (600k^2 K)^h(2h + 5)\mu$.

Combining (12.22), (12.23), we see that $x_{ri} = y(r)x'_{ri}$ is a non-trivial non-negative solution of the set of congruences (12.20). As for the size of the xs, we can combine (12.21) and (12.24) to obtain

$$\max x_{ri} \leqq C_6(k, h)m^{(1/k)+\mu+E}$$
$$\leqq m^{\lambda'}$$

from (12.9), (12.10).

We are left with the case where the condition $D(h, k, t, m, \mu)$ fails. This is the most innovative part of the entire proof. Restating our information, there are a divisor q of m; t coefficient vectors $\mathbf{a}_1, \ldots, \mathbf{a}_t$ (say); and an integer vector $\mathbf{v} = (v_1, \ldots, v_h)$, such that

$$q > m^\mu, \tag{12.25}$$

$$(v_1, \ldots, v_h, q) = 1, \tag{12.26}$$

and

$$\mathbf{v}\mathbf{a}_i \equiv 0 \pmod{q} \qquad (i = 1, \ldots, t). \tag{12.27}$$

We set all the variables, except the first $t = dw$, equal to zero, and write these variables as $x_{11}, \ldots, x_{dw}$. This gives restrictions of the forms $F_1, \ldots, F_h$ of the shape

$$G_j(\mathbf{x}) = G_j(x_{11}, \ldots, x_{dw}) = \sum_{u=1}^{d} (b_{u1}(j)x^k_{u1} + \cdots + b_{uw}(j)x^k_{uw})$$

$$(j = 1, \ldots, h) \tag{12.28}$$

An immediate consequence of (12.27) is that

$$v_1 G_1(\mathbf{x}) + \cdots + v_h G_h(\mathbf{x}) \equiv 0 \pmod{q} \tag{12.29}$$

for every integer vector **x**. To exploit this, we show that there is a divisor q' of q such that

$$q' \geqq q^{1/h}, \tag{12.30}$$

and some j, $1 \leqq j \leqq h$, for which

$$(v_j, q') = 1. \tag{12.31}$$

This is most easily seen by writing $q = p_1 \cdots p_a$, where the p_i are distinct primes. Because of (12.26), no p_i divides every one of $v_1, \ldots, v_h$; hence

$$\prod_{j=1}^{h} (v_j, q) \leqq p_1^{h-1} \cdots p_a^{h-1} = q^{h-1}.$$

Thus $(v_j, q) \leqq q^{1-(1/h)}$ for some j in $\{1, \ldots, h\}$, and (12.31) follows, where

$$q' = q/(v_j, q) \geqq q^{1/h}.$$

Suppose for simplicity of writing that the index j in (12.31) is $j = 1$. We remark that

$$G_1(\mathbf{x}) \equiv 0 \pmod{q'} \tag{12.32}$$

whenever $\mathbf{x} \in \mathbb{Z}^t$ and

$$G_2(\mathbf{x}) \equiv G_3(\mathbf{x}) \equiv \cdots \equiv G_h(\mathbf{x}) \equiv 0 \pmod{q'}. \tag{12.33}$$

This is a straightforward consequence of (12.29) and (12.31).

Now we are in a position to give a solution of $G_j(\mathbf{x}) \equiv 0 \pmod{m}$ $(j = 1, \ldots, h)$ with a suitably small **x**. By definition of $w = C_2(h, k, \lambda, \rho)$, for each $u \leqq d$ we may select non-negative integers

$$y_{u1}, \ldots, y_{uw},$$

so that

$$b_{u1}(j)y_{u1}^k + \cdots + b_{uw}(j)y_{uw}^k \equiv 0 \quad \left(\mathrm{mod} \, \frac{m}{q'}\right) \tag{12.34}$$

for $j = 1, \ldots, h$, and

$$0 < \max(y_{u1}, \ldots, y_{uw}) \leqq (m/q')^{\lambda+\rho}. \tag{12.35}$$

We denote the expression on the left-hand side of (12.34) by $f_u(j)$. Since $d = C_1(h - 1, k, \rho)$, we may now choose non-negative integers $z_1, \ldots, z_d$ such that

$$f_1(j)z_1^k + \cdots + f_d(j)z_d^k \equiv 0 \pmod{q'} \quad (j = 2, \ldots, h) \tag{12.36}$$

and

$$0 < \max(z_1, \ldots, z_d) \leqq (q')^{(1/k)+\rho}. \tag{12.37}$$

Let us write, in the usual way,

$$x_{ui} = z_u y_{ui} \qquad (u = 1, \ldots, d; i = 1, \ldots, w).$$

From (12.36), (12.28) it follows that (12.33) holds; so of course

$$G_1(\mathbf{x}) \equiv G_2(\mathbf{x}) \equiv \cdots \equiv G_h(\mathbf{x}) \equiv 0 \pmod{q'} \tag{12.38}$$

from (12.32). Moreover, by (12.34) we have

$$G_1(\mathbf{x}) \equiv \cdots \equiv G_h(\mathbf{x}) \equiv 0 \ \left(\bmod \frac{m}{q'} \right). \tag{12.39}$$

Since m is squarefree, we can combine (12.38), (12.39) to get

$$G_1(\mathbf{x}) \equiv G_2(\mathbf{x}) \equiv \cdots \equiv G_h(\mathbf{x}) \equiv 0 \pmod{m}.$$

As for the size of the x's, we can combine (12.35) and (12.37), giving

$$0 < \max (x_{11}, \ldots, x_{dv}) \leq (q')^{(1/k)+\rho}(m/q')^{\lambda+\rho}$$
$$\leq m^{\lambda+\rho}(q')^{-(\lambda-(1/k))}. \tag{12.40}$$

Now $q' \geq q^{1/h}$ and $q > m^{\mu}$ from (12.30), (12.25). Consequently the right-hand side of (12.40) is at most

$$m^{\lambda+\rho-(\lambda-(1/k))\mu/h} = m^{\lambda+\rho-2\rho} = m^{\lambda'}$$

from (12.5), (12.8). Thus there is always a nontrivial non-negative solution of (12.2) in integers not exceeding $m^{\lambda'}$, and Proposition 12.1 is proved for λ'. As explained earlier, this completes the proof of Theorem 12.1, apart from the task of proving Proposition 12.2.

12.3 Application of a 'discrete' circle method

In our proof of Proposition 12.2 we suppose that $m \geq C_8(h, k, \lambda, t)$,

$$s \geq t(1 + (6kh)/\mu). \tag{12.41}$$

Let $H = [m^{(1/k)+\mu}]$, $Z = [m^{(2h+5)\mu}]$. The number of solutions of (12.12) in natural numbers $x_1, \ldots, x_s, z_1, \ldots, z_h$ satisfying (12.13) is

$$W = m^{-h} \sum_{x_1,\ldots,x_s=1}^{H} \sum_{z_1,\ldots,z_h=1}^{Z} \sum_{u_1=1}^{m} \cdots \sum_{u_h=1}^{m} e_m \Big(\sum_{j=1}^{h} u_j(a_{j1}x_1^k + \cdots + a_{js}x_s^k - z_j) \Big).$$

After all,

$$\sum_{u=1}^{n} e_m(u(a_{j1}x_1^k + \cdots + a_{js}x_s^k - z_j)) = 0$$

whenever the jth congruence is violated, and so

$$\sum_{u_1=1}^{m} \cdots \sum_{u_h=1}^{m} e_m \Big(\sum_{j=1}^{h} u_j(a_{j1}x_1^k + \cdots + a_{js}x_s^k - z_j) \Big)$$

counts m^h or 0 according as the set of congruences holds, or not.

We can break up the multiple sum $m^h W$ in the following way. We have

$$m^h W = \sum_{x_1,\ldots,x_s=1}^{H} \sum_{z_1,\ldots,z_h=1}^{Z} \sum_{q\mid m} \sum_{\substack{v_1,\ldots,v_h=1\\(v_1,\ldots,v_h,q)=1}}^{q} e_q\Big(\sum_{j=1}^{h} v_j(a_{j1}x_1^k + \cdots + a_{js}x_s^k - z_j)\Big),$$

via a reduction of fractions $u_1/m, \ldots, u_h/m$ to their least common denominator. This can be written

$$m^h W = \sum_{q\mid m} S_q, \tag{12.42}$$

where

$$S_q = \sum_{x_1,\ldots,x_s=1}^{H} \sum_{z_1,\ldots,z_h=1}^{Z} \sum_{\substack{v_1,\ldots,v_h=1\\(v_1,\ldots,v_h,q)=1}}^{q} e_q\Big(\sum_{j=1}^{h} v_j(a_{j1}x_1^k + \cdots + a_{js}x_s^k - z_j)\Big).$$

$$\tag{12.43}$$

Evidently

$$S_1 = H^s Z^h. \tag{12.44}$$

This is actually the main term on the right-hand side of (12.42). For small $q > 1$ we can argue as in Section 11.6. For $1 \leq v < q$ we have

$$\Big|\sum_{z=1}^{Z} e_q(-vz)\Big| < q.$$

Since the condition $(v_1, \ldots, v_h, q) = 1$ in (12.43) ensures that $1 \leq v_j < q$ for at least one value of j, we can get the upper bound

$$|S_q| < H^s Z^{h-1} q^{h+1}$$

by summing over z_j first in (12.43). This tells us that

$$\sum_{\substack{q\mid m\\2\leq q\leq m^{2\mu}}} |S_q| < H^s Z^{h-1} (m^{2\mu})^{h+2}$$

$$\leq H^s Z^{h-1} (2Z) m^{-\mu} = H^s Z^h (2m^{-\mu}). \tag{12.45}$$

We claim that

$$\sum_{\substack{q\mid m\\q>m^{2\mu}}} |S_q| < H^s Z^h m^{-1}. \tag{12.46}$$

For suppose the contrary. Then

$$|S_q| \geq H^s Z^h m^{-2}$$

for some $q \mid m$, $q > m^{2\mu}$. It is an easy step to the inequality

$$\Big|\sum_{x_1,\ldots,x_s=1}^{H} e_q\Big(\sum_{j=1}^{h} v_j(a_{j1}x_1^k + \cdots + a_{js}x_s^k)\Big)\Big| \geq H^s m^{-2} q^{-h}$$

for some $\mathbf{v} = (v_1, \ldots, v_h)$ prime to q. That is,

$$\prod_{i=1}^{s} |T_i| \geq H^s m^{-2} q^{-h}, \tag{12.47}$$

where

$$T_i = \sum_{x=1}^{H} e_q\left(\sum_{j=1}^{h} v_i a_{ji} x^k\right)$$

$$= \sum_{x=1}^{H} e_q(\mathbf{v}\mathbf{a}_i x^k).$$

If, say, $|T_1| \geq |T_2| \geq \cdots \geq |T_s|$, then the left-hand side of (12.47) is

$$\leq H^{t-1} |T_t|^{s+1-t}.$$

It follows that for $i = 1, \ldots, t$ we have

$$|T_i| \geq |T_t| \geq H(m^2 q^h)^{-1/(s+1-t)} = P,$$

say. We have $P \geq H^{1-(1/(2K))}$, since

$$(m^2 q^h)^{1/(s+1-t)} \leq m^{(h+2)/(s+1-t)}$$

$$\leq m^{\mu/(2kt)} \leq H^{1/(2K)}$$

from (12.41).

By Theorem 5.1, then, for $i = 1, \ldots, t$ there is a natural number r_i having

$$r_i \leq H^{\mu/(2t)}(m^2 q^h)^{k/(s+1-t)}$$

$$\leq H^{\mu/(2t)} m^{\mu/(2t)} \leq m^{\mu/t},$$

$$\|(\mathbf{v}\mathbf{a}_i) r_i q^{-1}\| \leq H^{\mu/(2t)-k}(m^2 q^h)^{k/(s+1-t)} \leq m^{\mu/t} H^{-k}. \qquad (12.48)$$

Now

$$H^k m^{-\mu/t} \geq (\tfrac{1}{2} m^{(1/k)+\mu})^k m^{-\mu} > m.$$

The expression on the right-hand side of (12.48) is accordingly smaller than q^{-1}. We conclude that

$$q \mid r_i(\mathbf{v}\mathbf{a}_i) \qquad (i = 1, \ldots, t)$$

and indeed

$$q' \mid \mathbf{v}\mathbf{a}_i \qquad (i = 1, \ldots, t) \qquad (12.49)$$

where

$$q' = q/(q, r_1 \cdots r_t) \geq q/(r_1 \cdots r_t) \geq q m^{-\mu} > m^{\mu}. \qquad (12.50)$$

Since $\mathbf{v}$ is prime to q', the truth of (12.49) and (12.50) contradicts condition $D(h, k, t, m, \mu)$. Thus (12.46) must be true.

Combining (12.42), and (12.44)–(12.46), we get

$$m^h W > H^s Z^h (1 - 3m^{-\mu})$$

$$\gg H^s Z^h.$$

In particular, $W > 0$ and Proposition 12.2 is proved.

13
Small solutions of additive equations of odd degree

13.1 Preliminary reductions

Our object in the present chapter is to prove the following beautiful theorem of Schmidt (1979b).

Theorem 13.1. *Let k be an odd natural number. Let $s \geqq C_1(k, \varepsilon)$. Then given integers $a_1, \ldots, a_s$, the equation*

$$a_1 x_1^k + \cdots + a_s x_s^k = 0 \tag{13.1}$$

has a solution in integers $x_1, \ldots, x_s$ with

$$0 < \max (|x_1|, \ldots, |x_s|) \leqq A^\varepsilon. \tag{13.2}$$

Here

$$A = \max (1, |a_1|, \ldots, |a_s|). \tag{13.3}$$

The result is a rather easy consequence of Minkowski's linear forms theorem for $k = 1$, so we assume that $k \geqq 3$ in what follows. The only reason for assuming k to be odd is that then x^k can take both signs ($x \in \mathbb{Z}$). We could incorporate a result for even k if we allowed expressions $\pm x_i^k$ to replace x_i^k (see Schmidt's paper). We omit this 'bonus' for the sake of simplicity.

The theorem has important consequences for Diophantine inequalities involving forms of odd degree with real coefficients, as we shall see in Chapter 14.

Proof of Theorem 13.1. The proof is really quite similar to that of Theorem 11.1. Just one important new idea is required. However, we give the entire proof, occasionally referring to the arguments of Chapter 11 in order to avoid undue repetition.

Let $\Lambda = \Lambda_k$ be the set of numbers $\mu > 0$ which have the following property. *Whenever $a_1, \ldots, a_s$ are integers with (13.3) and $s \geqq C_2(k, \mu)$, there are integers $x_1, \ldots, x_s$ with (13.1) and*

$$0 < \max (|x_1|, \ldots, |x_s|) \leqq A^\mu. \tag{13.4}$$

Let λ be the greatest lower bound of Λ. By Theorem 11.1, $\lambda \leqq 1/k$.

Theorem 13.1 will follow if we can show that

$$\lambda = 0.$$

We will suppose that $\lambda > 0$ and ultimately reach a contradiction.
  The polynomial

$$g(\rho) = \lambda + k\lambda^2 - k\lambda\rho - k^2\lambda^2\rho - \rho$$

has $g(\lambda) = -k^2\lambda^3 < 0$. Hence we can pick ρ with

$$0 < \rho < \lambda \tag{13.5}$$

and $g(\rho) < 0$, i.e. with

$$\lambda + k\lambda^2 - k\lambda\rho - k^2\lambda^2\rho < \rho. \tag{13.6}$$

Pick $v > 0$ so small that

$$\rho + 8\lambda v < \lambda, \tag{13.7}$$

$$v < \tfrac{1}{5}, \tag{13.8}$$

$$v < \rho/10. \tag{13.9}$$

Finally pick μ with

$$\max{(\rho + 8\lambda v, \lambda - \tfrac{1}{2}\lambda v)} < \mu < \lambda. \tag{13.10}$$

We will show that $\mu \in \Lambda$, and this will be the desired contradiction.
  It suffices to show that (13.1) has a solution with (13.4) when
$s \ge C_2(k, \mu)$ and $A \ge C_3(k, \mu)$. (One can argue just as we did after the
statement of Proposition 11.2.) We consider only this case, of 'large A',
and of course we may also suppose that no a_i is zero.
  Pick τ with

$$\max{(\rho + 8\lambda v, \lambda - \tfrac{1}{2}\lambda v)} < \tau < \mu \tag{13.11}$$

and choose $\gamma > 0$ so small that

$$(1 + \gamma)\tau + (2\gamma)/k < \mu. \tag{13.12}$$

We now carry out an argument similar to that of the 'reduction' in
Section 11.2. If s is large in terms of γ, as we suppose, then we may
assume that $a_i/a_j \le A^\gamma$ $(i, j = 1, \ldots, t)$ where t is large. For suitable
natural numbers $p_1, \ldots, p_t$ we have

$$\tfrac{1}{2}A^{1+\gamma} \le b_i = a_i p_i^k \le A^{1+\gamma} = B \qquad (i = 1, \ldots, t).$$

Moreover, $p_i^k \le A^{2\gamma}$ $(i = 1, \ldots, t)$. Suppose we can show that

$$b_1 x_1^k + \cdots + b_t x_t^k = 0$$

is soluble in integers satisfying

$$|x_i| \le B^\tau \qquad (i = 1, \ldots, t).$$

Since

$$y_i = p_i x_i \leqq A^{(2\gamma)/k + (1+\gamma)\tau} \leqq A^\mu$$

by (13.12), we get the required type of solution of (13.1).

Hence it will suffice to show that *if* $a_1 x_1^k + \cdots + a_s x_s^k$ *is a form such that* $s \geqq C_4$, $A \geqq C_5$,

$$\tfrac{1}{2} A \leqq a_i \leqq A \qquad (i = 1, \ldots, s) \tag{13.13}$$

then (13.1) *is soluble in integers with*

$$0 < \max (|x_1|, \ldots, |x_s|) \leqq A^\tau. \tag{13.14}$$

Here C_4, C_5, and subsequent constants C_6, C_7, ... as well as constants implied by 'O', '$\ll$', and '$\gg$', depend on k, λ, ρ, ν, μ and τ, which of course have now been fixed.

Proposition 13.1. *If the condition* (13.13) *is satisfied, and if* $A \geqq C_5$ *and* $s \geqq C_6$, *then either* (13.1) *has a solution satisfying* (13.14), *or there are integers* $x_1, \ldots, x_s$, z *with*

$$a_1 x_1^k + \cdots + a_s x_s^k = z, \tag{13.15}$$

$$0 < \max (|x_1|, \ldots, |x_s|) \leqq A^\rho, \tag{13.16}$$

$$|z| \leqq A^{4\nu}. \tag{13.17}$$

This proposition is all that we need, as the reader who recalls Proposition 11.3 will easily believe. As for the details, observe that (by definition of λ) the constant $r = C_2(k, 2\lambda)$ is defined, and may be taken to be an integer; similarly we suppose $u = C_6$ is an integer. Now consider an equation (13.1) in $ru = C_6 C_2(k, 2\lambda)$ variables, with coefficients satisfying (13.13), $A \geqq C_5$. This can be written

$$\sum_{i=1}^r (a_{i1} x_{i1}^k + \cdots + a_{iu} x_{iu}^k) = 0. \tag{13.18}$$

To pick suitable solutions x_{ij}, we proceed as follows. We apply Proposition 13.1 to each of the forms $a_{i1} x_{i1}^k + \cdots + a_{iu} x_{iu}^k$ $(i = 1, \ldots, r)$. If any of these forms take on the value 0 at a point $\mathbf{x}$ with (13.14), we have finished. Otherwise, we can solve the equation

$$a_{i1} y_{i1}^k + \cdots + a_{iu} y_{iu}^k = z_i$$

for each $i = 1, \ldots, r$, with bounds

$$|y_{ij}| \leqq A^\rho, \qquad |z_i| \leqq A^{4\nu}$$

for the variables. Then, since $r = C_2(k, 2\lambda)$, we go on to solve the equation

$$z_1 w_1^k + \cdots + z_r w_r^k = 0$$

with

$$0 < \max_i |w_i| \leq (A^{4\nu})^{2\lambda}.$$

Of course $x_{ij} = y_{ij} w_i$ satisfies (13.18), and these values of x are bounded in modulus by

$$0 < \max_{i,j} |x_{ij}| \leq (A^{4\nu})^{2\lambda} A^\rho \leq A^\tau$$

according to (13.11). Thus (13.14) holds, and this suffices for the proof of Theorem 13.1.

We will now proceed to prove the proposition.

13.2 The circle method

We may suppose without loss of generality that s is even and that half the coefficients a_i in (13.15) are positive (say for $i \leq s/2$) and half are negative. Let N, H be the integer parts of $\frac{1}{2} A^\rho$, $A^{4\nu}$ respectively. Then

$$\tfrac{1}{4} A^\rho < N \leq \tfrac{1}{2} A^\rho, \tfrac{1}{2} A^{4\nu} < H \leq A^{4\nu}. \tag{13.19}$$

Let

$$N_i = [NA^{1/k} |a_i|^{-1/k}]. \tag{13.20}$$

Then

$$N \leq N_i \leq 2N. \tag{13.21}$$

Let Z be the number of solutions of (13.15) subject to

$$1 \leq x_i \leq N_i \quad (i = 1, \ldots, s) \quad \text{and} \quad 1 \leq z \leq H.$$

It suffices to prove that either (13.1) has a solution satisfying (13.14), or $Z > 0$.

We have, as in Section 11.4,

$$Z = \int_0^1 S_1(\alpha) \cdots S_s(\alpha) U(\alpha) \, d\alpha, \tag{13.22}$$

where

$$S_i(\alpha) = \sum_{x=1}^{N_i} e(\alpha a_i x^k)$$

and

$$U(\alpha) = \sum_{z=1}^{H} e(-\alpha z).$$

We define the *major arcs* $\mathcal{M}$ to be the union of the intervals of the type

$$\mathcal{M}(q, u) = \{\alpha : |\alpha - u/q| \leq A^{-1+\nu} N^{-k}\} \tag{13.23}$$

where

$$1 \leqq u \leqq q \leqq A^{v} \quad \text{and} \quad (q, u) = 1. \tag{13.24}$$

These intervals do not overlap, since their centres have mutual distances $\geqq A^{-2v} > 2A^{-1+v}$ by (13.8). The complement of the major arcs in

$$J = [A^{-1+v}N^{-k}, 1 + A^{-1+v}N^{-k}]$$

constitutes the *minor arcs*.

Lemma 13.1. *Suppose α lies in a minor arc and*

$$|S_1(\alpha) \cdots S_s(\alpha)U(\alpha)| \geqq HN^{s-k}A^{-2}. \tag{13.25}$$

Then (13.1) has a solution satisfying (13.14).

Proof. Define

$$\theta = 1/C_7 \tag{13.26}$$

where C_7 is sufficiently large. In particular, we choose C_7 so that $\theta < \lambda$. The quantity $C_2(k, \lambda + \theta)$ is well defined and may be taken to be an integer. Set

$$n = C_2(k, \lambda + \theta), \qquad h = n^2. \tag{13.27}$$

We may suppose C_6 to be so large that, taking $s = C_6$,

$$(k + 4/\rho)/(s - k + 1) < \theta.$$

Since, by (13.9), $2^h A < N^{2/\rho}$, we have

$$(2^h A^2 N^k)^{1/(s-h+1)} < N^{(k+4/\rho)/(s-h+1)} < N^\theta. \tag{13.28}$$

Now, if (13.25) holds, then we see that

$$|S_1(\alpha) \cdots S_s(\alpha)| \geq N^{s-k}A^{-2}. \tag{13.29}$$

If, say, $|S_1(\alpha)| \geqq \cdots \geqq |S_s(\alpha)|$, then the left-hand side of (13.29) is bounded by $|S_h(\alpha)|^{s-h+1}(2N)^{h-1}$, so that, by (13.28),

$$|S_i(\alpha)| \geqq |S_h(\alpha)| \geqq N(2^h N^k A^2)^{-1/(s-h+1)} > N^{1-\theta}$$

$(i = 1, \ldots, h)$. Now we may appeal to the case $M = 1$ of Theorem 5.1, since

$$N^{1-\theta} \geqq (2N)^{1-(1/K)+\theta}$$

from (13.26) and (13.19). This yields natural numbers $q_1, \ldots, q_h$ satisfying

$$q_i \leqq N^{2k\theta} \quad \text{and} \quad \|\alpha a_i q_i\| \leqq N^{-k+2k\theta} \quad (i = 1, \ldots, h). \tag{13.30}$$

It follows that there are integers $u_1, \ldots, u_h$ with

$$|\alpha a_i q_i^k - u_i| \leqq N^{-k+2k^2\theta} \quad (i = 1, \ldots, h). \tag{13.31}$$

Next, we observe that

$$|a_i q_i^k u_j - a_j q_j^k u_i| \leq |(\alpha a_j q_j^k - u_j) a_i q_i^k| + |(\alpha a_i q_i^k - u_i) a_j q_j^k|$$
$$\leq 2N^{-k+2k^2\theta} AN^{2k^2\theta} \tag{13.32}$$

for $i, j = 1, \ldots, h$. Unfortunately we cannot go on, as we did in Section 11.4, to deduce that the vectors

$$\mathbf{a}_i = (a_i q_i^k, u_i) \qquad (i = 1, \ldots, h). \tag{13.33}$$

are proportional. After all, in our present situation the sum length of $S_i(\alpha)$, which is about N, may well be much less than the $(1/k)$th powers of the coefficients a_j.

Instead, we proceed along the following lines. Write

$$\mathbf{a}_1 = r\mathbf{b}$$

where $\mathbf{b}$ is a primitive vector in $\mathbb{Z}^2$, say

$$\mathbf{b} = (q, u) \quad \text{with} \quad q > 0 \quad \text{and} \quad q, u \quad \text{coprime.} \tag{13.34}$$

Note that

$$r = a_1 q_1^k / q \geq a_1 / q,$$

so that

$$r \geq A/(2q). \tag{13.35}$$

Choose $\mathbf{c}$ so that $\mathbf{b}, \mathbf{c}$ becomes a basis for $\mathbb{Z}^2$. Then $\det(\mathbf{b}, \mathbf{c}) = 1$ and each a_i may be written as

$$\mathbf{a}_i = v_i \mathbf{b} + w_i \mathbf{c} \qquad (i = 1, \ldots, h) \tag{13.36}$$

with integers v_i, w_i. Note that w_i satisfies

$$|w_i| = \det(\mathbf{a}_i, \mathbf{b}) = r^{-1} \det(\mathbf{a}_i, \mathbf{a}_1)$$
$$\leq 2qA^{-1} \det(\mathbf{a}_i, \mathbf{a}_1) \leq 4qN^{-k+4k^2\theta} = M, \tag{13.37}$$

say. (The last inequality here comes from (13.32).) Since $q \leq a_1 q_1^k$, we have

$$M \leq 4AN^{-k+6k^2\theta} \tag{13.38}$$

from (13.30).

If $M < 1$, then 'all is well'—the argument is very like that of Section 11.4. It is the handling of the possibility that $M \geq 1$ that requires a rather nice innovation: namely, the use of blocks of variables to get rid of the $\mathbf{c}$ component in (13.36).

13.3 The minor arcs: a further analysis

Let us suppose for the present that $M \geq 1$. Recall from (13.27) that $h = n^2$. We now replace the indices $i = 1, \ldots, h$ by double indices j, l

where $1 \leq j$, $l \leq n$. So, for example, $a_1, \ldots, a_h$ are now written as $\mathbf{a}_{11}, \ldots, \mathbf{a}_{1n}, \ldots, \mathbf{a}_{n1}, \ldots, \mathbf{a}_{nn}$. We consider the equations

$$w_{j1} x_{j1}^k + \cdots + w_{jn} x_{jn}^k = 0 \tag{13.39}$$

for $j = 1, \ldots, n$. There are solutions $\mathbf{x}_j = (x_{j1}, \ldots, x_{jn})$ in integers satisfying

$$0 < \max_l |x_{jl}| \leq \max (1, M)^{\lambda + \theta} = M^{\lambda + \theta} \tag{13.40}$$

for $j = 1, \ldots, n$, because $n = C_2(k, \lambda + \theta)$ and (13.37) holds.

As a consequence of (13.39), the vectors

$$\mathbf{b}_j = x_{j1}^k \mathbf{a}_{j1} + \cdots + x_{jn}^k \mathbf{a}_{jn}$$

in $\mathbb{Z}^2$ are integer multiples of $\mathbf{b}$, and hence the first coordinate b_j of each $\mathbf{b}_j$ is divisible by q. We observe that

$$b_j = a_{j1} q_{j1}^k x_{j1}^k + \cdots + a_{jn} q_{jn}^k x_{jn}^k. \tag{13.41}$$

Now we consider a small solution $\mathbf{y} = (y_1, \ldots, y_n)$ of the equation

$$\frac{b_1}{q} y_1^k + \cdots + \frac{b_n}{q} y_n^k = 0 \tag{13.42}$$

with integer coefficients. There is, in fact, such a $\mathbf{y}$ in $\mathbb{Z}^n$ with

$$0 < \max_j |y_j| \leq \left(\max \left(1, \frac{|b_1|}{q}, \ldots, \frac{|b_n|}{q} \right) \right)^{\lambda + \theta}. \tag{13.43}$$

Looking carefully at (13.42), we find that we have produced an integral solution $z_{jl} = y_j q_{jl} x_{jl}$ of

$$a_{11} z_{11}^k + \cdots + a_{nn} z_{nn}^k = 0 \tag{13.44}$$

satisfying the bounds

$$0 < \max_{j,l} |z_{jl}| \leq \left(M \max \left(1, \frac{|b_1|}{q}, \ldots, \frac{|b_n|}{q} \right) \right)^{\lambda + \theta} N^{2k\theta}. \tag{13.45}$$

Here we appeal to (13.43), (13.30) and (13.40).

A little care is required to get a suitable power of A as an upper bound for the right-hand side of (13.45). In the first place,

$$|b_j| \leq n A N^{2k^2\theta} M^{k\lambda + k\theta}$$

from (13.41), (13.30), (13.40). Because we know that

$$q \leq a_1 q_1^k \leq A N^{2k^2\theta}, \qquad M \geq 1,$$

we may deduce that

$$\max \left(1, \frac{|b_1|}{q}, \ldots, \frac{|b_n|}{q} \right) \leq n A N^{2k^2\theta} M^{k\lambda + k\theta} q^{-1}.$$

Indeed, appealing to the definition (13.37) of M, we have

$$M \max \left(1, \frac{|b_1|}{q}, \ldots, \frac{|b_n|}{q}\right) \leq (Mq^{-1})nAN^{2k^2\theta}M^{k\lambda+k\theta}$$

$$= 4N^{-k+4k^2\theta}nAN^{2k^2\theta}M^{k\lambda+k\theta}$$

$$= 4nAN^{-k+6k^2\theta}M^{k\lambda+k\theta}$$

$$\leq 4nAN^{-k+6k^2\theta}4^{k\lambda+k\theta}A^{k\lambda+k\theta}N^{-k^2\lambda+6k^3\theta(\lambda+\theta)}.$$

(The last inequality is a deduction from (13.38).)

Tidying this up, the right-hand side of (13.45) is at most

$$(A^{1+k\lambda+k\theta}N^{-k-k^2\lambda+6k^3\theta(\lambda+\theta+1)})^{\lambda+\theta}N^{2k\theta} \leq A^{\lambda+k\lambda^2}N^{-k\lambda-k^2\lambda^2}A^{C_8\theta}$$

with a certain constant C_8 independent of θ. (Here we use the fact that $\theta < \lambda$.)

Recalling that $N \geq A^\rho/4$, we see that the last bound is at most

$$A^{\lambda+k\lambda^2-k\lambda\rho-k^2\lambda^2\rho+C_8\theta}4^{k\lambda+k^2\lambda^2} < A^\rho < A^\tau$$

because of (13.6), (13.11) and our hypothesis $A \geq C_5$. (We also need C_7 to be suitably large in (13.26).) Taking into account (13.44), this implies that (13.1) has a solution satisfying (13.14), and Lemma 13.1 is proved in the case $M \geq 1$.

We now turn to the case where $M < 1$. We revert to the original notation with indices $i = 1, \ldots, h$. We have $w_i = 0$ by (13.37), and hence each vector $\mathbf{a}_i$ $(i = 1, \ldots, h)$ is a multiple of $\mathbf{b}$. Therefore q divides each $a_i q_i^k$ $(i = 1, \ldots, h)$. Since $h = n^2 \geq n = C_2(k, \lambda + \theta)$, we choose integers $y_1, \ldots, y_h$ with

$$\frac{a_1 q_1^k}{q}y_1^k + \cdots + \frac{a_h q_h^k}{q}y_h^k = 0,$$

$$0 < \max(|y_1|, \ldots, |y_h|) \leq \left(\frac{AN^{2k^2\theta}}{q}\right)^{\lambda+\theta}.$$

Now if $q \geq A^v$, then we have solved (13.1) non-trivially in integers $q_1 y_1, \ldots, q_h y_h$ bounded in modulus by

$$N^{2k\theta}(AN^{2k^2\theta}/q)^{\lambda+\theta} \leq A^{\lambda+\theta}N^{2k\theta(2k\lambda+1)}q^{-\lambda}$$

$$\leq A^{\lambda+\theta(1+4k^2\rho\lambda+2k\rho)-v\lambda} < A^\tau,$$

in view of (13.11), (13.26). Again, we have a solution of (13.1) satisfying (13.14). On the other hand, suppose $q < A^v$; (13.31) yields

$$\left|\alpha - \frac{u}{q}\right| = \left|\alpha - \frac{u_1}{a_1 q_1^k}\right| \leq 2A^{-1}|\alpha a_1 q_1^k - u_1|$$

$$\leq 2A^{-1}N^{-k+2k^2\theta} < A^{-1+v}N^{-k},$$

so α lies in a major arc. We have shown that if (13.25) holds, then either (13.1) has a solution satisfying (13.14), or α lies in a major arc. Lemma 13.1 follows.

13.4 The major arcs

We need only carry on with the proof of Proposition 13.1 if we have the inequality

$$|S_1(\alpha) \cdots S_s(\alpha)U(\alpha)| < HN^{s-k}A^{-2} \qquad (13.46)$$

at every point of the minor arcs. (For if this is not the case, then Lemma 13.1 gives a solution of (13.1) satisfying (13.14).)

Assume that (13.46) holds throughout the minor arcs. Then, recalling (13.22),

$$Z = \left(\int_{\mathcal{M}} + \int_{\mathcal{N}\mathcal{M}} \right) S_1(\alpha) \cdots S_s(\alpha)U(\alpha)\,d\alpha$$

$$= \sum_{q \leq A^v} \sum_{\substack{u=1 \\ (u,q)=1}}^{q} \int_{\mathcal{M}(q,u)} S_1(\alpha) \cdots S_s(\alpha)U(\alpha)\,d\alpha$$

$$+ O(HN^{s-k}A^{-2}). \qquad (13.47)$$

Write

$$v(\beta) = k^{-1} \sum_{y=1}^{AN^k} y^{(1/k)-1} e(\beta y).$$

Let $\alpha = (u/q) + \beta$ be a point of a major arc. The argument leading to (11.11) applies here, and gives rise to the approximation

$$S_i(\alpha) = a_i^{-1/k} q^{-1} S_{q,a_iu} v(\beta) + O(qAN^k |\beta| + q)$$
$$= a_i^{-1/k} q^{-1} S_{q,a_iu} v(\beta) + O(A^{2v})$$

for $a_i > 0$. For negative a_i we work with $\overline{S_i(\alpha)}$ and reach the approximation

$$S_i(\alpha) = |a_i|^{-1/k} q^{-1} S_{q,a_iu} \overline{v(\beta)} + O(A^{2v}).$$

Since $A^{2v} < N$ from (13.9), (13.19), we can combine these estimates to get

$$S_1(\alpha) \cdots S_s(\alpha)e(-\alpha z) = |a_1 \cdots a_s|^{-1/k} q^{-s} S_{q,a_1u} \cdots S_{q,a_su}$$

$$\times e\left(-\frac{uz}{q}\right) e(-\beta z) |v(\beta)|^s + O(N^{s-1}A^{2v}) \qquad (13.48)$$

for $z = 1, \ldots, H$.

We replace $e(-\beta z)$ by 1 on the right-hand side of (13.48). Since

$e(-\beta z) - 1 = O(H\,|\beta|) = O(HA^{-1+\nu}N^{-k})$, this introduces an error of

$$O(A^{-s/k}(A^{1/k}N)^s HA^{-1+\nu}N^{-k}) = O(N^{s-k}A^{-1+5\nu}) = O(N^{s-1}A^{2\nu}).$$

(We used (13.8) for the last step.) Thus

$$S_1(\alpha) \cdots S_s(\alpha)e(-\alpha z) = |a_1 \cdots a_s|^{-1/k}q^{-s}S_{q,a_1 u} \cdots S_{q,a_s u}$$
$$\times e\left(-\frac{uz}{q}\right)|v(\beta)|^s + O(N^{s-1}A^{2\nu}). \quad (13.49)$$

We sum (13.49) over $z = 1, \ldots, H$ and integrate over $\mathcal{M}(q, u)$:

$$\int_{\mathcal{M}(q,u)} S_1(\alpha) \cdots S_s(\alpha)U(\alpha)\,d\alpha$$
$$= |a_1 \cdots a_s|^{-1/k}q^{-s}S_{q,a_1 u} \cdots S_{q,a_s u}\sum_{z=1}^{H} e\left(-\frac{uz}{q}\right)\int_{-A^{-1+\nu}N^{-k}}^{A^{-1+\nu}N^{-k}} |v(\beta)|^s\,d\beta$$
$$+ O(HN^{s-k-1}A^{-1+3\nu}).$$

We then sum over the distinct major arcs. Taking (13.47) into account, we find that

$$Z = |a_1 \cdots a_s|^{-1/k}\mathfrak{S}_1 J_1(A) + O(HN^{s-k-1}A^{-1+5\nu} + HN^{s-k}A^{-2}). \quad (13.50)$$

Here

$$\mathfrak{S}_1 = \sum_{q \leq A^\nu} \sum_{\substack{u=1 \\ (u,q)=1}}^{q} q^{-s}S_{q,a_1 u} \cdots S_{q,a_s u}\sum_{z=1}^{H} e\left(-\frac{uz}{q}\right)$$

and

$$J_1(A) = \int_{-A^{-1+\nu}N^{-k}}^{A^{-1+\nu}N^{-k}} |v(\beta)|^s\,d\beta.$$

13.5 Conclusion

The singular integral $J_1(A)$ can be estimated in the same way as $J(A)$ in Section 11.1. Let

$$J_1 = k^{-s} \sum_{m_1,\ldots,m_s} (m_1 \cdots m_s)^{(1/k)-1}$$

where the summation is restricted to integers satisfying

$$1 \leq m_1, \ldots, m_s \leq AN^k,$$
$$m_1 + \cdots + m_{s/2} = m_{(s/2)+1} + \cdots + m_s.$$

Then

$$J_1(A) - J_1 \ll \int_{A^{-1+\nu}N^{-k}}^{\infty} \beta^{-s/k}\,d\beta$$
$$\ll (A^{1-\nu}N^k)^{(s/k)-1}$$
$$\ll (AN^k)^{(s/k)-1}A^{-\nu}$$

since we suppose $s > 2k$. Moreover,

$$J_1 \gg (AN^k)^{(s/k)-1},$$

so that

$$J_1(A) \gg (AN^k)^{(s/k)-1}. \tag{13.51}$$

As regards the singular series $\mathfrak{S}_1$ we can repeat the calculation of Section 11.6 to obtain

$$\mathfrak{S}_1 = H + O\left(\sum_{q \leqq A^\nu} q^2\right) = H + O(A^{3\nu}) \gg H. \tag{13.52}$$

The term $|a_1 \cdots a_s|^{-1/k} \mathfrak{S}_1 J_1(A)$ on the right-hand side in (13.50) is seen to be

$$\gg HN^{s-k}A^{-1}$$

on combining the estimates (13.51), (13.52).

Now

$$HN^{s-k-1}A^{-1+5\nu} + HN^{s-k}A^{-2}$$

is of a smaller order of magnitude than $HN^{s-k}A^{-1}$, because

$$A^{5\nu} < (4N)^{1/2}$$

from (13.9), (13.19). Hence

$$Z \gg HN^{s-k}A^{-1}.$$

In particular, $Z > 0$, and Proposition 13.1 is proved.

14
Diophantine inequalities for forms of odd degree

14.1 Multilinear forms

We begin by recalling a well-known theorem of Birch (1957):

> If $\mathcal{G}(x_1, \ldots, x_s)$ is a form of odd degree k with integer coefficients, and if $s \geq C_1(k)$, then there is an integer point $\mathbf{x} = (x_1, \ldots, x_s) \neq \mathbf{0}$ with
>
> $$\mathcal{G}(\mathbf{x}) = 0.$$

The case $k = 1$ is trivial, and the case $k = 3$ had been settled a little before the general case by Lewis (1957) and by Davenport (unpublished). Later Davenport (1963) showed that one may take $C_1(3)$ equal to 16. This was regarded as very difficult to beat, but Heath-Brown (1983) eventually reduced 16 to 10 for *non-singular* cubic forms.

Forms with *real* coefficients are more difficult to deal with. We are going to prove the remarkable theorem of Schmidt (1980a).

Theorem 14.1. Let $\mathcal{F}(\mathbf{x}) = \mathcal{F}(x_1, \ldots, x_s)$ be a form of odd degree k with real coefficients, with $s \geq C_2(k)$. Then there is an integer point $\mathbf{x} \neq \mathbf{0}$ with

$$|\mathcal{F}(\mathbf{x})| < 1. \tag{14.1}$$

This contains the theorem of Birch stated above. In a sense, the method of proof is an elaboration of that of Birch, with the tool of Theorem 13.1 supplying the 'extra power'. In fact, Birch (1970) notes that he and Davenport knew in 1957 that if Theorem 13.1 were to be proved, it could be used to deduce an inequality (14.1) for real forms. This is not to say that the proof of Theorem 14.1 is easy!

As pointed out in Chapter 1, the case $k = 3$ of Theorem 14.1 was settled earlier by Pitman (1968) by methods which are special to cubic forms. She showed that one may take $C_2(3)$ equal to $(1314)^{256}$. This can be reduced considerably (Schmidt, unpublished). In principle, constants $C_2(k)$ could be computed by the methods of this chapter; but the values obtained would be astronomical.

In order to carry out our proof, which is inductive, we need to establish a strong form of (14.1). Given a vector $\mathbf{x} = (x_1, \ldots, x_s)$, put

$$|\mathbf{x}| = \max(|x_1|, \ldots, |x_s|)$$

(This convention is slightly different from the one used in earlier chapters.) Given a form $\mathscr{F}$, let

$$|\mathscr{F}|$$

be the maximum absolute value of its coefficients. We will show that:

Given odd k and given a positive number E, there is a $C_3 = C_3(k, E)$ with the following property. If $N \geqq 1$ is real and if $\mathscr{F}(\mathbf{x}) = \mathscr{F}(x_1, \ldots, x_s)$ is a form of degree k with real coefficients, with $s \geqq C_3$, then there is a non-zero integer point x with

$$|\mathbf{x}| \leqq N \tag{14.2}$$

and

$$|\mathscr{F}(\mathbf{x})| \ll N^{-E} |\mathscr{F}| \tag{14.3}$$

The constant implicit in $\ll$ depends on k, E. More generally, in what follows, the constants in $\ll$ will depend on E, k, h, $k_1, \ldots, k_h$, m, l.

Now (14.3) implies (14.1) if N is sufficiently large. The statement just formulated is still not strong enough for our inductive argument. We have to deal with simultaneous inequalities satisfied by several forms, and we have to find not one, but several linearly independent integer points.

A real form of degree k may be written

$$\mathscr{F}(x_1, \ldots, x_s) = \sum_{1 \leqq i_1, \ldots, i_k \leqq s} a(i_1, \ldots, i_k) x_{i_1} \cdots x_{i_k}$$

with $a(i_1, \ldots, i_k)$ a symmetric function of its arguments. With $\mathscr{F}$ we associate the 'multilinear form'

$$\hat{\mathscr{F}}(\mathbf{x}(1), \ldots, \mathbf{x}(k)) = \sum_{1 \leqq i_1, \ldots, i_k \leqq s} a(i_1, \ldots, i_k) x_{i_1}(1) \cdots x_{i_k}(k).$$

Here $\mathbf{x}(i) = (x_1(i), \ldots, x_s(i))$. Note that $\hat{\mathscr{F}}$ is linear in each vector $\mathbf{x}(i)$ ($1 \leqq i \leqq k$) and symmetric in the k vectors $\mathbf{x}(1), \ldots, \mathbf{x}(k)$, and that

$$\mathscr{F}(\mathbf{x}) = \hat{\mathscr{F}}(\mathbf{x}, \ldots, \mathbf{x}).$$

Our main result is

Theorem 14.2. *Given $h \geqq 1$, $m \geqq 1$ and odd numbers $k_1, \ldots, k_h$, and given a positive number E, there is a constant*

$$C_4 = C_4(k_1, \ldots, k_h, m, E)$$

as follows. If $N \geqq 1$ and if $\mathscr{F}_1, \ldots, \mathscr{F}_h$ are forms with real coefficients of respective degrees $k_1, \ldots, k_h$ in $\mathbf{x} = (x_1, \ldots, x_s)$ where $s \geqq C_4$, then there *are m linearly independent integer points* $\mathbf{x}(1), \ldots, \mathbf{x}(m)$ *with*

$$|\mathbf{x}(i)| \leqq N \qquad (1 \leqq i \leqq m) \tag{14.4}$$

and

$$|\mathscr{F}_j(\mathbf{x}(i_1), \ldots, \mathbf{x}(i_{k_j}))| \ll N^{-E} |\mathscr{F}_j| \tag{14.5}$$

for $1 \leqq j \leqq h$, $1 \leqq i_1, \ldots, i_{k_j} \leqq m$.

In particular it follows that

$$|\mathscr{F}_j(\mathbf{x}(i))| \ll N^{-E} |\mathscr{F}_j| \qquad (1 \leqq j \leqq h; 1 \leqq i \leqq m). \tag{14.6}$$

Let us see what this tells us about small solutions of homogeneous equations with *integer* coefficients. Suppose that $\mathscr{G}_1, \ldots, \mathscr{G}_h$ are forms in $\mathbb{Z}[x_1, \ldots, x_s]$ of respective odd degrees $k_1, \ldots, k_h$. Put

$$G = \max (1, |\mathscr{G}_1|, \ldots, |\mathscr{G}_h|) \tag{14.7}$$

$$\mathscr{F}_j(x) = G^{-1} \mathscr{G}_j(x) \qquad (1 \leqq j \leqq h).$$

Suppose that $s \geqq C_4(k_1, \ldots, k_h; m, \varepsilon^{-1}) = C_5(k_1, \ldots, k_h; m, \varepsilon)$ say. Apply Theorem 14.2 with $N = N_0 G^\varepsilon$, where $N_0 = N_0(k_1, \ldots, k_h; m, \varepsilon)$ is to be chosen in a moment. We obtain m linearly independent integer points $\mathbf{x}(1), \ldots, \mathbf{x}(m)$ with (14.4), (14.5). In particular we have $|\mathbf{x}(i)| \leqq N_0 G^\varepsilon$ ($i = 1, \ldots, m$), and for $1 \leqq j \leqq h$, $1 \leqq i_1, \ldots, i_{k_j} \leqq m$, we have

$$|\mathscr{G}_j(\mathbf{x}(i_1), \ldots, \mathbf{x}(i_{k_j}))| = G |\mathscr{F}_j(\mathbf{x}(i_1), \ldots, \mathbf{x}(i_{k_j}))|$$
$$\ll GN^{-1/\varepsilon} \ll N_0^{-1/\varepsilon}$$

so that

$$|\mathscr{G}_j(\mathbf{x}(i_1), \ldots, \mathbf{x}(i_{k_j})| < 1/k_j!$$

if N_0 was chosen large enough. Since $k_j! \mathscr{G}_j$ has integer coefficients, $\mathscr{G}_j(\mathbf{x}(i_1), \ldots, \mathbf{x}(i_{k_j})) = 0$. Thus we have

Theorem 14.3. *Given* $h \geqq 1$, $m \geqq 1$ *and odd numbers* $k_1, \ldots, k_h$, *there is a constant* $C_5 = C_5(k_1, \ldots, k_h; m, \varepsilon)$ *such that if* $\mathscr{G}_1, \ldots, \mathscr{G}_h$ *are forms of respective degrees* $k_1, \ldots, k_h$ *in* $\mathbb{Z}[x_1, \ldots, x_s]$, *where* $s \geqq C_5$, *then* $\mathscr{G}_1, \ldots, \mathscr{G}_h$ *vanish on an m-dimensional subspace which is spanned by integer points* $\mathbf{x}(1), \ldots, \mathbf{x}(m)$ *having*

$$|\mathbf{x}(i)| \ll G^\varepsilon \qquad (i = 1, \ldots, m).$$

Here G is given by (14.7).

We now turn to the proof of Theorem 14.2. We start by considering the simplest type of real forms.

14.2   Additive forms

Our task in this section and the next will be a proof of

Proposition 14.1.   *Given odd $k \geq 3$ and $E > 0$, there is a constant $C_6 = C_6(k, E)$ with the following property. Let $\mathcal{L}(\mathbf{x})$ be a nonzero additive form of degree k,*

$$\mathcal{L}(\mathbf{x}) = \lambda_1 x_1^k + \cdots + \lambda_s x_s^k,$$

where $s \geq C_6$. Then for real $N \geq 1$ there is a non-zero integer point $\mathbf{x} = (x_1, \ldots, x_s)$ with

$$|\mathbf{x}| \leq N \quad \text{and} \quad |\mathcal{L}(\mathbf{x})| < N^{-E} |\mathcal{L}|. \tag{14.8}$$

Proof.   We begin with simple reductions. Suppose we can prove the conclusion for $s \geq C_7(k, E)$ and $N \geq C_8(k, E)$. To deal with $1 \leq N < C_8(E)$, it will suffice to show that there is an $\mathbf{x}$ in $\mathbb{Z}^s$ with $|\mathbf{x}| = 1$ and

$$|\mathcal{L}(\mathbf{x})| < \mu |\mathcal{L}| \tag{14.9}$$

where $\mu = C_8(k, E)^{-E}$. If $s > \mu^{-1} + 1$, then there will be λ_i, λ_j with $i \neq j$ and

$$\|\lambda_i| - |\lambda_j\| < \mu |\mathcal{L}|;$$

so (14.9) will be true if we choose $x_i = 1$ and $x_j = \pm 1$, to get $|\lambda_i x_i^k + \lambda_j x_j^k| = \|\lambda_i| - |\lambda_j\|$, and other components of $\mathbf{x}$ equal to 0. It is now clear that *it will suffice to prove Proposition* 14.11 *for large values of N, say for $N \geq C_8(k, E)$.*

(The reader might like to try to produce an analogous argument for an *arbitrary* real form in place of $\mathcal{L}$. It is not so easy: see Baker and Schmidt (1980), Theorem 9. Since we are not going to tackle this here, we put up with the '$\ll$' in Theorems 14.2 and 14.3.)

Now choose $\gamma > 0$ so small that

$$2\gamma + (E + \gamma)/(E + \tfrac{1}{8}) < 1. \tag{14.10}$$

Proposition 14.1 is obvious if there is a λ_i with $|\lambda_i| < N^{-E} |\mathcal{L}|$. So we may suppose that $N^{-E} |\mathcal{L}| \leq |\lambda_i| \leq |\mathcal{L}|$ for each $|\lambda_i|$. Cover the interval $-E \leq \alpha \leq 0$ by a finite number of intervals of length γ. Given any integer t, and given s which is large as a function of t and E, one of these intervals will be such that at least t of the $|\lambda_i|$ are of the type N^{α_i} with $\alpha_i \in I$. Let's suppose this holds for $|\lambda_1|, \ldots, |\lambda_t|$. Put $H = N^\gamma \max(|\lambda_1|, \ldots, |\lambda_t|)$ and choose natural numbers $q_1, \ldots, q_t$ as large as possible, with

$$|\lambda_i| q_i^k \leq H \quad (i = 1, \ldots, t).$$

Thus $q_i^k \leqq N^{2\gamma}$ $(i = 1, \ldots, t)$. Since $H/|\lambda_i| \geqq N^\gamma$, we have

$$\tfrac{1}{2}H \leqq |\lambda_i| \, q_i^k \leqq H \qquad (i = 1, \ldots, t)$$

if N is sufficiently large.

Now, of course, we consider the form

$$\mathcal{M}(\mathbf{y}) = \lambda_1 q_1^k y_1^k + \cdots + \lambda_t q_t^k y_t^k,$$

which has

$$|\mathcal{M}| \leqq H.$$

Suppose Proposition 14.1 is true for this form with

$$N_0 = N^{(E+\gamma)/(E+(1/8))} \qquad E_0 = E + \tfrac{1}{8}$$

in place of N, E. Then we can solve

$$|\mathcal{M}(\mathbf{y})| < N_0^{-(E+(1/8))} |\mathcal{M}|, \qquad 0 < |\mathbf{y}| \leqq N_0.$$

Hence

$$\begin{aligned}
|\mathcal{L}(q_1 y_1, \ldots, q_t y_t, 0, \ldots, 0)| &= |\mathcal{M}(\mathbf{y})| \\
&< N_0^{-(E+(1/8))} H \\
&\leqq N^{-(E+\gamma)} N^\gamma |\mathcal{L}| = N^{-E} |\mathcal{L}|,
\end{aligned}$$

while all $q_i y_i$ are bounded in modulus by

$$N^{2\gamma/k} N_0 \leqq N$$

from (14.10).

Summing up, it is clear that if Proposition 14.3 is true with $E + (1/8)$ in place of E for forms $\mathcal{L}$ with

$$\tfrac{1}{2} |\mathcal{L}| \leqq |\lambda_i| \leqq |\mathcal{L}| \qquad (i = 1, \ldots, s), \tag{14.11}$$

then it is true with E for general forms. By homogeneity we may replace (14.11) by

$$|\mathcal{L}| = 1 \quad \text{and} \quad \tfrac{1}{2} \leqq |\lambda_i| \leqq 1 \qquad (i = 1, \ldots, s). \tag{14.12}$$

It will now suffice to prove the following two statements:

(i) *The conclusion of Proposition 14.1 is true for $0 < E \leqq \tfrac{1}{2}$ for forms $\mathcal{L}$ with (14.12), provided only that $s \geqq C_9(k, E)$;*

(ii) *The conclusion of Proposition 14.1 is true for E for forms $\mathcal{L}$ with (14.12), provided only that $s \geqq C_9(k, E)$ and that Proposition 14.1 is true for $E - \tfrac{1}{4}$ for general diagonal forms.*

We shall give a (simultaneous) proof of (i) and (ii) by analytic means in the next section.

14.3 The Davenport–Heilbronn circle method

In this section we use a variant of the circle method which first appears in a paper of Davenport and Heilbronn (1946). There are many later uses of the 'Davenport–Heilbronn circle method': see Davenport (1977) and Vaughan (1981) for references; also the papers of Baker (1982a) and Baker and Harman (1982b), (1984a).

In our particular situation, we may suppose that s is even, and that λ_i is positive for $i \leq s/2 = t$ (say) and negative for $i > t$. Under these conditions we wish to estimate the number Z of solutions of

$$|\mathcal{L}(\mathbf{x})| < N^{-E} \quad |\mathcal{L}| = N^{-E} \tag{14.13}$$

in integer points $\mathbf{x} = (x_1, \ldots, x_s)$ in $\mathbb{Z}^s$ subject to

$$1 \leq x_i \leq N \quad (i = 1, \ldots, s). \tag{14.14}$$

As the reader might expect from past experience, we shall show that *either* $Z > 0$, *or* the form $\mathcal{L}$ has certain redeeming features which enable us to get the conclusion of Proposition 14.1 without estimating Z.

Put

$$\theta = (4k + E)^{-1}(k + 2E + 2)^{-1}, \tag{14.15}$$

$$m = 1 \text{ in the case (i)},$$

$$m = C_6(k, E - \tfrac{1}{4}) \text{ in the case (ii)}. \tag{14.16}$$

Let n be such that any non-trivial equation

$$\mathcal{G}(\mathbf{x}) = a_1 x_1^k + \cdots + a_n x_n^k = 0$$

over $\mathbb{Z}$ has a solution with

$$0 < |\mathbf{x}| \leq |\mathcal{G}|^{\theta}. \tag{14.17}$$

Set

$$h = mn, \tag{14.18}$$

and choose φ so small that

$$10Kh\varphi < 1. \tag{14.19}$$

Now let $s = C_9(k, E)$ be so large that

$$(2k + 4E + 2)(s - h + 1)^{-1} < \varphi. \tag{14.20}$$

We will assume N to be large.

Lemma 14.1. *We have, for real Q,*

$$\int_{-\infty}^{\infty} e(\alpha Q)\left(\frac{\sin \pi \alpha}{\pi \alpha}\right)^2 d\alpha = \begin{cases} 1 - |Q| & \text{if} \quad |Q| \leq 1 \\ 0, & \text{if} \quad Q > 1. \end{cases}$$

Proof. We recall from Section 2.2 that

$$\int_{-\infty}^{\infty} \left(\frac{\sin \pi \alpha}{\pi \alpha}\right)^2 d\alpha = 1.$$

Hence

$$\int_{-\infty}^{\infty} \left(\frac{\sin \pi \gamma \alpha}{\pi \alpha}\right)^2 d\alpha = |\gamma|$$

for any real γ. This gives

$$\int_{-\infty}^{\infty} e(\alpha Q)\left(\frac{\sin \pi \alpha}{\pi \alpha}\right)^2 d\alpha = \int_{-\infty}^{\infty} \cos 2\pi \alpha Q \left(\frac{\sin \pi \alpha}{\pi \alpha}\right)^2 d\alpha$$

$$= \frac{1}{2}\int_{-\infty}^{\infty} \frac{\sin^2 \pi \alpha (Q+1) + \sin^2 \pi \alpha (Q-1) - 2\sin^2 \pi \alpha Q}{(\pi \alpha)^2} d\alpha$$

$$= \frac{1}{2}\{|Q+1| + |Q-1| - 2|Q|\}$$

and the lemma follows. (Alternatively, we can use (9.45) in conjunction with Fourier's inversion formula.)

Instead of working with $((\sin \pi \alpha)/\pi \alpha)^2$ we could use a 'repeated smoothing' as in Sections 2.1, 9.3, and work with $((\sin \pi \alpha)/\pi \alpha)^r$, where r is large. See Davenport (1956), Lemma 1. One can in fact get an analogue of the more refined analysis in Section 2.2; see Baker and Harman (1982*b*). However, here and throughout we can afford to be very wasteful, on account of the strength of Theorem 13.1.

Substituting $N^E Q$ for Q and writing $\beta = N^E \alpha$ we get

$$N^E \int_{-\infty}^{\infty} e(\beta Q)\left(\frac{\sin \pi \beta N^{-E}}{\pi \beta}\right)^2 d\beta = \begin{cases} 1 - N^E |Q| & \text{if} \quad |Q| < N^{-E} \\ 0 & \text{if} \quad |Q| \geq N^{-E}. \end{cases}$$

$$(14.21)$$

Now define

$$S(\alpha) = \sum_{x=1}^{N} e(\alpha x^k), \qquad I(\alpha) = \int_0^N e(\alpha x^k) \, dx.$$

Substituting $\mathcal{L}(\mathbf{x})$ for Q into (14.21) and taking the sum over $\mathbf{x}$ subject to (14.14) we obtain (writing α instead of β)

$$N^E \int_{-\infty}^{\infty} S(\lambda_1 \alpha) \cdots S(\lambda_s \alpha)\left(\frac{\sin \pi \alpha N^{-E}}{\pi \alpha}\right)^2 d\alpha$$

$$= \sum_{x_1=1}^{N} \cdots \sum_{x_s=1}^{N} \max(0, 1 - N^E |\mathcal{L}(\mathbf{x})|). \quad (14.22)$$

Similarly, by taking integrals instead of sums, we obtain

$$N^E \int_{-\infty}^{\infty} I(\lambda_1\alpha) \cdots I(\lambda_s\alpha) \left(\frac{\sin \pi\alpha N^{-E}}{\pi\alpha}\right)^2 d\alpha$$

$$= \int_0^N \cdots \int_0^N \max\left(0, 1 - N^E |\mathscr{L}(\mathbf{x})|\right) d\mathbf{x}. \quad (14.23)$$

The right-hand side of (14.22) is a lower bound for Z. The general idea now will be to show that the right-hand side of (14.23) is large (which is easy), and to show that the left-hand sides of (14.22) and (14.23) differ little (which is tricky).

Lemma 14.2. *The right-hand side of* (14.22) *is*

$$\gg N^{s-E-k}.$$

Proof. Make the substitution $y_i = N^E |\lambda_i| x_i^k$ $(i = 1, \ldots, s)$. Then

$$dx_i = N^{-E/k} |\lambda_i|^{-1/k} k^{-1} y_i^{1/k-1} dy_i,$$

and the integral in question becomes

$$N^{-Es/k} |\lambda_1 \cdots \lambda_s|^{-1/k} k^{-s} \int_0^{|\lambda_1| N^{E+k}} \cdots \int_0^{|\lambda_s| N^{E+k}}$$

$$\times \frac{\max\left(0, 1 - |y_1 + \cdots + y_t - y_{t+1} - \cdots - y_{2t}|\right)}{(y_1 \cdots y_s)^{1-(1/k)}} d\mathbf{y} \quad (14.24)$$

because of our special convention on the signs of $\lambda_1, \ldots, \lambda_s$. The domain

$$\frac{1}{4t} N^{E+k} \leq y_i \leq \frac{1}{2t} N^{E+k} \qquad (i = 1, \ldots, t),$$

$$\frac{1}{16t} N^{E+k} \leq y_i \leq \frac{1}{8t} N^{E+k} \qquad (i = t+1, \ldots, 2t-1),$$

$$|y_1 + \cdots + y_t - y_{t+1} - \cdots - y_{2t}| \leq \tfrac{1}{2}$$

is contained in the domain of integration of the integral (14.24). The volume of this domain is

$$\gg N^{(E+k)(s-1)}$$

and the integrand in the domain is

$$\gg N^{(E+k)s((1/k)-1)},$$

so that the integral (14.24) is $\gg N^{(E+k)((s/k)-1)}$, and the right-hand side of 14.23) is

$$\gg N^{-Es/k} N^{(E+k)((s/k)-1)} = N^{s-E-k}.$$

Lemma 14.3.   *We have*

$$\int_{|\alpha|<(1/(2k))N^{1-k}} S(\lambda_1\alpha)\cdots S(\lambda_s\alpha)\left(\frac{\sin\pi\alpha N^{-E}}{\pi\alpha}\right)^2 d\alpha \gg N^{s-k-2E}.$$

$$(14.25)$$

Proof.   We note that

$$|I(\lambda_j\alpha)| = \left|\int_0^N e(\lambda_j\alpha x^k)\,dx\right|$$

$$= k^{-1}|\lambda_j\alpha|^{-1/k}\left|\int_0^{|\lambda_j|\alpha N^k} e(t)t^{(1/k)-1}\,dt\right|.$$

The integral on the right is bounded as a function of the upper limit of integration. This gives the first estimate in

$$I(\lambda_j\alpha) \ll \min(|\alpha|^{-1/k}, N):$$

$$(14.26)$$

the second estimate is obvious. Thus

$$\int_{|\alpha|\geq(1/(2k))N^{1-k}} I(\lambda_1\alpha)\cdots I(\lambda_s\alpha)\left(\frac{\sin\pi\alpha N^{-E}}{\pi\alpha}\right)^2 d\alpha$$

$$\ll N^{-2E}\int_{|\alpha|\geq(1/(2k))N^{1-k}} |\alpha|^{-s/k}\,d\alpha$$

$$\ll N^{(k-1)((s/k)-1)-2E}.$$

This is of smaller order of magnitude than N^{s-k-2E}. Thus by the preceding lemma,

$$\int_{|\alpha|<(1/(2k))N^{1-k}} I(\lambda_1\alpha)\cdots I(\lambda_s\alpha)\left(\frac{\sin\pi\alpha N^{-E}}{\pi\alpha}\right)^2 d\alpha \gg N^{s-k-2E}.$$

$$(14.27)$$

It remains to compare the integral here with the one in (14.25). Now

$$S(\lambda_j\alpha) = \sum_{x=1}^N e(\lambda_j\alpha x^k) = \sum_{x=1}^N e(f(x)),$$

say, where

$$|f'(x)| = k\,|\lambda_j\alpha x^{k-1}| \leq kN^{k-1}\,|\alpha| \leq \tfrac{1}{2}$$

in our interval of integration. Further, $f''(x)$ is of constant sign. Thus by van der Corput's lemma (Vaughan (1981); Lemma 4.2) we have

$$S(\lambda_j\alpha) = I(\lambda_j\alpha) + O(1)$$

$$(14.28)$$

(it is worth comparing this with Lemma 4.4, with $q = 1$, and with equation (11.11)).

From (14.26) and (14.28) we get

$$|S(\lambda_1\alpha) \cdots S(\lambda_s\alpha) - I(\lambda_1\alpha) \cdots I(\lambda_s\alpha)| \ll \min(|\alpha|^{-1/k}, N)^{s-1}$$

in our interval of integration. The left-hand sides of (14.25), (14.27) have a difference

$$\ll N^{-2E+s-1} \int_{|\alpha|<N^{-k}} d\alpha + N^{-2E} \int_{|\alpha|\geq N^{-k}} |\alpha|^{-(s-1)/k} \, d\alpha$$
$$\ll N^{-2E+s-k-1},$$

whence Lemma 14.3 follows.

Lemma 14.4. *We have*

$$\int_{|\alpha|>N^{k+2E+1}} |S(\lambda_1\alpha) \cdots S(\lambda_s\alpha)| \left(\frac{\sin \pi\alpha N^{-E}}{\pi\alpha}\right)^2 d\alpha \leq N^{s-k-2E-1}.$$

Proof. The left-hand side is not greater than

$$N^s \int_{N^{k+2E+1}}^{\infty} \frac{d\alpha}{\alpha^2} = N^{s-k-2E-1}.$$

Suppose for a moment that

$$\int_{(1/(2k))N^{1-k}<|\alpha|\leq N^{k+2E+1}} S(\lambda_1\alpha) \cdots S(\lambda_s\alpha) \left(\frac{\sin \pi\alpha N^{-E}}{\pi\alpha}\right)^2 d\alpha$$
$$\leq 2N^{s-k-2E-1}. \quad (14.29)$$

It then follows from Lemmas 14.3 and 14.4 that the same integral extended over the real line is $\gg N^{s-k-2E}$, and in view of (14.22) it follows that

$$Z \gg N^{s-k-E}.$$

Thus, if (14.29) is true, then $Z > 0$.

We may thus suppose that (14.29) is false. There is an α in

$$\frac{1}{2k} N^{1-k} < |\alpha| \leq N^{k+2E+1} \quad (14.30)$$

with

$$|S(\lambda_1\alpha) \cdots S(\lambda_s\alpha)| > N^{s-2k-4E-2}.$$

If, say, $|S(\lambda_1\alpha)| \geq \cdots \geq |S(\lambda_s\alpha)|$, then the left-hand side here is

$$\leq |S(\lambda_h\alpha)|^{s-h+1} N^{h-1},$$

and we have

$$|S(\lambda_i\alpha)| \geq N^{1-(2k+4E+2)/(s-h+1)} \geq N^{1-\varphi}$$

for $i = 1, \ldots, h$, by (14.20).

In view of (14.19), we may apply Theorem 5.1 with $M = 1$, $P = N^{1-\varphi}$. We obtain natural numbers $q_1, \ldots, q_h$ with

$$q_i \leqq N^{2k\varphi} \quad \text{and} \quad \|\alpha\lambda_i q\| \leqq N^{-k+2k\varphi} \quad (i = 1, \ldots, h). \quad (14.31)$$

Setting $q = q_1 \cdots q_h$ we have

$$q \leqq N^{2kh\varphi} \quad \text{and} \quad \|\alpha\lambda_i q\| \leqq N^{-k+2kh\varphi} \quad (i = 1, \ldots, h). \quad (14.32)$$

With suitable integers $g_1, \ldots, g_h$ we have

$$|\alpha\lambda_i q - g_i| \leqq N^{-k+2kh\varphi} \quad (i = 1, \ldots, h), \quad (14.33)$$

whence

$$\lambda_i = \frac{g_i}{\alpha q} + \mu_i \quad (i = 1, \ldots, h)$$

with, for large N,

$$|\mu_i| \leqq |\alpha\lambda_i|^{-1} N^{-k+2kh\varphi} \leqq 4kN^{-1+2kh\varphi} \leqq N^{-4/5} \quad (i = 1, \ldots, h)$$

by (14.33), (14.11), (14.30) and (14.19).

Thus, with $\mathbf{x} = (x_1, \ldots, x_s) = (y_1, \ldots, y_h, 0, \ldots, 0) = (\mathbf{y}, \mathbf{0})$,

we have

$$\mathscr{L}(\mathbf{x}) = (\alpha q)^{-1} \mathscr{G}(\mathbf{y}) + \mathscr{M}(\mathbf{y}), \quad (14.34)$$

where $\mathscr{G}$ has integer coefficients $g_1, \ldots, g_h$ and where

$$|\mathscr{M}| \leqq N^{-4/5}. \quad (14.35)$$

From (14.30), (14.32), (14.33) we obtain

$$|g_i| \leqq 2N^{k+2E+1+2kh\varphi} \quad (i = 1, \ldots, h)$$

and therefore for large N,

$$|\mathscr{G}| < N^{k+2E+2}. \quad (14.36)$$

Now we recall that there are two cases to deal with. Suppose first that (i) $0 \leqq E \leqq \frac{1}{2}$. In view of the definition of

$$h = n$$

(see the lines preceding (14.18)) there is a non-zero integer point $\mathbf{y}$ with $\mathscr{G}(\mathbf{y}) = 0$ and with

$$|\mathbf{y}| \leqq \max(1, |\mathscr{G}|^\theta)$$
$$< N^{(k+2E+2)\theta} \leqq N^{1/(4k)}$$

by (14.36) and (14.15). With $\mathbf{x} = (\mathbf{y}, \mathbf{0})$, we have for large N,

$$|\mathcal{L}(\mathbf{x})| = |\mathcal{M}(\mathbf{y})| \leqq h\,|\mathcal{M}|\,|\mathbf{y}|^k$$
$$\leqq hN^{-4/5}N^{1/4} < N^{-1/2} \leqq N^{-E}$$

by (14.34), (14.35).

We have to be a bit more subtle in case

(ii) *Proposition* 14.1 *holds for* $E - \frac{1}{4}$ *in place of* E. We have $h = mn$ by (14.18); write

$$\mathbf{y} = (\mathbf{y}_1, \ldots, \mathbf{y}_m),\ \mathcal{G}(\mathbf{y}) = \mathcal{G}_1(\mathbf{y}_1) + \cdots + \mathcal{G}_m(\mathbf{y}_m),$$
$$\mathcal{M}(\mathbf{y}) = \mathcal{M}_1(\mathbf{y}_1) + \cdots + \mathcal{M}_m(\mathbf{y}_m)$$

where each $\mathbf{y}_i$ has n components. By definition of n and (14.15), there are nonzero $\mathbf{y}_1, \ldots, \mathbf{y}_m$ in $\mathbb{Z}^n$ with

$$\mathcal{G}_i(\mathbf{y}_i) = 0 \quad \text{and} \quad |\mathbf{y}_i| \leqq \max(1, |\mathcal{G}|^\theta) < N^{(k+2E+2)\theta}$$
$$< \min(N^{1/(4E)}, N^{1/(4k)}) \qquad (i = 1, \ldots, m). \tag{14.37}$$

Setting $\mathbf{x} = z_1\mathbf{y}_1 + \cdots + z_m\mathbf{y}_m$, we have

$$\mathcal{L}(\mathbf{x}) = \mathcal{L}(z_1\mathbf{y}_1 + \cdots + z_m\mathbf{y}_m)$$
$$= \mathcal{M}_1(\mathbf{y}_1)z_1^k + \cdots + \mathcal{M}_m(\mathbf{y}_m)z_m^k = \mathcal{N}(\mathbf{z}),$$

say.

Now it is a question of choosing $\mathbf{z}$ to make $\mathcal{N}(z)$ small. Since $m = C_6(k, E - \frac{1}{4})$ we can in fact find nonzero $\mathbf{z}$ in $\mathbb{Z}^m$ with

$$|\mathbf{z}| \leqq N^{1-1/(4E)},$$
$$|\mathcal{N}(\mathbf{z})| \leqq N^{-(1-(1/(4E))(E-(1/4))}\,|\mathcal{N}|. \tag{14.38}$$

Of course $\mathbf{x}$ has

$$0 < |\mathbf{x}| \leqq N$$

on combining this with (14.37). Moreover, we have

$$|\mathcal{N}| \leqq m\,|\mathcal{M}|\,N^{1/4} \leqq mN^{-4/5}N^{1/4} < N^{-1/2}$$

from (14.37), (14.35). From (14.38), then, we have

$$|\mathcal{L}(\mathbf{x})| = |\mathcal{N}(\mathbf{z})| < N^{-E+(1/2)}N^{-1/2} = N^{-E}.$$

This completes the proof of assertion (ii).

14.4   An induction on the degree

Theorem 14.2 will be proved by induction on the (odd) values of

$$k = \max(k_1, \ldots, k_h). \tag{14.39}$$

The case $k = 1$ is proved by the usual box principle. We may suppose that $|\mathscr{F}_1| = \cdots = |\mathscr{F}_h| = 1$. Given $N \geq 1$, consider the points $\mathbf{z}$ in $\mathbb{Z}^s$ with $0 \leq z_i \leq N$. There are

$$([N] + 1)^s$$

such points $\mathbf{z}$. For each of them, the point

$$\mathbf{p}(\mathbf{z}) = (\mathscr{F}_1(\mathbf{z}), \ldots, \mathscr{F}_h(\mathbf{z}))$$

lies in the h-dimensional cube C: $0 \leq p_1, \ldots, p_h \leq sN$. Let r be the integer in

$$([N] + 1)^{s/h} - 1 \leq r < ([N] + 1)^{s/h}$$

and divide C into r^h subcubes of side sN/r. Since $r^h < ([N] + 1)^s$, two points $\mathbf{p}(\mathbf{z})$, $\mathbf{p}(\mathbf{z}')$ will lie in the same subcube. Thus $\mathbf{x} = \mathbf{z} - \mathbf{z}'$ has $0 < |\mathbf{x}| \leq N$, and for $i = 1, \ldots, h$,

$$|\mathscr{F}_i(\mathbf{x})| \leq sN/r \leq sN(([N] + 1)^{s/h} - 1)^{-1} \ll N^{-E}$$

if, say,

$$s \geq C_{10}(h, E) = C_4(\underbrace{1, 1, \ldots, 1}_{h}; 1, E).$$

So Theorem 14.2 is true for $k = m = 1$. More generally, it is true for $k = 1$ with $C_4(1, \ldots, 1; m, E) = mC_4(1, \ldots, 1; 1, E)$.

Before commencing the induction we introduce some useful notation. Let $\mathscr{F}(\mathbf{x})$ be a real form, and suppose a form $\mathscr{G}$ can be obtained from $\mathscr{F}$ by writing

$$\mathscr{G}(y_1, \ldots, y_t) = \mathscr{F}(y_1\mathbf{x}_1 + \cdots + y_t\mathbf{x}_t) \tag{14.40}$$

for certain linearly independent integer points $\mathbf{x}_1, \ldots, \mathbf{x}_t$. We write

$$\mathscr{F} \to \mathscr{G},$$

and define

$$\psi(\mathscr{F}, \mathscr{G}) = \min \max (|\mathbf{x}_1|, \ldots, |\mathbf{x}_t|) \tag{14.41}$$

where the minimum is over $\mathbf{x}_1, \ldots, \mathbf{x}_t$ with (14.40). It is useful to note that when (14.40) holds, we have the identity

$$\mathscr{G}(\mathbf{w}(1), \ldots, \mathbf{w}(k)) = \mathscr{F}(w_1(1)\mathbf{x}_1 + \cdots + w_t(1)\mathbf{x}_t, \ldots,$$
$$w_1(k)\mathbf{x}_1 + \cdots + w_t(k)\mathbf{x}_t) \tag{14.42}$$

in the k vectors

$$\mathbf{w}(1) = (w_1(1), \ldots, w_t(1)), \ldots, \mathbf{w}(k) = (w_1(k), \ldots, w_t(k)).$$

(The right-hand side is linear in each $\mathbf{w}(i)$, and agrees with $\mathscr{G}(\mathbf{w}(1))$ when $\mathbf{w}(i) = \mathbf{w}(1)$ for $i = 1, \ldots, k$. As it is also symmetric in the $\mathbf{w}(i)$ the identity (14.42) is evidently correct.)

By substituting suitable vectors $\mathbf{w}(i)$ into (14.42) we see that, whenever

$\mathcal{F} \to \mathcal{G}$, we have

$$|\mathcal{G}| \ll |\mathcal{F}| \, \psi(\mathcal{F}, \mathcal{G})^k \tag{14.43}$$

provided that $s(\mathcal{F})$ (the number of variables of the form $\mathcal{F}$) is bounded.

Suppose that $\mathcal{F} \to \mathcal{G}$ and $\mathcal{G} \to \mathcal{H}$. Then besides an identity (14.40) in $y_1, \ldots, y_t$ we have (say)

$$\mathcal{H}(z_1, \ldots, z_l) = \mathcal{G}(z_1 \mathbf{y}_1 + \cdots + z_l \mathbf{y}_l)$$

with linearly independent $\mathbf{y}_1, \ldots, \mathbf{y}_l$ in $\mathbb{Z}^s$. Writing $\mathbf{y}_i = (y_1^{(i)}, \ldots, y_t^{(i)})$, and setting

$$\mathbf{x}^{(i)} = y_1^{(i)} \mathbf{x}_1 + \cdots + y_t^{(i)} \mathbf{x}_t,$$

we easily verify that

$$\mathcal{H}(z_1, \ldots, z_l) = \mathcal{F}(z_1 \mathbf{x}^{(1)} + \cdots + z_l \mathbf{x}^{(l)}).$$

Since $\mathbf{x}^{(1)}, \ldots, \mathbf{x}^{(l)}$ are linearly independent, we find that $\mathcal{F} \to \mathcal{H}$ and that

$$\psi(\mathcal{F}, \mathcal{H}) \leqq t \psi(\mathcal{F}, \mathcal{G}) \psi(\mathcal{G}, \mathcal{H}) \ll \psi(\mathcal{F}, \mathcal{G}) \psi(\mathcal{G}, \mathcal{H}) \tag{14.44}$$

as long as $t = s(\mathcal{G})$ is bounded.

If $s(\mathcal{F}) \geqq m$, write

$$\Omega_m(\mathcal{F}, N) = \min \max_{1 \leqq i_1, \ldots, i_k \leqq m} |\mathcal{F}(\mathbf{y}(i_1), \ldots, \mathbf{y}(i_k))|,$$

where the minimum is over all sets of m linearly independent points $\mathbf{y}(1), \ldots, \mathbf{y}(m)$ with $|\mathbf{y}(i)| \leqq N$ $(i = 1, \ldots, m)$. Clearly

$$\Omega_m(\mathcal{F}, N) \leqq \Omega_m(\mathcal{F}, N') \qquad \text{for } N' > N. \tag{14.45}$$

Suppose that $\mathcal{F} \to \mathcal{G}$, and choose $\mathbf{x}_1, \ldots, \mathbf{x}_t$ in (14.40) so that (14.41) holds with these vectors. For $m \leqq s(\mathcal{G})$, we may select $\mathbf{y}(1), \ldots, \mathbf{y}(m)$, linearly independent with $|\mathbf{y}(i)| \leqq N$, such that

$$\Omega_m(\mathcal{G}, N) = \max_{1 \leqq i_1, \ldots, i_k \leqq m} |\mathcal{G}(\mathbf{y}(i_1), \ldots, \mathbf{y}(i_k))|.$$

Then by (14.42), the m linearly independent vectors

$$\mathbf{x}(i) = y_1(i) \mathbf{x}_1 + \cdots + y_t(i) \mathbf{x}_t \qquad (i = 1, \ldots, m)$$

satisfy

$$|\mathcal{F}(\mathbf{x}(i_1), \ldots, \mathbf{x}(i_k))| = |\mathcal{G}(\mathbf{y}(i_1), \ldots, \mathbf{y}(i_k))| \leqq \Omega_m(\mathcal{G}, N).$$

Since $|\mathbf{x}(i)| \leqq Nt \max(|\mathbf{x}_1|, \ldots, |\mathbf{x}_t|) = Ns(\mathcal{G}) \psi(\mathcal{F}, \mathcal{G})$, it can be seen that

$$\Omega_m(\mathcal{F}, Ns(\mathcal{G}) \psi(\mathcal{F}, \mathcal{G})) \leqq \Omega_m(\mathcal{G}, N) \tag{14.46}$$

whenever $\mathcal{F} \to \mathcal{G}$ and $s(\mathcal{G}) \geqq m$.

In the next section we will need the identity

$$\mathscr{F}(\mathbf{a} + \mathbf{b}) = \hat{\mathscr{F}}(\mathbf{a} + \mathbf{b}, \ldots, \mathbf{a} + \mathbf{b})$$

$$= \sum_{u=0}^{k} \binom{k}{u} \hat{\mathscr{F}}(\underbrace{\mathbf{a}, \ldots, \mathbf{a}}_{u}, \underbrace{\mathbf{b}, \ldots, \mathbf{b}}_{k-u}). \tag{14.47}$$

The last equality is reached by using the linearity of $\hat{\mathscr{F}}$ in each argument and the symmetry property of $\hat{\mathscr{F}}$.

14.5 The case of a single form

Suppose now that $k > 1$ is odd and that Theorem 14.2 has already been proved for forms of odd degree less than k. Our next aim will be to prove the theorem for a single form $\mathscr{F}$ of degree k. It is in this section and the next that the argument resembles that of Birch (1957). The ideas can be traced to Brauer (1945) (who did not need to avoid equations of even degree, because he was working in a p-adic field). We have already seen this sort of reasoning in a simple form in Section 10.3.

Lemma 14.5. *Suppose* $k \geqq 1$, $l \geqq 1$, $E > 0$. *Now if* $\mathscr{F}$ *is a form of degree* k *with* $s(\mathscr{F}) \geqq C_{11}(k, l, E)$, *and if* $N \geqq 1$, *then there is a form* $\mathscr{G}$ *with* $\mathscr{F} \to \mathscr{G}$ *and*

$$\psi(\mathscr{F}, \mathscr{G}) \leqq N, \tag{14.48}$$

and where $\mathscr{G}$ *is a form in* $l + 1$ *variables, of the type*

$$\mathscr{G} = \lambda y^{k} + \mathscr{F}_{1}(z_{1}, \ldots, z_{l}) + \mathscr{M}(y, z_{1}, \ldots, z_{l}) \tag{14.49}$$

with

$$|\mathscr{M}| \ll N^{-E} |\mathscr{F}|.$$

Here the constant in $\ll$ *depends on* k, l, E.

Proof.   The construction of $\mathscr{G}$ is in two steps.

Let $\mathbf{e}(1), \ldots, \mathbf{e}(h)$ be the first h unit vectors in $\mathbb{R}^{s}$. Consider the forms

$$\mathscr{H}_{p_{1}, \ldots, p_{k-u}}(\mathbf{x}) = \hat{\mathscr{F}}(\underbrace{\mathbf{x}, \ldots, \mathbf{x}}_{u}, \underbrace{\mathbf{e}(p_{1}), \ldots, \mathbf{e}(p_{k-u})}_{k-u}) \tag{14.50}$$

where u takes all the odd values from 1 to $k - 2$, and $p_{1}, \ldots, p_{k-u}$ all the values from 1 to h. The number of the forms (14.50) is less than kh^{k}, and each is of odd degree $\leqq k - 2$ in $\mathbf{x}$. Each form (14.50) has

$$|\mathscr{H}_{p_{1}, \ldots, p_{k-u}}| \leqq |\mathscr{F}|.$$

So by the part of Theorem 14.2 which we already know, we see that for $s \geqq C_{12}(k, h, E)$ there is an integer point $\mathbf{x}_{0} \neq \mathbf{0}$ with

$$|\mathbf{x}_{0}| \leqq N$$

and

$$|\mathcal{H}_{p_1,\ldots,p_{k-u}}(\mathbf{x}_0)| \ll N^{-E-k}|\mathcal{F}| \tag{14.51}$$

for all the forms (14.50). The constant in $\ll$ here depends on k, h, E. We may choose $\mathbf{x}_0$ so that the number $v(\mathbf{x}_0)$ of its non-zero components does not exceed $C_{12}(k, h, E)$.

In the second step, write $\mathbf{x} = (x_1, \ldots, x_h, 0, \ldots, 0)$ and consider the forms

$$\mathcal{L}_v(\mathbf{x}) = \mathcal{F}(\underbrace{\mathbf{x}_0, \ldots, \mathbf{x}_0}_{k-v}, \underbrace{\mathbf{x}, \ldots, \mathbf{x}}_{v}) \tag{14.52}$$

where v takes the odd values from 1 to $k-2$. There are fewer than k such forms, and each is of odd degree less than k. Each of the forms (14.52) has

$$|\mathcal{L}_v| \leqq (v(\mathbf{x}_0)|\mathbf{x}_0|)^k |\mathcal{F}|.$$

Again using the part of Theorem 14.2 which is known, we see that for $h \geqq C_{13}(k, l, E)$ there will be linearly independent integer points $\mathbf{x}_1, \ldots, \mathbf{x}_{l+1}$ with

$$|\mathbf{x}_i| \leqq N \qquad (i = 1, \ldots, l+1)$$

and with

$$|\mathcal{L}_v(\mathbf{x}_{i_1}, \ldots, \mathbf{x}_{i_v})| \ll N^{-E-k}|\mathbf{x}_0|^k |\mathcal{F}|(1 \leqq i_1, \ldots, i_v \leqq l+1) \tag{14.53}$$

for odd v in $1 \leqq v \leqq k-2$.

Now set $h = C_{13}(k, l, E)$. We can carry out both steps if $s = C_{12}(k, h, E) = C_{12}(k, C_{13}(k, l, E), E)$. We may suppose without loss of generality that $\mathbf{x}_0, \mathbf{x}_1, \ldots, \mathbf{x}_l$ are linearly independent.

Writing

$$\mathcal{G}(y, z_1, \ldots, z_l) = \mathcal{F}(y\mathbf{x}_0 + z_1\mathbf{x}_1 + \cdots + z_l\mathbf{x}_l)$$

we have $\mathcal{F} \to \mathcal{G}$ and

$$\psi(\mathcal{F}, \mathcal{G}) \leqq \max(|\mathbf{x}_0|, |\mathbf{x}_1|, \ldots, |\mathbf{x}_l|) \leqq N.$$

Recalling (14.47), $\mathcal{G}$ is of the form (14.49) with

$$\lambda = \mathcal{F}(\mathbf{x}_0), \qquad \mathcal{F}_1(z_1, \ldots, z_l) = \mathcal{F}(z_1\mathbf{x}_1 + \cdots + z_l\mathbf{x}_l)$$

and

$$\mathcal{M}(y, z_1, \ldots, z_l)$$
$$= \sum_{u=1}^{k-1} \binom{k}{u} \mathcal{F}(\underbrace{y\mathbf{x}_0, \ldots, y\mathbf{x}_0}_{u}, \underbrace{z_1\mathbf{x}_1+\cdots+z_l\mathbf{x}_l, \ldots, z_1\mathbf{x}_1+\cdots+z_l\mathbf{x}_l}_{k-u})$$
$$= \sum_{u=1}^{k-1} \mathcal{M}_u,$$

say. Now if u is odd, then

$$|\mathcal{M}_u| \ll N^{-E-k} |\mathcal{F}| \left(\max_{1 \leq j \leq l} |\mathbf{x}_j|\right)^{k-u} \ll N^{-E} |\mathcal{F}|$$

by (14.51). On the other hand, if u is even, so that $k - u = v$, say, is odd, then by (14.53)

$$|\mathcal{M}_u| \ll N^{-E-k} |\mathbf{x}_0|^k |\mathcal{F}|$$
$$\ll N^{-E} |\mathcal{F}|.$$

It follows that

$$|\mathcal{M}| \ll N^{-E} |\mathcal{F}|.$$

This completes the proof of Lemma 14.5.

Lemma 14.5 is the starting point for the following lemma, which we prove by induction on t.

Lemma 14.6. *Suppose $k \geq 1$, $l \geq 1$, $t \geq 1$, $E > 0$. Now if $N \geq 1$ and if $\mathcal{F}$ is a form of degree k with $s(\mathcal{F}) \geq C_{14}(k, l, t, E)$, then there is a form $\mathcal{G}_t$ with $\mathcal{F} \to \mathcal{G}_t$ and*

$$\psi(\mathcal{F}, \mathcal{G}_t) \leq N, \tag{14.54}$$

and with $\mathcal{G}_t$ being a form in $l + t$ variables, of the type

$$\mathcal{G}_t = \lambda_1 y_1^k + \cdots + \lambda_t y_t^k + \mathcal{F}_t(z_1, \ldots, z_l) + \mathcal{M}_t(y_1, \ldots, y_t, z_1, \ldots, z_l) \tag{14.55}$$

where

$$|\mathcal{M}_t| \ll N^{-E} |\mathcal{F}|. \tag{14.56}$$

The constant in $\ll$ depends on k, l, t and E.

Proof. The case $t = 1$ is Lemma 14.5. By the case $t - 1$ we see that if $s(\mathcal{F}) \geq C_{14}(k, h, t - 1, 3E + 3k)$, then $\mathcal{F} \to \mathcal{G}_{t-1}$ with

$$\psi(\mathcal{F}, \mathcal{G}_{t-1}) \leq N^{1/3}, \tag{14.57}$$

and with

$$\mathcal{G}_{t-1} = \lambda_1 y_1^k + \cdots + \lambda_{t-1} y_{t-1}^k + \mathcal{F}_{t-1}(z_1, \ldots, z_h)$$
$$+ \mathcal{M}_{t-1}(y_1, \ldots, y_{t-1}, z_1, \ldots, z_h)$$

and

$$|\mathcal{M}_{t-1}| \ll N^{-E-k} |\mathcal{F}|.$$

From (14.57) we get

$$|\mathcal{G}_{t-1}| \ll \psi(\mathcal{F}, \mathcal{G}_{t-1})^k |\mathcal{F}| \ll N^{k/3} |\mathcal{F}|,$$

so that also

$$|\mathscr{F}_{t-1}| \ll N^{k/3} |\mathscr{F}|.$$

The constant in $\ll$ may depend on $s(\mathscr{F})$ here.

Now if $h \geq C_{11}(k, l, 3E + 3k)$ then by Lemma 14.5 we have $\mathscr{F}_{t-1} \to \mathscr{G}$ where

$$\psi(\mathscr{F}_{t-1}, \mathscr{G}) \leq N^{1/3} \tag{14.58}$$

and with

$$\mathscr{G} = \lambda_t y_t^k + \mathscr{F}_t(z_1, \ldots, z_l) + \mathscr{M}(y_t, z_1, \ldots, z_l)$$

and

$$|\mathscr{M}| \ll N^{-E-k} |\mathscr{F}_{t-1}| \ll N^{-E} |\mathscr{F}|.$$

Say $\mathscr{G}(y_t, z_1, \ldots, z_l) = \mathscr{F}_{t-1}(y_t \mathbf{x}_0 + z_1 \mathbf{x}_1 + \cdots + z_l \mathbf{x}_l)$ with

$$|\mathbf{x}_i| \leq \psi(\mathscr{F}_{t-1}, \mathscr{G}) \leq N^{1/3} \qquad (i = 0, 1, \ldots, l).$$

Here $\mathbf{x}_0, \mathbf{x}_1, \ldots, \mathbf{x}_l$ are points with h components. Put

$$\mathscr{G}_t(y_1, \ldots, y_t, z_1, \ldots, z_l) = \mathscr{G}_{t-1}(y_1, \ldots, y_{t-1}, y_t \mathbf{x}_0 + z_1 \mathbf{x}_1 + \cdots + z_l \mathbf{x}_l);$$

this makes sense since $\mathscr{G}_{t-1}$ has $t - 1 + h$ variables. From (14.44), (14.57), (14.58) we get

$$\psi(\mathscr{F}, \mathscr{G}_t) \ll \psi(\mathscr{F}, \mathscr{G}_{t-1}) \psi(\mathscr{G}_{t-1}, \mathscr{G}_t)$$
$$\ll N^{2/3},$$

so that $\psi(\mathscr{F}, \mathscr{G}_t) \leq N$ *if N is large.* Further,

$$\mathscr{G}_t = \lambda_1 y_1^k + \cdots + \lambda_t y_t^k + \mathscr{F}_t(z_1, \ldots, z_l) + \mathscr{M}_t(y_1, \ldots, y_t, z_1, \ldots, z_l)$$

where

$$\mathscr{M}_t = \mathscr{M}(y_t, z_1, \ldots, z_l) + \mathscr{M}_{t-1}(y_1, \ldots, y_{t-1}, y_t \mathbf{x}_0 + z_1 \mathbf{x}_1 + \cdots + z_l \mathbf{x}_l).$$

Here

$$|\mathscr{M}_t| \ll |\mathscr{M}| + |\mathscr{M}_{t-1}| \left(\max_i |\mathbf{x}_i| \right)^k$$
$$\ll N^{-E} |\mathscr{F}| + N^{-E-k} |\mathscr{F}| N^{k/3} \ll N^{-E} |\mathscr{F}|.$$

To summarize: setting $h = C_{11}(k, l, 3E + 3k)$ and

$$s = s(\mathscr{F}) = C_{14}(k, h, t - 1, 3E + 3k),$$

the assertion holds with a constant in (14.56) which depends on k, l, t, E, provided that N is large as a function of k, l, t, E. It is then clear that the assertion holds for $N \geq 1$, at the possible cost of making the constant in (14.56) larger. It is further clear that the assertion is also true for larger values of s, with a constant in (14.56) which depends on k, l, t, E but is independent of s.

The following is an obvious

Corollary. *Suppose $k \geq 1$, $t \geq 1$, $E > 0$. If $N \geq 1$ and if $\mathscr{F}$ is a form of degree k with $s(\mathscr{F}) \geq C_{15}(k, t, E)$, then*

$$\mathscr{F} \to \lambda_1 y_1^k + \cdots + \lambda_t y_t^k + \mathscr{M}_t(y_1, \ldots, y_t) = \mathscr{G}_t, \tag{14.59}$$

say, where

$$\psi(\mathscr{F}, \mathscr{G}_t) \leq N$$

and

$$|\mathscr{M}_t| \ll N^{-E} |\mathscr{F}|.$$

Now choose $t = mC_6(k, 3E + 3k)$, where C_6 is the constant of Proposition 14.1, and let $\mathscr{F}$ be a form with $s(\mathscr{F}) = C_{15}(k, t, 3E + 3k)$. There is for $N \geq 1$ a form $\mathscr{G}_t$ with (14.59) and

$$\psi(\mathscr{F}, \mathscr{G}_t) \leq N^{1/3}, \qquad |\mathscr{M}_t| \ll N^{-E-k} |\mathscr{F}|. \tag{14.60}$$

Here $|\mathscr{G}_t| \ll \psi(\mathscr{F}, \mathscr{G}_t)^k |\mathscr{F}| \ll N^k |\mathscr{F}|$, so that the additive form

$$\mathscr{L}(\mathbf{y}) = \lambda_1 y_1^k + \cdots + \lambda_t y_t^k$$

has

$$|\mathscr{L}| \ll N^k |\mathscr{F}|.$$

By Proposition 14.1, and by our choice of t, there are integer points $\mathbf{y}(i) = (y_1(i), \ldots, y_t(i))$, such that $y_l(i) y_l(j) = 0$ for all l and $i \neq j$, and

$$0 < |\mathbf{y}(i)| \leq N^{1/3} \quad \text{and} \quad |\mathscr{L}(\mathbf{y}(i))| \leq N^{-E-k} |\mathscr{L}| \ll N^{-E} |\mathscr{F}|$$

$$(i = 1, \ldots, m). \tag{14.61}$$

We have

$$|\mathscr{L}(\mathbf{y}(i_1), \ldots, \mathbf{y}(i_k))| \ll N^{-E} |\mathscr{F}| \qquad (1 \leq i_1, \ldots, i_k \leq m)$$

since the left-hand side is zero unless $i_1 = \cdots = i_k$. On the other hand,

$$|\hat{\mathscr{M}}_t(\mathbf{y}(i_1), \ldots, \mathbf{y}(i_k))| \ll N^{k/3} |\mathscr{M}_t| \ll N^{-E} |\mathscr{F}| \qquad (1 \leq i_1, \ldots, i_k \leq m).$$

In other words,

$$\Omega_m(\mathscr{G}_t, N^{1/3}) \ll N^{-E} |\mathscr{F}|.$$

For large N we have

$$N^{1/3} s(\mathscr{G}_t) \psi(\mathscr{F}, \mathscr{G}_t) \leq N$$

by (14.60), so that, by (14.45), (14.46),

$$\Omega_m(\mathscr{F}, N) \leq \Omega_m(\mathscr{F}, N^{1/3} s(\mathscr{G}_t) \psi(\mathscr{F}, \mathscr{G}_t))$$
$$\leq \Omega_m(\mathscr{G}_t, N^{1/3}) \ll N^{-E} |\mathscr{F}|.$$

This proves Theorem 14.2 for our particular k and for $h = 1$.

14.6 Conclusion

We finally proceed to prove the main theorem for h forms $\mathscr{F}_1, \ldots, \mathscr{F}_h$ of respective degrees $k_1, \ldots, k_h$, and with our given value of k in (14.39). The case $h = 1$ was carried out in the preceding section. We proceed by induction on h. By the case $h - 1$ we can choose

$$n = C_4(k_2, \ldots, k_h; m; 3E + 3k).$$

Suppose that $s = C_4(k_1; n, 3E + 3k)$. Then

$$\Omega_n(\mathscr{F}_1, N^{1/3}) \ll N^{-E-k} |\mathscr{F}_1|$$

so that there are n independent integer points $\mathbf{z}(1), \ldots, \mathbf{z}(n)$ with $|\mathbf{z}(i)| \leqq N^{1/3}$ and

$$|\mathscr{F}_1(\mathbf{z}(i_1), \ldots, \mathbf{z}(i_{k_1}))| \ll N^{-E-k} |\mathscr{F}_1| \qquad (1 \leqq i_1, \ldots, i_{k_1} \leqq n).$$

Define new forms

$$\mathscr{G}_j(y_1, \ldots, y_n) = \mathscr{F}_j(y_1 \mathbf{z}(1) + \cdots + y_n \mathbf{z}(n)) \qquad (j = 1, \ldots, h).$$

Then, by use of (14.42), we see that

$$|\mathscr{G}_1| \ll N^{-E-k} |\mathscr{F}_1| \tag{14.62}$$

and

$$|\mathscr{G}_j| \ll N^k |\mathscr{F}_j| \qquad (j = 2, \ldots, h). \tag{14.63}$$

By our choice of n, there are m linearly independent points $\mathbf{y}(1), \ldots, \mathbf{y}(m)$ in $\mathbb{Z}^n$ with $|\mathbf{y}(j)| \leqq N^{1/3}$ and

$$|\mathscr{G}_j(\mathbf{y}(i_1), \ldots, \mathbf{y}(i_{k_j}))| \ll N^{-E-k} |\mathscr{G}_j| \ll N^{-E} |\mathscr{F}_j|$$
$$(2 \leqq j \leqq h; 1 \leqq i_1, \ldots, i_{k_j} \leqq m). \tag{14.64}$$

On the other hand by (14.62),

$$|\mathscr{G}_1(\mathbf{y}(i_1), \ldots, \mathbf{y}(i_{k_1}))| \ll N^{-E} |\mathscr{F}_1| \qquad (1 \leqq i_1, \ldots, i_{k_1} \leqq m). \tag{14.65}$$

Put

$$\mathbf{x}(i) = y_1(i)\mathbf{z}(1) + \cdots + y_n(i)\mathbf{z}(n) \qquad (i = 1, \ldots, m).$$

The $\mathbf{x}(i)$ are linearly independent, and $|\mathbf{x}(i)| \ll N^{2/3}$, whence $|\mathbf{x}(i)| \leqq N$ if N is large. Moreover, by the identity (14.42),

$$\mathscr{F}_j(\mathbf{x}(i_1), \ldots, \mathbf{x}(i_{k_j})) = \mathscr{G}_j(\mathbf{y}(i_1), \ldots, \mathbf{y}(i_{k_j}))$$
$$(1 \leqq j \leqq h; 1 \leqq i_1, \ldots, i_{k_j} \leqq m).$$

Hence by (14.64) and (14.65)

$$|\mathscr{F}_j(\mathbf{x}(i_1), \ldots, \mathbf{x}(i_{k_j}))| \ll N^{-E} |\mathscr{F}_j| \qquad (1 \leqq j \leqq h; 1 \leqq i_1, \ldots, i_{k_j} \leqq m).$$

This completes the proof of Theorem 14.2.

15
Exponential sums: forms with integer coefficients

15.1 Introduction

Let

$$\mathscr{F} = (\mathscr{F}^{(k)}, \ldots, \mathscr{F}^{(2)}) \tag{15.1}$$

be a system of forms in $\mathbf{x} = (x_1, \ldots, x_s)$, with the subsystem $\mathscr{F}^{(d)}$ $(2 \leq d \leq k)$ consisting of $r_d \geq 0$ forms of degree d with rational coefficients. The number of integer points with $|\mathbf{x}| \leq P$ is $\sim (2P)^s$. When we substitute such a point into a form of degree d, we will in general obtain a value of the order of magnitude of P^d, and hence the 'probability' for the form to vanish should be about P^{-d}. Therefore the probability for $\mathscr{F}$ to vanish simultaneously should be about P^{-R}, where

$$R = \sum_{d=2}^{k} d r_d, \qquad r = \sum_{d=2}^{k} r_d.$$

Let $z_P(\mathscr{F})$ denote the number of integer points in $|\mathbf{x}| \leq P$ with $\mathscr{F}(\mathbf{x}) = 0$. When $s > R$, one might naïvely expect $z_P(\mathscr{F})$ to have the order of magnitude of P^{s-R}.

Experience with the simplest cases suggests that a formula

$$z_P(\mathscr{F}) = \mu P^{s-R} + O(P^{s-R-\gamma}) \tag{15.2}$$

might be true, with $\gamma = \gamma(\mathscr{F}) > 0$. Here $\mu = \mu(\mathscr{F})$ is an infinite product of 'local densities' of zeros of $\mathscr{F}$. Thus μ vanishes if the system

$$\mathscr{F}(\mathbf{x}) = \mathbf{0}$$

has no non-trivial solution in $\mathbb{R}^s$, or if the system of congruences

$$\mathscr{F}(\mathbf{x}) \equiv \mathbf{0} \pmod{p^l}$$

has only the trivial solution for some power p^l. For more details of μ see Schmidt (1985), which is the principal source for Chapters 15 and 16.

An example where formula (15.2) does hold, with $\mu > 0$, is that of a single form

$$\mathscr{F} = a_1 X_1^d + \cdots + a_s X_s^d,$$

where $d \geq 18$, $s > d^2$ and $a_1 a_2 \cdots a_s \neq 0$, and where either d is odd or not all the coefficients are of the same sign (Davenport and Lewis (1963)).

Now if $\mathscr{F}$ is a form of degree $d > 1$ over $\mathbb{Q}$, write $h(\mathscr{F})$ for the least number h such that $\mathscr{F}$ 'splits into h products', that is,

$$\mathscr{F} = \mathscr{A}_1 \mathscr{B}_1 + \cdots + \mathscr{A}_h \mathscr{B}_h$$

with forms $\mathscr{A}_i$, $\mathscr{B}_i$ of positive degrees and with rational coefficients. When $\mathscr{F} = (\mathscr{F}_1, \ldots, \mathscr{F}_r)$ consists of forms of equal degree $d > 1$, write

$$h(\mathscr{F}) = \min h(\mathscr{F}),$$

with the minimum taken over forms $\mathscr{F}$ of the rational pencil of $\mathscr{F}$, i.e.

$$\mathscr{F} = c_1 \mathscr{F}_1 + \cdots + c_r \mathscr{F}_r = \mathbf{c} \mathscr{F},$$

where $\mathbf{c} \neq \mathbf{0}$ has rational components. The quantity $h(\mathscr{F})$, and its analogues with different fields in place of $\mathbb{Q}$, will play an important role in the last few chapters of the book. To see its significance in the present context, we cite a result of Schmidt (1982e) that a system of r quadratic forms with $h(\mathscr{F}) > 2r^2 + 3r$ satisfies an asymptotic formula (15.2). If $h(\mathscr{F}) > 4r^2 + 4r^3$, and the system possesses a real non-singular zero (i.e. a zero with $\partial \mathscr{F}_j / \partial x_i$ ($1 \leq j \leq r$, $1 \leq i \leq s$) of rank r) then $\mu(\mathscr{F}) > 0$. Again, (15.2) holds, with $\mu(\mathscr{F}) > 0$, for a system of r cubic forms with $h(\mathscr{F}) > C_1 r^4$ (Schmidt, (1982d)).

It is, in fact, not unreasonable to impose a lower bound on the size of $h(\mathscr{F})$ in pursuit of a formula (15.2). To illustrate this, let

$$\mathscr{F} = \mathscr{A}_1^D + \cdots + \mathscr{A}_h^D,$$

where D is even and where the system of forms $\mathscr{A} = (\mathscr{A}_1, \ldots, \mathscr{A}_h)$ of degree d does obey (15.2). Here

$$z_P(\mathscr{F}) = z_P(\mathscr{A}) \sim \mu P^{s - dh},$$

whereas $R(\mathscr{F}) = Dd$ may be larger or smaller than dh. The trouble seems to be that $h(\mathscr{F})$ is small, namely $h(\mathscr{F}) \leq h$.

The most successful method up to now in proving formula (15.2) is the Hardy–Littlewood circle method. One of the highlights is Schmidt's proof (1985) of formula (15.2) under a suitable lower bound on each $h(\mathscr{F}^{(d)})$. We quote his main result, although we are not going to give the whole proof.

Theorem 15.1. *There is a function $\chi(d)$ such that a system $\mathscr{F}$ as in (15.1) with*

$$h(\mathscr{F}^{(d)}) \geq \chi(d) r_d k R \tag{15.3}$$

for each d with $2 \leq d \leq k$, $r_d \neq 0$, satisfies (15.2). For instance, one may *take* $\chi(2) = 2$, $\chi(3) = 32$, $\chi(4) = 1152$, *and in general*

$$\chi(d) < 2^{4d} \cdot d!$$

Write $\upsilon(\mathbf{r}) = \upsilon(r_k, \ldots, r_2)$ *for the least number such that a system* (15.1) *has a non-trivial p-adic zero for each prime p. If, firstly, we replace R by* $\upsilon(\mathbf{r})$ *on the right-hand side of* (15.3); *and, secondly, suppose that*

$$\dim V_\mathbb{R} \geqq s - r$$

where $V_\mathbb{R}$ *is the manifold of real zeros of* $\mathscr{F}$, *then we have* $\mu(\mathscr{F}) > 0$. *This last condition is always satisfied if all the forms of* $\mathscr{F}$ *are of odd degree.*

Here $\dim V_\mathbb{R}$ denotes dimension in a topological sense; for the rest of the book we will use 'dimension' in the sense of algebraic geometry. Although we do not need to do so here, the two concepts can be effectively related (Schmidt (1982d)).

We mention briefly the previous work on which Schmidt was able to draw. A paper of Tartakovsky (1935) discusses such systems of equations by the Hardy–Littlewood method, but his work appears to be incomplete and in any case strong conditions are imposed on the forms. Davenport wrote a powerful series of papers on the case of a single cubic form (Davenport (1959), (1962b), (1963)), finally showing that the condition $s \geqq 16$ suffices to ensure that the equation

$$\mathscr{F}(x_1, \ldots, x_s) = 0$$

has a non-trivial solution in integers. Davenport assumed throughout that $\mathscr{F}(\mathbf{x})$ did not represent 0 for integral $\mathbf{x} \neq \mathbf{0}$, and obtained a contradiction by deriving (15.2) for $z_P(\mathscr{F})$.

Davenport used the formula

$$\int_0^1 S(\alpha)\,d\alpha$$

to count zeros of $\mathscr{F}$ in $P\mathscr{B}$, where

$$S(\alpha) = \sum_{\mathbf{x} \in P\mathscr{B}} e(\alpha \mathscr{F}(x)),$$

$\mathscr{B}$ being a box in $\mathbb{R}^s$. The counterpart of this formula in Theorem 15.1 (assuming forms with integer coefficients) is

$$\sum_{\mathbf{x} \in P\mathscr{B}:\, \mathscr{F}(\mathbf{x})=0} 1 = \int_{U_h} S(\boldsymbol{\alpha})\,d\boldsymbol{\alpha}$$

with

$$S(\alpha) = \sum_{x \in P\mathscr{B}} e(\alpha \mathscr{F}(x)). \qquad (15.4)$$

Here $\alpha = (\alpha^{(k)}, \ldots, \alpha^{(2)})$ denotes a point of $\mathbb{R}^s$ with r components, $\alpha^{(d)}$ having r_d components; $\alpha^{(d)} = (\alpha_1^{(d)}, \ldots, \alpha_{r_d}^{(d)})$.

Davenport's methods for estimating $S(\alpha)$ in the first of the three papers (where he supposed $s \geqq 32$) were applied by Birch (1962). Birch proved a result like Theorem 15.1 for a system of forms of equal degree d, with a hypothesis on $V^*(\mathscr{F})$, the set of x in $\mathbb{C}^s$ for which the matrix $\partial \mathscr{F}_j / \partial x_i$ $(1 \leqq j \leqq r, 1 \leqq i \leqq s)$ has rank less than r. Birch assumed in place of (15.3) that

$$s > \dim V^* + (d-1)2^{d-1} r(r+1).$$

As we shall explain below, Birch's theorem can be deduced as a special case from results in Schmidt (1985).

The formulation of Theorem 15.1 allowed Schmidt to deduce the previously unattainable lower bound

$$z_P(\mathscr{F}) > C_0(\mathscr{F}) P^{s - C_2}$$

for a system $\mathscr{F}$ of r forms of odd degree $\leqq k$ in $s > C_2(k, r)$ variables. The reason is that, if $h(\mathscr{F})$ is small, then for some $\mathscr{F}$ in the rational pencil of $\mathscr{F}$, the equation $\mathscr{F} = 0$ is implied by a system of equations

$$\mathscr{A}_1 = \cdots = \mathscr{A}_h = 0$$

of lower degree than $\mathscr{F}$. Now an inductive strategy becomes visible; we refer to Schmidt (1985) for the details.

The task we set ourselves in Chapters 15 and 16 is an exposition of Schmidt's estimates of the exponential sums (15.4). This can act as a jumping-off point for the reader who wants to study the applications described above, for which we do not have space here. It also acts as a starting point for our study of sums

$$\sum_{x \in \mathbb{F}_p^s} e_p(\mathscr{B}(x)),$$

where the polynomial $\mathscr{B}$ has coefficients in the field $\mathbb{F}_p$, in Chapter 17. There, and in Chapter 18, we shall give arithmetic applications—to small solutions of congruences.

In the present chapter, the exponential sums are estimated in terms of a certain invariant $g(\mathscr{F})$. In Chapter 16, which is essentially algebraic in nature, we will derive a relation between $g(\mathscr{F})$ and $h(\mathscr{F})$, and more generally between $g(\mathscr{F})$ and $h(\mathscr{F})$ in the case of systems of forms. Chapter 15 might be considered a relatively natural strengthening of Birch (1962), but Chapter 16 is going to involve some rather startling manoeuvres.

15.2 Algebraic geometry

At this stage it would be as well to make a few rudimentary remarks about algebraic geometry; proofs can be found in Schmidt (1976), Chapter VI, Lang (1958), or Hartshorne (1977). Let k, Ω be fields such that $k \subset \Omega$, the transcendence degree of Ω over k is infinite, and Ω is algebraically closed. We call k the *ground field*, and Ω the *universal domain*. For example, we may take $k = \mathbb{Q}$, $\Omega = \mathbb{C}$. Or $k = \mathbb{F}_q$, the finite field of q elements, and (with $X_1, X_2, \ldots$ 'variables') take $\Omega = \mathbb{F}_q(X_1, X_2, \ldots)$, i.e. the algebraic closure of $\mathbb{F}_q(X_1, X_2, \ldots)$.

Consider Ω^n, the space of n-tuples of elements in Ω. Suppose $\mathscr{I}$ is an ideal in $k[X_1, X_2, \ldots, X_n] = k[\mathbf{X}]$. Let $A(\mathscr{I})$ be the set of $\mathbf{x} \in \Omega^n$ having $f(\mathbf{x}) = 0$ for every $f(\mathbf{X}) \in \mathscr{I}$. Every set $A(\mathscr{I})$ defined this way is called an *algebraic set* (or k-algebraic set). Because of Hilbert's basis theorem, $A(\mathscr{I})$ can also be characterized as the set of $\mathbf{x}$ in Ω^n with $f_1(\mathbf{x}) = \cdots = f_m(\mathbf{x}) = 0$. Here $f_1, \ldots, f_m$ is a basis of $\mathscr{I}$.

The family of algebraic sets, which includes $\varnothing$ and Ω^n, is closed under the operations of finite union and arbitrary intersections; so it is convenient to introduce the *Zariski topology* in Ω^n by defining the closed sets as the algebraic sets.

Suppose S is an algebraic set. We call S *reducible* if $S = S_1 \cup S_2$, where S_1 and S_2 are algebraic sets, and $S \neq S_1, S_2$. Otherwise, we call S *irreducible*. It turns out that the irreducible sets can be characterized as those sets which are the (Zariski-topology) closure of a single point:

$$S = \{\mathbf{x}\}^-.$$

An irreducible set is referred to as a *variety* (more precisely, a k-variety). If V is a variety, a point $\mathbf{x}$ in V is called a *generic point* of V if $V = \{\mathbf{x}\}^-$.

Lemma 15.1. *Let S be a non-empty algebraic set. We may write*

$$S = V_1 \cup \cdots \cup V_t \tag{15.5}$$

where $V_1, \ldots, V_t$ are varieties with $V_i \not\subseteq V_j$ if $i \neq j$. This decomposition is unique in the obvious sense.

Proof. See Hartshorne (1977), Chapter I, Proposition 1.5.

The V_i in the unique representation of S in (15.5) are called the *irreducible components* of S.

Let $\mathbf{x} \in \Omega^n$. The *transcendence degree* of $\mathbf{x}$ over k is the maximum number of algebraically independent components of $\mathbf{x}$ over k. This is clearly equal to the transcendence degree of $k(\mathbf{x}) = k(x_1, \ldots, x_n)$ over k. It turns out that any of the generic points of a variety V have the same transcendence degree, d say. We define the *dimension* of V to be d and the *codimension* to be $n - d$. As for an algebraic set, we define its dimension to be the maximum dimension of its irreducible components.

15.3   Manifolds $\mathcal{M}$ and invariants g

With each form $\mathcal{F}(\mathbf{X})$ of degree d we associate the unique multilinear form $\mathcal{F}(\mathbf{X}_1 | \mathbf{X}_2 | \cdots | \mathbf{X}_d)$ with

$$\mathcal{F}(\mathbf{X} | \mathbf{X} | \cdots | \mathbf{X}) = (-1)^d d! \, \mathcal{F}(\mathbf{X}).$$

(This differs by a factor $(-1)^d d!$ from the multilinear form $\hat{\mathcal{F}}$ in Chapter 14.) Suppose $\mathcal{F} = (\mathcal{F}_1, \ldots, \mathcal{F}_r)$ is a system of forms with complex coefficients and of equal degree $d > 1$. The *complex pencil* of $\mathcal{F}$ consists of forms

$$\boldsymbol{\alpha}\mathcal{F} = \alpha_1 \mathcal{F}_1 + \cdots + \alpha_r \mathcal{F}_r$$

with nonzero $\boldsymbol{\alpha} \in \mathbb{C}^r$. Let $\mathbf{e}_1, \ldots, \mathbf{e}_s$ be the basis vectors. We associate with $\mathcal{F}$ the set $\mathcal{M} = \mathcal{M}(\mathcal{F})$ of $(d-1)$-tuples $(\mathbf{x}_1, \ldots, \mathbf{x}_{d-1}) \in \mathbb{C}^{s(d-1)}$ for which the matrix

$$(m_{ij}) = \mathcal{F}_j(\mathbf{x}_1 | \cdots | \mathbf{x}_{d-1} | \mathbf{e}_i) \qquad (1 \le i \le s, \, 1 \le j \le r) \qquad (15.6)$$

has rank $< r$. The set $\mathcal{M}$ is algebraic in $\mathbb{C}^{s(d-1)}$ (take a ground field as small as possible containing the coefficients of $\mathcal{F}$). After all, $\mathcal{M}$ is the set of points in $\mathbb{C}^{s(d-1)}$ where all $r \times r$ subdeterminants of our matrix vanish simultaneously.

Another way of looking at $\mathcal{M}$ is that it consists of the $(d-1)$-tuples $(\mathbf{x}_1, \ldots, \mathbf{x}_{d-1})$ for which some form $\mathcal{F}$ of the complex pencil has

$$\mathcal{F}(\mathbf{x}_1 | \cdots | \mathbf{x}_{d-1} | \mathbf{Z}) = 0 \qquad (15.7)$$

identically in $\mathbf{Z}$. The set $\mathcal{M}$ depends only on the complex pencil of $\mathcal{F}$, i.e. it is invariant under substitutions $\mathcal{F} \to T\mathcal{F}$ where T is a non-singular linear map of $\mathbb{C}^r$.

The set $V^* = V^*(\mathcal{F})$ mentioned in Section 15.1 consists of $\mathbf{x}$ in $\mathbb{C}^s$ for which the matrix $\partial \mathcal{F}_j / \partial x_i$ $(1 \le i \le s, \, 1 \le j \le r)$ has rank less than r, i.e. for which the matrix

$$\mathcal{F}_j(\mathbf{x} | \cdots | \mathbf{x} | \mathbf{e}_i) \qquad (1 \le i \le s, \, 1 \le j \le r)$$

has rank less than r. Hence V^* is the intersection of $\mathcal{M}$ with the diagonal $\mathbf{x}_1 = \mathbf{x}_2 = \cdots = \mathbf{x}_{d-1}$. This diagonal has codimension $s(d-2)$, and hence V^*, interpreted in this way as a submanifold of $\mathbb{C}^{s(d-1)}$, has codimension

$$\le \operatorname{codim} \mathcal{M} + s(d-2)$$

(Lang (1958), Ch. II, Section 7). Hence if V^* is interpreted as a submanifold of $\mathbb{C}^s$, we get

$$\operatorname{codim} V^* \le \operatorname{codim} \mathcal{M}. \qquad (15.8)$$

Now we introduce the quantity $g(\mathcal{F})$ for a system $\mathcal{F}$ of forms with *rational* coefficients. An integer $(d-1)$-tuple $(\mathbf{x}_1, \ldots, \mathbf{x}_{d-1})$ now lies in

$\mathcal{M}$ precisely if there is a form $\mathcal{F}$ of the *rational* pencil with (15.7). We write $g = g(\mathcal{F})$ for the largest real number such that

$$z_P(\mathcal{M}) = \sum_{\mathbf{x} \in \mathcal{M}, |\mathbf{x}| \leq P} 1 \ll P^{s(d-1)-g+\varepsilon} \qquad (15.9)$$

holds for each $\varepsilon > 0$. Since it is easily seen that $z_P(\mathcal{M}) \ll P^{\dim \mathcal{M}}$, we have

$$\operatorname{codim} \mathcal{M} \leqq g. \qquad (15.10)$$

The number g is invariant under substitutions $\mathcal{F} \to T\mathcal{F}$ where T is a non-singular linear map of $\mathbb{Q}^r$. It is easily seen to be invariant also under substitutions $\mathcal{F}(\mathbf{X}) \to \mathcal{F}(\tau(\mathbf{X}))$ where τ is a non-singular linear map of $\mathbb{Q}^s$.

Given $\boldsymbol{\alpha} = (\boldsymbol{\alpha}^{(d)}, \ldots, \boldsymbol{\alpha}^{(2)})$ as in (15.4), we define

$$\|\boldsymbol{\alpha}\| = \max_{2 \leqq j \leqq d} \max_{1 \leqq i \leqq r_d} \|\alpha_i^{(j)}\|.$$

We are now ready to state two results along lines at least slightly familiar from earlier chapters; if $S(\boldsymbol{\alpha})$ is large, then $\boldsymbol{\alpha}$ has a good rational approximation.

Theorem 15.2. *Let $\mathcal{F} = (\mathcal{F}_1, \ldots, \mathcal{F}_r)$ be a system of forms of equal degree $d \geqq 2$, with rational coefficients. Define $S(\boldsymbol{\alpha}) = S(\boldsymbol{\alpha}, \mathcal{B}, \mathcal{F})$ as in (15.4), where $\mathcal{B}$ is a box in $\mathbb{R}^r$. Let*

$$0 < \Omega < g(\mathcal{F})/(2^{d-1}(d-1)r). \qquad (15.11)$$

Let $\Delta > 0$ and let $P > C_3(\mathcal{F}, \Omega, \mathcal{B}, \Delta)$. Then each $\boldsymbol{\alpha}$ in $\mathbb{R}^r$ satisfies at least one of the following two alternatives. Either

(i)
$$|S(\boldsymbol{\alpha})| \leqq P^{s-\Delta\Omega},$$
or

(ii) *there is a natural number $q = q(\boldsymbol{\alpha}) \leqq P^{\Delta}$ with*

$$\|q\boldsymbol{\alpha}\| \leqq P^{-d+\Delta}.$$

This should be compared with the case $L = 1$ of Lemma 3.10, where $r = s = 1$. The same conclusion is reached there with a hypothesis

$$0 < \Delta \leqq 1 \qquad (0 < \Omega < 2^{-d+1}).$$

We shall use Weyl's inequality, as we did there, though in a slightly different guise.

Schmidt (1985) shows that one can reach an asymptotic formula (15.2) for a given system $\mathcal{F}$ of forms of equal degree provided that the conclusion of Theorem 15.2 is true with some $\Omega > r + 1$. This will be the case if

$$g > 2^{d-1}(d-1)r(r+1).$$

By (15.8) and (15.10), this is certainly true if

$$\text{codim } V^* > 2^{d-1}(d-1)r(r+1),$$

so that Birch's theorem does indeed follow from Schmidt's work.

The analogue of Theorem 15.2 for systems of forms of differing degrees is a bit more complicated. Let $\mathscr{F} = (\mathscr{F}^{(k)}, \ldots, \mathscr{F}^{(2)})$ where $\mathscr{F}^{(d)}$ consists of r_d forms of degree d with rational coefficients. For each d with $r_d > 0$ put

$$g_d = g(\mathscr{F}^{(d)}). \tag{15.12}$$

Further set

$$\gamma_d = g_d^{-1} 2^{d-1}(d-1)r_d. \tag{15.13}$$

when $r_d > 0$, $g_d > 0$; and $\gamma_d = 0$ when $r_d = 0$, and $\gamma_d = +\infty$ when $r_d > 0$, $g_d = 0$. Finally set

$$\tau = \gamma_2 + 4\gamma_3 + \cdots + 4^{k-1}\gamma_k. \tag{15.14}$$

Theorem 15.3. *Let $\mathscr{F}$, τ be as above. Define $S(\alpha) = S(\alpha, \mathscr{B}, \mathscr{F})$ as in* (15.4), *where $\mathscr{B}$ is a box in $\mathbb{R}^r$. Let*

$$0 < \Omega < \tau^{-1}. \tag{15.15}$$

Let $0 < \Delta \leqq 1$ and let $P > C_4(\mathscr{F}, \Omega, \mathscr{B}, \Delta)$. Then each α in $\mathbb{R}^r$ satisfies at least one of the alternatives (i), (ii) *of Theorem 15.2.*

15.4 Weyl's inequality

Given a real function $\mathscr{F}(\mathbf{X})$ on $\mathbb{Z}^s$, define

$$\mathscr{F}_d(\mathbf{X}_1, \ldots, \mathbf{X}_d) = \sum_{\varepsilon_1=0}^{1} \cdots \sum_{\varepsilon_d=0}^{1} (-1)^{\varepsilon_1 + \cdots + \varepsilon_d} \mathscr{F}(\varepsilon_1 \mathbf{X}_1 + \cdots + \varepsilon_d \mathbf{X}_d).$$

Then $\mathscr{F}_d$ is symmetric in its d arguments, and $\mathscr{F}_d(\mathbf{X}_1, \ldots, \mathbf{X}_{d-1}, \mathbf{0}) = 0$. We have $\mathscr{F}_1(\mathbf{X}_1) = \mathscr{F}(\mathbf{0}) - \mathscr{F}(\mathbf{X}_1)$, and

$$\mathscr{F}_2(\mathbf{X}_1, \mathbf{X}_2) = \mathscr{F}(\mathbf{0}) - \mathscr{F}(\mathbf{X}_1) - \mathscr{F}(\mathbf{X}_2) + \mathscr{F}(\mathbf{X}_1 + \mathbf{X}_2)$$
$$= \mathscr{F}_1(\mathbf{X}_1) + \mathscr{F}_1(\mathbf{X}_2) - \mathscr{F}_1(\mathbf{X}_1 + \mathbf{X}_2).$$

For $d \geqq 2$ and for given $\mathbf{X}_3, \ldots, \mathbf{X}_{d+1}$ we apply this formula to

$$\mathscr{G}(\mathbf{X}) = \sum_{\varepsilon_3=0}^{1} \cdots \sum_{\varepsilon_{d+1}=0}^{1} (-1)^{\varepsilon_3 + \cdots + \varepsilon_{d+1}} \mathscr{F}(\mathbf{X} + \varepsilon_3 \mathbf{X}_3 + \cdots + \varepsilon_{d+1} \mathbf{X}_{d+1}),$$

obtaining

$$\mathscr{F}_{d+1}(\mathbf{X}_1, \mathbf{X}_2, \mathbf{X}_3, \ldots, \mathbf{X}_{d+1}) = \mathscr{F}_d(\mathbf{X}_1, \mathbf{X}_3, \ldots, \mathbf{X}_{d+1})$$
$$+ \mathscr{F}_d(\mathbf{X}_2, \mathbf{X}_3, \ldots, \mathbf{X}_{d+1})$$
$$- \mathscr{F}_d(\mathbf{X}_1 + \mathbf{X}_2, \mathbf{X}_3, \ldots, \mathbf{X}_{d+1}).$$

This relation remains valid if the roles of $\mathbf{X}_1, \ldots, \mathbf{X}_{d+1}$ are interchanged. For fixed $\mathbf{x}_1, \ldots, \mathbf{x}_{d-1}$ set $\mathscr{G}(\mathbf{X}) = \mathscr{F}_d(\mathbf{x}_1, \ldots, \mathbf{x}_{d-1}, \mathbf{X})$. We obtain

$$\mathscr{F}_{d+1}(\mathbf{x}_1, \ldots, \mathbf{x}_{d-1}, \mathbf{X}_d, \mathbf{X}_{d+1}) = -\mathscr{G}_2(\mathbf{X}_d, \mathbf{X}_{d+1}). \qquad (15.16)$$

Given a finite set A of integer points in $\mathbb{R}^s$, write $A - \mathbf{x}$ for the set of points $\mathbf{a} - \mathbf{x}$ with $\mathbf{a} \in A$. Define $A^D = \bigcup_{\mathbf{x} \in A} (A - \mathbf{x})$,

$$A(\mathbf{x}_1, \ldots, \mathbf{x}_t) = \bigcap_{\varepsilon_1=0}^{1} \cdots \bigcap_{\varepsilon_t=0}^{1} (A - \varepsilon_1 \mathbf{x}_1 - \cdots - \varepsilon_t \mathbf{x}_t).$$

Then $A(\mathbf{x}) = A \cap (A - \mathbf{x})$, and for $t \geq 2$,

$$A(\mathbf{x}_1, \ldots, \mathbf{x}_t) = A(\mathbf{x}_1, \ldots, \mathbf{x}_{t-1}) \cap (A(\mathbf{x}_1, \ldots, \mathbf{x}_{t-1}) - \mathbf{x}_t).$$

Lemma 15.2. *Let $\mathscr{F}$ be defined on $\mathbb{Z}^s$, and put*

$$S = \sum_{\mathbf{x} \in A} e(\mathscr{F}(\mathbf{x})).$$

Then for each $d \geq 2$,

$$|S|^{2^{d-1}} \leq |A^D|^{2^{d-1}-d} \sum_{\mathbf{x}_1 \in A^D} \cdots \sum_{\mathbf{x}_{d-1} \in A^D} \left| \sum_{\mathbf{x}_d \in A(\mathbf{x}_1, \ldots, \mathbf{x}_{d-1})} e(\mathscr{F}_d(\mathbf{x}_1, \ldots, \mathbf{x}_d)) \right|.$$

This lemma is the appropriate version of Weyl's inequality for the present chapter. Naturally, we prove it by induction on d.

Proof. In

$$|S|^2 = \sum_{\mathbf{x} \in A} \sum_{\mathbf{y} \in A} e(\mathscr{F}(\mathbf{x}) - \mathscr{F}(\mathbf{y}))$$

set $\mathbf{x}_1 = \mathbf{x} - \mathbf{y}$, $\mathbf{x}_2 = \mathbf{y}$. Then $\mathbf{x}_1 \in A^D$ and $\mathbf{x}_2 \in A \cap (A - \mathbf{x}_1) = A(\mathbf{x}_1)$. We note that

$$\mathscr{F}(\mathbf{x}) - \mathscr{F}(\mathbf{y}) = \mathscr{F}(\mathbf{x}_1 + \mathbf{x}_2) - \mathscr{F}(\mathbf{x}_2) = \mathscr{F}_2(\mathbf{x}_1, \mathbf{x}_2) - \mathscr{F}_1(\mathbf{x}_1).$$

Thus

$$|S|^2 = \sum_{\mathbf{x}_1 \in A^D} e(-\mathscr{F}_1(\mathbf{x}_1)) \sum_{\mathbf{x}_2 \in A(\mathbf{x}_1)} e(\mathscr{F}_2(\mathbf{x}_1, \mathbf{x}_2))$$

$$\leq \sum_{\mathbf{x}_1 \in A^D} \left| \sum_{\mathbf{x}_2 \in A(\mathbf{x}_1)} e(\mathscr{F}_2(\mathbf{x}_1, \mathbf{x}_2)) \right|,$$

which is the case $d = 2$ of the lemma.

For the step from d to $d+1$, we square the inequality of the lemma, and use Cauchy's inequality, to obtain

$$|S|^{2^d} \le |A^D|^{2^d - 2d + d - 1} \sum_{\mathbf{x}_1 \in A^D} \cdots \sum_{\mathbf{x}_{d-1} \in A^D}$$

$$\times \left| \sum_{\mathbf{x}_d \in A(\mathbf{x}_1, \ldots, \mathbf{x}_{d-1})} e(\mathscr{F}_d(\mathbf{x}_1, \ldots, \mathbf{x}_d)) \right|^2. \quad (15.17)$$

Denote the sum over $\mathbf{x}_d$ on the inside by S_d. Then S_d is like S, except that $\mathscr{F}(\mathbf{X})$ is replaced by $\mathscr{G}(\mathbf{X}) = \mathscr{F}_d(\mathbf{x}_1, \ldots, \mathbf{x}_{d-1}, \mathbf{X})$, and A is replaced by $A(\mathbf{x}_1, \ldots, \mathbf{x}_{d-1})$. By applying the case $d = 2$ of the lemma, and taking into account (15.16) as well as the relation $A(\mathbf{x}_1, \ldots, \mathbf{x}_{d-1})^D \subseteq A^D$, we get

$$|S_d|^2 \le \sum_{\mathbf{x}_d \in A^D} \left| \sum_{\mathbf{x}_{d+1} \in A(\mathbf{x}_1, \ldots, \mathbf{x}_d)} e(\mathscr{F}_{d+1}(\mathbf{x}_1, \ldots, \mathbf{x}_{d+1})) \right|.$$

By substituting this into (15.17) we get the desired result.

Lemma 15.3. *Suppose $\mathscr{F}$ is a form of degree $j > 0$. Then*
  (a) $\mathscr{F}_d = 0$ *when $d > j$.*
  (b) $\mathscr{F}_j(\mathbf{X}_1, \ldots, \mathbf{X}_j)$ *is multilinear.*
  (c) *When $1 \le d < j$, then*

$$\mathscr{F}_d(\mathbf{X}_1, \ldots, \mathbf{X}_d) = \sum_{l=1}^{j-d+1} \mathscr{G}_{d,l}(\mathbf{X}_1, \ldots, \mathbf{X}_d),$$

where $\mathscr{G}_{d,l}$ is a form of degree l in $\mathbf{X}_d$, and a form of total degree $j - l$ in $\mathbf{X}_1, \ldots, \mathbf{X}_{d-1}$.

Proof. Since $\mathscr{F}_d(\mathbf{0}, \mathbf{X}_2, \ldots, \mathbf{X}_d) = 0$, each monomial occurring in $\mathscr{F}_d(\mathbf{X}_1, \ldots, \mathbf{X}_d)$ has some component of $\mathbf{X}_1$ as a factor. The same is true for $\mathbf{X}_2, \ldots, \mathbf{X}_d$. Since the total degree is j, each monomial has a degree between 1 and $j - d + 1$ in each of $\mathbf{X}_1, \ldots, \mathbf{X}_d$. Now (a), (b), and (c) follow.

Now if $\mathscr{F}$ is a form of degree d, then by (b), $\mathscr{F}_d(\mathbf{X}_1, \ldots, \mathbf{X}_d)$ is symmetric and multilinear. Taking in particular $\mathscr{F}(X) = X^d$, we find that

$$\sum_{\varepsilon_1 = 0}^{1} \cdots \sum_{\varepsilon_d = 0}^{1} (-1)^{\varepsilon_1 + \cdots + \varepsilon_d} (\varepsilon_1 X_1 + \cdots + \varepsilon_d X_d)^d = \rho(d) X_1 \cdots X_d$$

$$(15.18)$$

with a numerical multiplier $\rho(d)$. Evidently $\rho(d) = (-1)^d d!$ if we consider how $X_1 \cdots X_d$ may arise on the left-hand side.

Each of the 2^d summands in the definition of $\mathscr{F}_d(\mathbf{X}, \ldots, \mathbf{X})$ is a multiple of $\mathscr{F}(\mathbf{X})$, so that $\mathscr{F}_d(\mathbf{X}, \ldots, \mathbf{X}) = \sigma(d) \mathscr{F}(\mathbf{X})$ with a numerical factor $\sigma(d)$. Comparing with (15.18) with $X_1 = X_2 = \cdots = X_d$, we see that

$\sigma(d) = \rho(d) = (-1)^d d!$. Thus

$$\mathcal{F}_d(\mathbf{X}_1, \ldots, \mathbf{X}_d) = \mathcal{F}(\mathbf{X}_1 \mid \cdots \mid \mathbf{X}_d) \qquad (15.19)$$

where the right-hand side is the multilinear form of the last section.

As in Chapter 14, write $|\mathcal{F}|$ for the maximum absolute value of the coefficient of $\mathcal{F}$ in $\mathbb{R}[x_1, \ldots, x_s]$. Write $\|\mathcal{F}\|$ for the maximum of $\|f\|$ over the coefficients f of $\mathcal{F}$.

Lemma 15.4. *Suppose $\mathcal{F}$ is a form of degree j. Then*

$$\|\mathcal{F}_d\| \leq 2^d d^j \|\mathcal{F}\|.$$

Proof. We have

$$\|\mathcal{F}(\mathbf{X}_1 + \cdots + \mathbf{X}_p)\| \leq p^j \|\mathcal{F}\|.$$

Since $\mathcal{F}_d$ consists of 2^d summands, each of the form

$$\pm \mathcal{F}(\varepsilon_1 \mathbf{X}_1 + \cdots + \varepsilon_d \mathbf{X}_d),$$

our (rather crude) estimate for $\|\mathcal{F}_d\|$ follows.

We write $\mathscr{C}$ for the cube $\{\mathbf{x} \in \mathbb{R}^s; |\mathbf{x}| \leq 1\}$.

Lemma 15.5. *Suppose*

$$\mathcal{F}(\mathbf{X}) = \mathcal{F}^{(0)} + \mathcal{F}^{(1)}(\mathbf{X}) + \cdots + \mathcal{F}^{(k)}(\mathbf{X}),$$

where $\mathcal{F}^{(j)}$ $(0 \leq j \leq k)$ is a form of degree j with real coefficients. Then

(A) $\mathcal{F}_k(\mathbf{X}_1, \ldots, \mathbf{X}_k) = \mathcal{F}^{(k)}(\mathbf{X}_1 \mid \cdots \mid \mathbf{X}_k)$.
(B) *Suppose that $1 \leq d < k$ and that $\|\mathcal{F}^{(j)}\| \leq cP^{\theta-j}$ for $d < j \leq k$, where $\theta \geq 0$, $P > 1$. Also suppose (if $d > 1$) that $\mathbf{x}_1, \ldots, \mathbf{x}_{d-1}$ are integer points in $P\mathscr{C}$. Then*

$$\mathcal{F}_d(\mathbf{x}_1, \ldots, \mathbf{x}_{d-1}, \mathbf{X}) = \mathcal{F}^{(d)}(\mathbf{x}_1 \mid \cdots \mid \mathbf{x}_{d-1} \mid \mathbf{X}) + \sum_{l=1}^{k-d+1} \mathcal{G}_d^{(l)}(\mathbf{x}_1, \ldots, \mathbf{x}_{d-1}, \mathbf{X}),$$

where $\mathcal{G}_d^{(l)}(\mathbf{X}) = \mathcal{G}_d^{(l)}(\mathbf{x}_1, \ldots, \mathbf{x}_{d-1}, \mathbf{X})$ is a form in $\mathbf{X}$ of degree l with

$$\|\mathcal{G}_d^{(l)}\| \ll P^{\theta-l} \qquad (1 \leq l \leq k - d + 1).$$

The implied constant depends on k, s and c.

Proof. (A) follows from Lemma 15.3 and from (15.19). As for (B), in the notation of Lemma 15.3(c) we have

$$\mathcal{F}_d = \mathcal{F}_d^{(d)} + \sum_{j=d+1}^{k} \mathcal{F}_d^{(j)}$$

$$= \mathcal{F}_d^{(d)} + \sum_{j=d+1}^{k} \sum_{l=1}^{j-d+1} \mathcal{G}_{d,l}^{(j)}$$

$$= \mathcal{F}_d^{(d)} + \sum_{l=1}^{k-d+1} \mathcal{G}_d^{(l)},$$

where

$$\mathcal{G}_d^{(l)} = \sum_{j=\max(d+1,\,l+d-1)}^{k} \mathcal{G}_{d,l}^{(j)}.$$

Since $\|\mathcal{G}_{d,l}^{(j)}\| \leq \|\mathcal{F}_d^{(j)}\| \ll \|\mathcal{F}^{(j)}\|$, and since $\mathcal{G}_{d,l}^{(j)}$ is a form of degree $j - l$ in $\mathbf{X}_1, \ldots, \mathbf{X}_{d-1}$, and $|\mathbf{x}_1|, \ldots, |\mathbf{x}_{d-1}|$ are bounded by P, we see that

$$\|\mathcal{G}_d^{(l)}\| \ll \sum_{j=d+1}^{k} P^{j-l} \|\mathcal{F}^{(j)}\| \ll \sum_{j=d+1}^{k} P^{j-l} P^{\theta-j}$$

$$\ll P^{\theta-l} \qquad (1 \leq l \leq k - d + 1).$$

15.5 Predominantly linear exponential sums

We now follow a strategy which can be traced back to Davenport (1959) and Birch (1962), although they needed only 'purely linear' sums in their work.

Lemma 15.6. *Suppose* $\mathcal{G}(\mathbf{X}) = \mathcal{G}^{(0)} + \mathcal{G}^{(1)}(\mathbf{X}) + \cdots + \mathcal{G}^{(m)}(\mathbf{X})$, *where* $\mathcal{G}^{(j)}$ *is a form of degree* j *with* $\|\mathcal{G}^{(j)}\| \leq Q^{-j}$ $(j = 1, \ldots, m)$ *with some given* $Q > 1$. *Suppose that* $0 < \gamma < 1$ *and that* $M_1, \ldots, M_s$ *lie in* $1 \leq M_i \leq Q^{1-\gamma}$ $(1 \leq i \leq s)$. *Given* $\boldsymbol{\beta} = (\beta_1, \ldots, \beta_s)$, *put*

$$S = \sum_{\mathbf{x}:1 \leq x_i \leq M_i} e(\boldsymbol{\beta}\mathbf{x} + \mathcal{G}(\mathbf{x})).$$

Then

$$S \ll \prod_{i=1}^{s} \min(M_i, \|\beta_i\|^{-1}).$$

The implied constant depends only on m, s, *and* γ.

Proof. We may suppose that the constant term $\mathcal{G}^{(0)} = 0$. We may further suppose without loss of generality that $|\mathcal{G}^{(j)}| \leq Q^{-j}$ $(j = 1, \ldots, m)$. For the vectors $\mathbf{x}$ of the sum,

$$|\mathcal{G}^{(j)}(\mathbf{x})| \ll Q^{-j} Q^{(1-\gamma)j} = Q^{-\gamma j} \qquad (j = 1, \ldots, m),$$

so that $|\mathcal{G}(\mathbf{X})| \ll Q^{-\gamma}$. Let l be an integer with

$$s < l\gamma < s + 1,$$

and put

$$\mathcal{H}(\mathbf{X}) = \sum_{n=0}^{l} (2\pi i \mathcal{G}(\mathbf{X}))^n / n!$$

Then

$$e(\mathcal{G}(\mathbf{X})) = \mathcal{H}(\mathbf{x}) + O(\mathcal{G}(\mathbf{x})^l)$$
$$= \mathcal{H}(\mathbf{x}) + O(Q^{-l\gamma}) = \mathcal{H}(\mathbf{x}) + O(Q^{-s}),$$

and therefore

$$S = \sum_{\mathbf{x}:1 \leq x_i \leq M_i} e(\boldsymbol{\beta}\mathbf{x}) \mathcal{H}(\mathbf{x}) + O(1). \tag{15.20}$$

We now write

$$\mathcal{H}(\mathbf{X}) = 1 + \mathcal{H}^{(1)}(\mathbf{X}) + \cdots + \mathcal{H}^{(ml)}(\mathbf{X}), \tag{15.21}$$

where $\mathcal{H}^{(j)}$ is a form of degree j. Our hypothesis implies that

$$|\mathcal{H}^{(j)}| \ll Q^{-j} \qquad (1 \leq j \leq ml). \tag{15.22}$$

Now, using a geometric progression as in (3.4),

$$\sum_{x=1}^{M} e(\beta x) x^t = \sum_{x=1}^{M} (x^t - (x-1)^t)(e(\beta x) + e(\beta(x+1)) + \cdots + e(\beta M))$$

$$\ll \sum_{x=1}^{M} x^{t-1} \min(M, \|\beta\|^{-1})$$

$$\ll M^t \min(M, \|\beta\|^{-1}).$$

Hence for a monomial $\mathcal{M}(\mathbf{X}) = X_1^{j_1} \cdots X_s^{j_s}$ of total degree $j_1 + \cdots + j_s = j$, we have

$$\sum_{\mathbf{x}:1 \leq x_i \leq M_i} e(\boldsymbol{\beta}\mathbf{x}) \mathcal{M}(\mathbf{x}) \ll Q^j \prod_{i=1}^{s} \min(M_i, \|\beta_i\|^{-1}).$$

This, together with (15.20)–(15.22), gives the desired result.

Lemma 15.7. *Suppose* $\mathcal{G}(\mathbf{X}) = \mathcal{G}^{(0)} + \mathcal{G}^{(1)}(\mathbf{X}) + \cdots + \mathcal{G}^{(m)}(\mathbf{X})$ *where* $\mathcal{G}^{(j)}$ *is a form of degree* j. *Suppose* $0 \leq \theta < \frac{1}{4}$, $P > 1$, *and suppose there is a natural number* q *with*

$$q \leq P^\theta \qquad and \qquad \|q \mathcal{G}^{(j)}\| \leq c P^{\theta - j} \quad (j = 1, \ldots, m)$$

where c *is a constant. Given* $\boldsymbol{\beta}$ *and given a box* $\mathcal{B}$ *with sides at most* 1, *write*

$$S = \sum_{\mathbf{x} \in P\mathcal{B}} e(\boldsymbol{\beta}\mathbf{x} + \mathcal{G}(\mathbf{x})).$$

Then

$$S \ll P^{2\theta s + \varepsilon} \prod_{i=1}^{s} \min(P^{1-2\theta}, \|q\beta_i\|^{-1}),$$

with a constant in $\ll$ *which depends only on* m, s, c, ε.

Proof. Choose $\delta = \varepsilon/(2\varepsilon + 2s)$ so that $0 < \delta < \min(\frac{1}{2}, \varepsilon/(2s))$. Put

$$Q = P^{1-2\theta-\delta}, \qquad M = Q^{1-\delta}.$$

The box $P\mathcal{B}$ may be split into $\ll (P/(Mq))^s$ boxes with sides $\leq Mq$. In each such small box write $\mathbf{x} = \mathbf{a} + q\mathbf{y}$, where $\mathbf{a}$ runs through a residue system modulo q, and $\mathbf{y}$ runs through a box with sides $\leq M$. Given the small box, and given the residue class $\mathbf{a}$, we have $\mathbf{x} = \mathbf{b} + q\mathbf{z}$, where $\mathbf{z}$ runs through a box $\mathscr{B}(\mathbf{b})$ of the type $1 \leq z \leq M_i$ with $M_i \leq M$ $(i = 1, \ldots, s)$. Put

$$S(\mathbf{b}) = \sum_{\mathbf{z} \in \mathscr{B}(\mathbf{b})} e(q\boldsymbol{\beta}\mathbf{z} + \mathscr{G}(\mathbf{b} + q\mathbf{z})).$$

Since the number of possibilities for $\mathbf{b}$ is

$$\ll (P/(Mq))^s q^s = (P/M)^s \ll P^{2\theta s + 2\delta s},$$

it will suffice for us to show that

$$S(\mathbf{b}) \ll \prod_{i=1}^{s} \min (P^{1-2\theta}, \|q\beta_i\|^{-1}).$$

Now

$$\mathscr{G}^{(l)}(\mathbf{X} + \mathbf{Y}) = \mathscr{H}^{(l)}(\mathbf{X}, \mathbf{Y}) = \mathscr{H}_l^{(0)}(\mathbf{X}, \mathbf{Y}) + \cdots + \mathscr{H}_l^{(l)}(\mathbf{X}, \mathbf{Y}),$$

say, where each term is a form of total degree l in $\mathbf{X}$, $\mathbf{Y}$, and where $\mathscr{H}_l^{(j)}$ is of degree j in $\mathbf{Y}$ and of degree $l - j$ in $\mathbf{X}$. Clearly

$$\|q\mathscr{H}_l^{(j)}\| \leq \|q\mathscr{H}^{(l)}\| \leq 2^l \|q\mathscr{G}^{(l)}\|.$$

For fixed $\mathbf{b}$ we have

$$\mathscr{G}^{(l)}(\mathbf{b} + q\mathbf{Z}) = \mathscr{H}_l^{(0)}(\mathbf{b}) + q\mathscr{H}_l^{(1)}(\mathbf{b}, \mathbf{Z}) + \cdots + q^l\mathscr{H}_l^{(l)}(\mathbf{b}, \mathbf{Z})$$
$$= \mathscr{R}_l^{(0)} + \mathscr{R}_l^{(1)}(\mathbf{Z}) + \cdots + \mathscr{R}_l^{(l)}(\mathbf{Z})$$

say, where $\mathscr{R}_l^{(j)}$ is a form of degree j. Since $|\mathbf{b}| \ll P$, we obtain for $j = 1, \ldots, l$ that

$$\|\mathscr{R}_l^{(j)}\| \ll q^{j-1} |\mathbf{b}|^{l-j} \|q\mathscr{G}^{(l)}\|$$
$$\ll q^j P^{l-j} P^{\theta - l}$$
$$= P^\theta (P/q)^{-j} \ll P^\theta P^{j(\theta-1)}$$
$$\ll Q^{-j(1+\delta)}$$

after a short calculation.

Now

$$\mathscr{G}(\mathbf{b} + q\mathbf{Z}) = \mathscr{R}^{(0)} + \mathscr{R}^{(1)}(\mathbf{Z}) + \cdots + \mathscr{R}^{(m)}(\mathbf{Z})$$

with $\mathscr{R}^{(j)} = \mathscr{R}_j^{(j)} + \mathscr{R}_{j+1}^{(j)} + \cdots + \mathscr{R}_m^{(j)}$ $(j = 0, 1, \ldots, m)$, so that

$$\|\mathscr{R}^{(j)}\| \leq Q^{-j}$$

if P and hence Q is large. Lemma 15.6 yields

$$S(\mathbf{b}) \ll \prod_{i=1}^{s} \min (M, \|q\beta_i\|^{-1}).$$

15.6   Exponential sums and multilinear inequalities

Lemma 15.8. *Suppose* $\mathscr{F}(\mathbf{X}) = \mathscr{F}^{(0)} + \mathscr{F}^{(1)}(\mathbf{X}) + \cdots + \mathscr{F}^{(k)}(\mathbf{X})$, *where* $\mathscr{F}^{(j)}$ *is a form of degree j with real coefficients. Let* $\mathscr{B}$ *be a box with sides* $\leqq 1$, *let* $P > 1$, *and put*

$$S = \sum_{\mathbf{x} \in P\mathscr{B}} e(\mathscr{F}(\mathbf{x})).$$

Now let $2 \leqq d \leqq k$ *and* $\varepsilon > 0$. *Suppose that either* $d = k$, *and put* $\theta = 0$ *and* $q = 1$. *Or else, suppose that* $2 \leqq d < k$, *that* $0 \leqq \theta < \frac{1}{4}$, *and that there is a natural number* q *with*

$$q \leqq P^{\theta} \qquad and \qquad \|q\mathscr{F}^{(j)}\| \leqq cP^{\theta - j} \quad for \ d < j \leqq k.$$

Then

$$|S|^{2^{d-1}} \ll P^{(2^{d-1} - d + 2\theta)s + \varepsilon} \sum \prod_{i=1}^{s} \min\left(P^{1-2\theta}, \|q\mathscr{F}^{(d)}(\mathbf{x}_1 \mid \cdots \mid \mathbf{x}_{d-1} \mid \mathbf{e}_i)\|^{-1}\right),$$

where the sum $\sum$ *is over* $(d-1)$-*tuples of integer points* $\mathbf{x}_1, \ldots, \mathbf{x}_{d-1}$ *in* $P\mathscr{C}$, *and where* $\mathbf{e}_1, \ldots, \mathbf{e}_s$ *are the basis vectors. The implied constant depends only on c, s, k and* ε.

Proof. Our hypothesis on $\mathscr{B}$ implies that $(P\mathscr{B})^D \subset P\mathscr{C}$. Lemma 15.2 gives

$$|S|^{2^{d-1}} \ll P^{(2^{d-1} - d)s} \sum_{\mathbf{x}_1 \in P\mathscr{C}} \cdots \sum_{\mathbf{x}_{d-1} \in P\mathscr{C}}$$

$$\times \left| \sum_{\mathbf{x}_d \in (P\mathscr{B})(\mathbf{x}_1, \ldots, \mathbf{x}_{d-1})} e(\mathscr{F}_d(\mathbf{x}_1, \ldots, \mathbf{x}_{d-1}, \mathbf{x}_d)) \right|. \quad (15.23)$$

In the case when $d = k$,

$$\mathscr{F}_d(\mathbf{x}_1, \ldots, \mathbf{x}_{d-1}, \mathbf{X}) = \boldsymbol{\beta}\mathbf{X},$$

with

$$\beta_i = \mathscr{F}_d^{(d)}(\mathbf{x}_1, \ldots, \mathbf{x}_{d-1}, \mathbf{e}_i) = \mathscr{F}^{(d)}(\mathbf{x}_1 \mid \cdots \mid \mathbf{x}_{d-1} \mid \mathbf{e}_i) \qquad (i = 1, \ldots, s).$$
$$(15.24)$$

Hence a bound

$$\ll \prod_{i=1}^{s} \min\left(P, \|\beta_i\|^{-1}\right)$$

holds for the inner sum in (15.23). In the case when $2 \leqq d < k$, Lemma 15.5 tells us that

$$\mathscr{F}_d(\mathbf{x}_1, \ldots, \mathbf{x}_{d-1}, \mathbf{X}) = \boldsymbol{\beta}\mathbf{X} + \sum_{l=1}^{k-d+1} \mathscr{G}_d^{(l)}(\mathbf{X}),$$

where $\boldsymbol{\beta}$ is given by (15.24) and where $\mathscr{G}_d^{(l)}$ is a form of degree l with

$$\|q\mathscr{G}_d^{(l)}\| \ll P^{\theta - l} \qquad (l = 1, \ldots, k - d + 1).$$

Now Lemma 15.7 gives a bound

$$\ll P^{2\theta s+\varepsilon} \prod_{i=1}^{s} \min\left(P^{1-2\theta}, \|q\beta_i\|^{-1}\right)$$

for the inner sum in (15.23), and Lemma 15.8 follows.

Lemma 15.9. *Make all the assumptions of the preceding lemma. Suppose further that*

$$|S| \geqq P^{s-H}, \tag{15.25}$$

where $H > 0$. Then the number N of $(d-1)$-tuples of integer points $\mathbf{x}_1, \ldots, \mathbf{x}_{d-1}$ in $P\mathscr{C}$ with

$$\|q\mathscr{F}^{(d)}(\mathbf{x}_1 | \cdots | \mathbf{x}_{d-1} | \mathbf{e}_i)\| < P^{-1+2\theta} \qquad (i = 1, \ldots, s) \tag{15.26}$$

satisfies

$$N \gg P^{s(d-1)-2^{d-1}H-\varepsilon} \tag{15.27}$$

with a constant in $\gg$ depending only on c, s, k and ε.

Proof. Let $N_0(\mathbf{x}_1, \ldots, \mathbf{x}_{d-2})$ be the number of points $\mathbf{x}_{d-1}$ in $P\mathscr{C}$ with (15.26). Then $N = N_0$ when $d = 2$, and

$$N = \sum_{\mathbf{x}_1 \in P\mathscr{C}} \cdots \sum_{\mathbf{x}_{d-2} \in P\mathscr{C}} N_0(\mathbf{x}_1, \ldots, \mathbf{x}_{d-2}) \tag{15.28}$$

when $d > 2$. It will be convenient to set

$$J = [P^{1-2\theta}] + 1. \tag{15.29}$$

For any set of integer points $\mathbf{x}_1, \ldots, \mathbf{x}_{d-2}$ and any integers $a_1, \ldots, a_s$ with $0 \leqq a_i < J$ the inequalities

$$J^{-1}a_i \leqq \{q\mathscr{F}^{(d)}(\mathbf{x}_1 | \cdots | \mathbf{x}_{d-2} | \mathbf{x}_{d-1} | \mathbf{e}_i)\} < J^{-1}(a_i + 1) \qquad (1 \leqq i \leqq s)$$

cannot hold for more than $N_0(\mathbf{x}_1, \ldots, \mathbf{x}_{d-2})$ integer points $\mathbf{x}_{d-1}$ lying in a prescribed box of side P; for if $\mathbf{x}'_{d-1}$ is one solution of these inequalities, and if $\mathbf{x}_{d-1}$ denotes the general solution, then

$$\|q\mathscr{F}^{(d)}(\mathbf{x}_1 | \cdots | \mathbf{x}_{d-2} | \mathbf{x}_{d-1} - \mathbf{x}'_{d-1} | \mathbf{e}_i)\| < J^{-1} \qquad (i = 1, \ldots, s)$$

and $\mathbf{x}_{d-1} - \mathbf{x}'_{d-1} \in P\mathscr{C}$. Thus the number of possibilities for $\mathbf{x}_{d-1}$ is indeed at most $N_0(\mathbf{x}_1, \ldots, \mathbf{x}_{d-2})$.

Dividing the cube $P\mathscr{C}$ into 2^s cubes of side P, we obtain

$$\sum_{\mathbf{x}_{d-1} \in P\mathscr{C}} \prod_{i=1}^{s} \min\left(J, \|q\mathscr{F}^{(d)}(\mathbf{x}_1 | \cdots | \mathbf{x}_{d-1} | \mathbf{e}_i)\|^{-1}\right)$$

$$\ll N_0(\mathbf{x}_1, \ldots, \mathbf{x}_{d-2}) \sum_{a_1=0}^{J} \cdots \sum_{a_s=0}^{J} \left(\prod_{i=1}^{s} \min\left(J, \max\left(\frac{J}{a_i}, \frac{J}{|J-a_i-1|}\right)\right)\right)$$

$$\ll N_0(\mathbf{x}_1, \ldots, \mathbf{x}_{d-2}) J^s (\log J)^s.$$

In conjunction with (15.28), (15.29) and the preceding lemma, this gives

$$|S|^{2^{d-1}} \ll NP^{(2^{d-1}-d+1)s+2\varepsilon}.$$

Since $\varepsilon > 0$ is arbitrary here, the hypothesis (15.25) yields (15.27). This completes the proof of Lemma 15.9.

15.7   An application of the geometry of numbers

Lemma 15.10.   *Let*

$$\mathcal{L}_i(\mathbf{X}) = \lambda_{i1}X_1 + \cdots + \lambda_{is}X_s \qquad (i = 1, \dots, s)$$

be linear forms with $\lambda_{ij} = \lambda_{ji}$ $(1 \le i, j \le s)$. Given $A > 1$ and $Z > 0$, let $N(Z)$ be the number of integer points $\mathbf{x}$ with

$$|\mathbf{x}| \le ZA \quad and \quad \|\mathcal{L}_i(\mathbf{X})\| \le ZA^{-1} \qquad (i = 1, \dots, s). \tag{15.30}$$

Then for $0 < Z_1 \le Z_2 < 1$ we have

$$N(Z_1) \gg (Z_1/Z_2)^s N(Z_2), \tag{15.31}$$

with an implied constant which depends only on s.

Proof. This is a lemma of Davenport (1959) and is a fairly easy deduction from Lemmas 9.1, 9.2. We use the notation given there, so that

$$\left. \begin{aligned} \zeta_j(x, \dots, x_{2s}) &= A(\mathcal{L}_j(x_1, \dots, x_s) - x_{s+j}) \\ \zeta_{s+j}(x_1, \dots, x_{2s}) &= A^{-1}x_j, \end{aligned} \right\} \qquad (j = 1, \dots, s)$$

and the lattice

$$\Pi = \{(\zeta_1(\mathbf{x}), \dots, \zeta_{2s}(\mathbf{x})): \mathbf{x} \in \mathbb{Z}^{2s}\}$$

has successive minima $\pi_1, \dots, \pi_{2s}$ with respect to the unit ball. Moreover,

$$\pi_s \gg 1 \tag{15.32}$$

from (9.6). Write $\mathcal{L}(\mathbf{x}) = (\mathcal{L}_1(\mathbf{x}), \dots, \mathcal{L}_s(\mathbf{x}))$.

   With each Z with

$$\sqrt{(2s)}Z \ge \pi_1, \qquad Z < 1$$

we may associate an integer $l = l(Z)$ with $1 \le l \le 2s$,

$$\pi_l \le \sqrt{(2s)}Z < \pi_{l+1}$$

(the upper bound is omitted if $l = 2s$). Just as in the proof of Lemma 9.2,

$$N(Z) \ll (\pi_1 \cdots \pi_l)^{-1}Z^l. \tag{15.33}$$

We need the inequality in the reverse direction, namely

$$N(Z) \gg (\pi_1 \cdots \pi_l)^{-1} Z^l. \tag{15.34}$$

Let $(\mathbf{u}^{(j)}, \mathbf{t}^{(j)}) = (u_1^{(j)}, \ldots, u_s^{(j)}, t_1^{(j)}, \ldots, t_s^{(j)})$ $(j = 1, \ldots, 2s)$ be linearly independent points in $\mathbb{Z}^{2s}$ such that

$$(\zeta_1(\mathbf{u}^{(j)}, \mathbf{t}^{(j)}), \ldots, \zeta_{2s}(\mathbf{u}^{(j)}, \mathbf{t}^{(j)}))$$

lies in the closed sphere of centre $\mathbf{0}$ and radius π_j. Thus

$$|\mathbf{u}^{(j)}| \leq \pi_j A, \qquad |\mathcal{L}(\mathbf{u}^{(j)}) - \mathbf{t}^{(j)}| \leq \pi_j A^{-1} \tag{15.35}$$

for $j = 1, \ldots, 2s$. Consider the points $\mathbf{u}$ in $\mathbb{Z}^s$ given by

$$\mathbf{u} = m_1 \mathbf{u}^{(1)} + \cdots + m_l \mathbf{u}^{(l)} \tag{15.36}$$

where $m_1, \ldots, m_l$ take all integer values satisfying

$$|m_j| < (1/(2l)) Z \pi_j^{-1} \qquad (j = 1, \ldots, l). \tag{15.37}$$

By (15.35), every such point satisfies

$$|\mathbf{u}| < \tfrac{1}{2} Z A, \qquad |\mathcal{L}(\mathbf{u}) - (m_1 \mathbf{t}^{(1)} + \cdots + m_l \mathbf{t}^{(l)})| < \tfrac{1}{2} Z A^{-1}.$$

Thus $N(Z)$ is at least equal to the number of distinct points $\mathbf{u}$ so obtained. If two of the points $\mathbf{u}$ coincided, then their difference, which is given by (15.36) and (15.37) without the factor $\tfrac{1}{2}$ in (15.37), would be $\mathbf{0}$. This would imply

$$|m_1 \mathbf{t}^{(1)} + \cdots + m_l \mathbf{t}^{(l)}| < Z A^{-1}$$

and so

$$m_1 \mathbf{t}^{(1)} + \cdots + m_l \mathbf{t}^{(l)} = \mathbf{0}.$$

But this, together with $\mathbf{u} = \mathbf{0}$, would contradict the linear independence of $(\mathbf{u}^{(1)}, \mathbf{t}^{(1)}), \ldots, (\mathbf{u}^{(2s)}, \mathbf{t}^{(2s)})$. Thus the points $\mathbf{u}$ constructed above are distinct, and (15.34) follows.

Now to deduce (15.31). Suppose first that $\sqrt{(2s)} Z_1 \geq \pi_1$; we determine $l(Z_1) = l$ as above, and similarly define $l(Z_2) = h$, say. Then $h \geq l$. By (15.33) applied to Z_2 and (15.34) applied to Z_1, we have

$$\frac{N(Z_2)}{N(Z_1)} \ll \frac{Z_2^h}{Z_1^l \pi_{l+1} \cdots \pi_h} \ll \left(\frac{Z_2}{Z_1}\right)^h \ll \left(\frac{Z_2}{Z_1}\right)^s.$$

(The last inequality is obvious if $h \leq s$. If $h > s$, then we omit the second inequality and use (15.32):

$$Z_2^h < Z_2^s, \qquad Z_1^l \pi_{l+1} \cdots \pi_h \gg Z_1^l \pi_{l+1} \cdots \pi_s \geq Z_1^s,$$

so that (15.31) does hold.)

Now suppose that $\sqrt{(2s)} Z_2 \geq \pi_1$, $\sqrt{(2s)} Z_1 < \pi_1$. Then $N(Z_1) = 1$, and with h as above,

$$\frac{N(Z_2)}{N(Z_1)} \ll \frac{Z_2^h}{\pi_1 \cdots \pi_h} \ll \left(\frac{Z_2}{Z_1}\right)^s$$

much as before. Finally, if $\sqrt{(2s)}Z_2 < \pi_1$, then $N(Z_2) = N(Z_1) = 1$ and (15.31) is obvious. This completes the proof of Lemma 15.10.

Corollary 1. *Suppose that $1 < R \leq P < J$, and let*

N be the number of $\mathbf{x}$ with $|\mathbf{x}| \leq P$, $\|\mathcal{L}_i(\mathbf{x})\| \leq J^{-1}$ $(i = 1, \ldots, s)$,

N' be the number of $\mathbf{x}$ with $|\mathbf{x}| \leq R$, $\|\mathcal{L}_i(\mathbf{x})\| \leq J^{-1}RP^{-1}$ $(i = 1, \ldots, s)$.

Then

$$N' \gg (R/P)^s N. \tag{15.38}$$

Proof. Apply the lemma with $A = (PJ)^{1/2}$, $Z_2 = (P/J)^{1/2}$, $Z_1 = R/A$.

Corollary 2. *Suppose that $6 < R \leq J \leq P$. Define N as in Corollary 1, and let*

N'' be the number of $\mathbf{x}$ with $|\mathbf{x}| \leq R$, $\|\mathcal{L}_i(\mathbf{x})\| \leq J^{-2}R$ $(i = 1, \ldots, s)$.

Then

$$N'' \gg (R/P)^s N.$$

Proof. Divide the cube $|\mathbf{x}| \leq P$ into $O((P/J)^s)$ cubes of side $\leq J/3$. One of these subcubes will contain $\gg (J/P)^s N$ points $\mathbf{x}$ with $\|\mathcal{L}_i(\mathbf{x})\| \leq J^{-1}$ $(i = 1, \ldots, s)$. If $\mathbf{x}^*$ is a fixed one of these points, then $\mathbf{y} = \mathbf{x} - \mathbf{x}^* \in \frac{1}{3}J\mathscr{C}$ and $\|\mathcal{L}_i(\mathbf{y})\| \leq (J/2)^{-1}$. By Corollary 1, applied with $\frac{1}{6}R$, $\frac{1}{3}J$, $\frac{1}{2}J$ in place of R, P, J, we find that the number of $\mathbf{x}$ with $|\mathbf{x}| \leq R$ and $\|\mathcal{L}_i(\mathbf{x})\| \leq J^{-2}R$ $(i = 1, \ldots, s)$ is

$$\gg (R/(2J))^s (J/P)^s N \gg (R/P)^s N.$$

Lemma 15.11. *Make the same assumptions as in Lemma 15.9. Suppose $\eta > 0$, and*

$$\eta + 4\theta \leq 1. \tag{15.39}$$

Then the number $N(\eta)$ of $(d-1)$-tuples

$$\mathbf{x}_1, \ldots, \mathbf{x}_{d-1} \text{ in } P^\eta \mathscr{C}$$

with

$$\|q\mathscr{F}^{(d)}(\mathbf{x}_1 | \cdots | \mathbf{x}_{d-1} | \mathbf{e}_i)\| \leq P^{-d+4\theta+(d-1)\eta}$$ $(i = 1, \ldots, s)$

is

$$\gg P^{s(d-1)\eta - 2^{d-1}H - \varepsilon}$$

with the constant in $\gg$ dependent only on c, s, k, η and ε.

Proof. We may assume P is so large that $P^\eta > 12$. For fixed $\mathbf{x}_2, \ldots, \mathbf{x}_{d-1}$, let

$$\mathscr{L}_i(\mathbf{X}) = q\mathscr{F}^{(d)}(\mathbf{X} \mid \mathbf{x}_2 \mid \cdots \mid \mathbf{x}_{d-1} \mid \mathbf{e}_i).$$

Put $J = P^{1-2\theta}$, $R = \frac{1}{2}P^\eta$, and let $N_1(\mathbf{x}_2, \ldots, \mathbf{x}_{d-1})$ be the number of

$$\mathbf{x}_1 \in P\mathscr{C} \quad \text{with} \quad \|\mathscr{L}_i(\mathbf{x}_1)\| \leq J^{-1} \quad (i = 1, \ldots, s).$$

By Lemma 15.9,

$$\sum_{\mathbf{x}_2 \in P\mathscr{C}} \cdots \sum_{\mathbf{x}_{d-1} \in P\mathscr{C}} N_1(\mathbf{x}_2, \ldots, \mathbf{x}_{d-1}) \gg P^{s(d-1)-2^{d-1}H-\varepsilon}.$$

Let $N_1''(\mathbf{x}_2, \ldots, \mathbf{x}_{d-1})$ be the number of

$$\mathbf{x}_1 \in P^\eta\mathscr{C} \quad \text{with} \quad \|\mathscr{L}_i(\mathbf{x}_1)\| \leq J^{-2}R = \frac{1}{2}P^{-2+4\theta+\eta} \quad (i = 1, \ldots, s).$$

We infer from Corollary 2 that

$$N_1''(\mathbf{x}_2, \ldots, \mathbf{x}_{d-1}) \gg P^{(\eta-1)s}N_1(\mathbf{x}_2, \ldots, \mathbf{x}_{d-1}).$$

Therefore the number of $(d-1)$-tuples $\mathbf{x}_1, \ldots, \mathbf{x}_{d-1}$ with

$$\mathbf{x}_1 \in P^\eta\mathscr{C}, \mathbf{x}_2 \in P\mathscr{C}, \ldots, \mathbf{x}_{d-1} \in P\mathscr{C}$$

and with

$$\|q\mathscr{F}^{(d)}(\mathbf{x}_1 \mid \cdots \mid \mathbf{x}_{d-1} \mid \mathbf{e}_i)\| \leq \frac{1}{2}P^{-2+4\theta+\eta} \quad (i = 1, \ldots, s)$$

is

$$\gg P^{s(d-2+\eta)-2^{d-1}H-\varepsilon}.$$

Next, for fixed $\mathbf{x}_1, \mathbf{x}_3, \ldots, \mathbf{x}_{d-1}$, let

$$\mathscr{L}_i(\mathbf{X}) = q\mathscr{F}^{(d)}(\mathbf{x}_1 \mid \mathbf{X} \mid \mathbf{x}_3 \mid \cdots \mid \mathbf{x}_{d-1} \mid \mathbf{e}_i) \quad (i = 1, \ldots, s).$$

Put $J = 2P^{2-4\theta-\eta}$, $R = P^\eta$ and let $N_2(\mathbf{x}_1, \mathbf{x}_3, \ldots, \mathbf{x}_{d-1})$ be the number of

$$\mathbf{x}_2 \in P\mathscr{C} \quad \text{with} \quad \|\mathscr{L}_i(\mathbf{x}_2)\| \leq J^{-1} \quad (i = 1, \ldots, s).$$

We have just seen that

$$\sum_{\mathbf{x}_1 \in P^\eta\mathscr{C}} \sum_{\mathbf{x}_3 \in P\mathscr{C}} \cdots \sum_{\mathbf{x}_{d-1} \in P\mathscr{C}} N_2(\mathbf{x}_1, \mathbf{x}_3, \ldots, \mathbf{x}_{d-1}) \gg P^{s(d-2+\eta)-2^{d-1}H-\varepsilon}.$$

Let $N_2'(\mathbf{x}_1, \mathbf{x}_3, \ldots, \mathbf{x}_{d-1})$ be the number of

$$\mathbf{x}_2 \in P^\eta\mathscr{C} \quad \text{with} \quad \|\mathscr{L}_i(\mathbf{x}_2)\| \leq J^{-1}RP^{-1} = \frac{1}{2}P^{-3+4\theta+2\eta} \quad (i = 1, \ldots, s).$$

We infer from Corollary 1 that

$$N_2'(\mathbf{x}_1, \mathbf{x}_3, \ldots, \mathbf{x}_{d-1}) \gg P^{(\eta-1)s}N_2(\mathbf{x}_1, \mathbf{x}_3, \ldots, \mathbf{x}_{d-1}).$$

Hence the number of $(d-1)$-tuples

$$\mathbf{x}_1 \in P^\eta\mathscr{C}, \mathbf{x}_2 \in P^\eta\mathscr{C}, \mathbf{x}_3 \in P\mathscr{C}, \ldots, \mathbf{x}_{d-1} \in P\mathscr{C}$$

with

$$\|q\mathscr{F}^{(d)}(\mathbf{x}_1\,|\,\cdots\,|\,\mathbf{x}_{d-1}\,|\,\mathbf{e}_i)\| \leqq \tfrac{1}{2}P^{-3+4\theta+2\eta} \qquad (i=1,\ldots,s)$$

is

$$\gg P^{s(d-3+2\eta)-2^{d-1}H-\varepsilon}.$$

Continuing with this process, considering $\mathbf{x}_3,\ldots,\mathbf{x}_{d-1}$ in turn, and applying Corollary 1 with $R=P^\eta$ each time, we finally reach the desired conclusion.

15.8 Systems of forms

Let $\mathscr{F}=(\mathscr{F}^{(k)},\ldots,\mathscr{F}^{(2)})$ be a system of forms as in (15.1), and let

$$\boldsymbol{\alpha}=(\boldsymbol{\alpha}^{(k)},\ldots,\boldsymbol{\alpha}^{(2)})\in\mathbb{R}^r,$$

where $\boldsymbol{\alpha}^{(d)}$ has r_d components. Further let $\mathscr{M}_d=\mathscr{M}(\mathscr{F}^{(d)})$ $(2\leqq d\leqq k)$ be the algebraic set of Section 15.3.

Lemma 15.12. *Suppose that $H>0$, $\varepsilon>0$ and that $2\leqq d\leqq k$. Suppose that either $d=k$, in which case set $\theta=0$ and $q=1$. Or else, suppose that $2\leqq d<k$ with $r_d>0$, that $0\leqq\theta<\tfrac{1}{4}$, and there is a natural number q with*

$$q\leqq P^\theta \qquad \text{and} \qquad \|q\boldsymbol{\alpha}^{(j)}\|\leqq cP^{\theta-j} \quad \text{for } d<j\leqq k. \qquad (15.40)$$

Given a box $\mathscr{B}$ with sides $\leqq1$, and given $P>1$, define $S(\boldsymbol{\alpha})$ by (15.4).

Let $\eta>0$ with (15.39); then one of the following three alternatives must hold. Either

(i) $|S(\boldsymbol{\alpha})|\leqq P^{s-H}$, *or*
(ii) *there is a natural number n with*

$$n\ll P^{r_d(d-1)\eta} \qquad \text{and} \qquad \|nq\boldsymbol{\alpha}^{(d)}\|\ll P^{-d+4\theta+r_d(d-1)\eta}, \qquad \text{or}$$

(iii) $z_R(\mathscr{M}_d)\gg R^{(d-1)s-2^{d-1}(H/\eta)-\varepsilon}$

holds with $R=P^\eta$. The constants in $\ll$ and $\gg$ here depend only on c, s, k, $r_k,\ldots,r_2$, η, ε and $\mathscr{F}$, and hence only on c, $\mathscr{F}$, η, ε.

Proof. We have $\boldsymbol{\alpha}\mathscr{F}=\mathscr{F}^{(2)}+\cdots+\mathscr{F}^{(k)}$ with $\mathscr{F}^{(d)}=\boldsymbol{\alpha}^{(d)}\mathscr{F}^{(d)}$. In the case when $2\leqq d<k$, the hypothesis (15.40) implies that $\|q\mathscr{F}^{(j)}\|\ll P^{\theta-j}$ for each j in $d<j\leqq k$, with a constant in $\ll$ which depends only on c and $\mathscr{F}^{(j)}$. We may apply Lemma 15.11 when (i) fails. The number $N(\eta)$ of integer $(d-1)$-tuples $\mathbf{x}_1,\ldots,\mathbf{x}_{d-1}$ in $R\mathscr{C}$ with

$$\|q\boldsymbol{\alpha}^{(d)}\mathscr{F}^{(d)}(\mathbf{x}_1\,|\,\cdots\,|\,\mathbf{x}_{d-1}\,|\,\mathbf{e}_i)\|\leqq P^{-d+4\theta+(d-1)\eta} \qquad (i=1,\ldots,s)$$

$$(15.41)$$

satisfies

$$N(\eta)\gg R^{s(d-1)-2^{d-1}(H/\eta)-\varepsilon}.$$

Suppose $\mathscr{F}^{(d)} = (\mathscr{F}_1^{(d)}, \ldots, \mathscr{F}_{r_d}^{(d)})$. Given $\mathbf{x}_1, \ldots, \mathbf{x}_{d-1}$ as above, form the matrix

$$(m_{ij}) = (\mathscr{F}_j^{(d)}(\mathbf{x}_1 | \cdots | \mathbf{x}_{d-1} | \mathbf{e}_i)) \qquad (1 \leq i \leq s, \ 1 \leq j \leq r_d).$$

Now if this matrix has rank less than r_d for each of the $(d-1)$-tuples counted by $N(\eta)$, then clearly alternative (iii) holds. Hence we may suppose that at least one of these matrices has rank r_d. We may suppose without loss of generality that the matrix with $1 \leq i \leq r_d$ is non-singular. Write n for the absolute value of the determinant of this submatrix. We have

$$m_{ij} \ll R^{d-1},$$

and hence

$$n \ll R^{r_d(d-1)} = P^{r_d(d-1)\eta}.$$

From (15.41) we have

$$q \sum_{j=1}^{r_d} \alpha_j^{(d)} m_{ij} = b_i + \rho_i \qquad (1 \leq i \leq s),$$

where the b_i are integers and the ρ_i are bounded in modulus by the right-hand side of (15.41). Let $a_1, \ldots, a_{r_d}$ be the solution of the system of linear equations

$$\sum_{j=1}^{r_d} a_j m_{ij} = n b_i \qquad (1 \leq i \leq r_d). \tag{15.42}$$

Then

$$\sum_{j=1}^{r_d} (qn\alpha_j^{(d)} - a_j) m_{ij} = n\rho_i \qquad (1 \leq i \leq r_d). \tag{15.43}$$

Cramer's rule applied to (15.42) shows that the a_j are integers, and applied to (15.43) it shows that

$$\|qn\alpha_j^{(d)}\| \leq |qn\alpha_j^{(d)} - a_j| \ll R^{(d-1)(r_d-1)} P^{-d+4\theta+(d-1)\eta} = P^{-d+4\theta+(d-1)r_d\eta}.$$

The proof of Lemma 15.12 is complete.

For d with $r_d > 0$, define $g_d = g(\mathscr{F}^{(d)})$ and γ_d as in Section 15.3, and put

$$\gamma_d' = 2^{d-1}/g_d = \gamma_d/((d-1)r_d). \tag{15.44}$$

Let us suppose ε is sufficiently small. We apply Lemma 15.12 with ε^3 in place of ε. The third alternative of Lemma 15.12 may not happen for large P if $(2^{d-1}H/\eta) + \varepsilon^3 < g_d - \varepsilon^3$. In particular it may not happen with $\eta = H\gamma_d' + \varepsilon^{5/2}$. The condition (15.39) is fulfilled when $4\theta + H\gamma_d' < 1$.

Corollary. *Let* $\mathscr{F} = (\mathscr{F}^{(k)}, \ldots, \mathscr{F}^{(2)})$, $\alpha = (\alpha^{(k)}, \ldots, \alpha^{(2)})$ *and* P *be as* *above. Suppose that either $d = k$, in which case set $\theta = 0$, $q = 1$. Or suppose that $2 \leqq d < k$, that $r_d > 0$, that $0 \leqq \theta < \frac{1}{4}$, and that there is a natural number q with* (15.40). *Suppose that $\varepsilon > 0$, and that $H > 0$ satisfies*

$$4\theta + H\gamma'_d < 1. \tag{15.45}$$

Then either

(i) $|S(\boldsymbol{\alpha})| \leqq P^{s-H}$, *or*

(ii) *there is a natural number n with*

$$n \ll P^{H\gamma_d + \varepsilon^2} \qquad and \qquad \|nq\boldsymbol{\alpha}^{(d)}\| \ll P^{-d + 4\theta + H\gamma_d + \varepsilon^2}.$$

The constant in $\ll$ depends only on c, $\mathcal{F}$, H, η and ε.

In particular, when $H\gamma'_k < 1$ and when

$$|S(\boldsymbol{\alpha})| > P^{s-H},$$

there is an n_k with

$$n_k \ll P^{H\gamma_k + \varepsilon^2} \qquad and \qquad \|n_k\boldsymbol{\alpha}^{(k)}\| \ll P^{-k + H\gamma_k + \varepsilon^2}. \tag{15.46}$$

Suppose now that $r_{k-1} > 0$ and that $4H\gamma_k + H\gamma'_{k-1} < 1 - \varepsilon$. Assuming ε is sufficiently small, we apply the corollary with $d = k - 1$, $\theta = H\gamma_k + \varepsilon^2$ and $q = n_k$. We infer that there is a natural number n_{k-1} with

$$n_{k-1} \ll P^{H\gamma_{k-1} + \varepsilon^2} \qquad and \qquad \|n_k n_{k-1}\boldsymbol{\alpha}^{(k-1)}\| \ll P^{-(k-1) + 4H\gamma_k + H\gamma_{k-1} + \delta}$$

$$\tag{15.47}$$

where δ is a multiple of ε^2. In the case when $r_{k-1} = 0$ we have $\gamma_{k-1} = 0$, and (15.47) is trivially satisfied with $n_{k-1} = 1$. Now when $4^2 H\gamma_k + 4H\gamma_{k-1} + H\gamma'_{k-2} < 1$, the argument may be repeated. Ultimately we obtain

Lemma 15.13. *Put*

$$\tau_d = \gamma_d + 4\gamma_{d+1} + \cdots + 4^{k-d}\gamma_k \qquad (2 \leqq k \leqq d).$$

Suppose that $\varepsilon > 0$, and that $H > 0$ with

$$H\tau_2 < 1. \tag{15.48}$$

Given $\mathcal{F}$, $\boldsymbol{\alpha}$ and P as above, we have either

(i) $|S(\boldsymbol{\alpha})| \leqq P^{s-H}$, *or*

(ii) *there are natural numbers n_k, n_{k-1}, $\ldots$, n_2 with*

$$n_d \leqq P^{H\gamma_d + \varepsilon}$$

and

$$\|n_k \cdots n_d\boldsymbol{\alpha}^{(d)}\| \leqq P^{-d + H\tau_d + \varepsilon} \qquad (2 \leqq d \leqq k). \tag{15.49}$$

The implied constants depend only on H, $\mathcal{F}$ and ε.

Proof of Theorem 15.2. We may suppose without loss of generality that $\mathscr{B}$ has sides $\leqq 1$. We apply the corollary to Lemma 15.12 with $d = k$. We suppose (15.11) to hold, so that

$$\Omega\gamma_d < 1.$$

We set $H = \Delta\Omega$, so that $H\gamma_d + \varepsilon < \Delta$ when $\varepsilon > 0$ is small. Thus when P is large, say when $P \geqq C_5(\mathscr{F}, \Omega, \Delta)$, then provided (15.45) holds, either

(i) $|S(\boldsymbol{\alpha})| \leqq P^{s-\Delta\Omega}$, or
(ii) there is a q with

$$q \leqq P^\Delta \quad \text{and} \quad \|q\boldsymbol{\alpha}\| \leqq P^{-d+\Delta}.$$

When (15.45) fails, we have $H\gamma_d' \geqq 1$, that is, $\Omega\Delta\gamma_d' \geqq 1$. This implies that

$$\Delta > (d-1)r.$$

But in this case (ii) is always true by Dirichlet's rational approximation theorem, since

$$\Delta(r+1)/r > (d-1)(r+1) \geqq d, \quad \text{that is,} \quad -\Delta/r < -d + \Delta.$$

Proof of Theorem 15.3. We apply Lemma 15.13 with $H = \Delta\Omega$. In view of (15.15) we have $\Omega\tau < 1$. Setting $q = n_k n_{k-1} \cdots n_2$ we have

$$q \ll P^{H\tau+\varepsilon} \quad \text{and} \quad \|q\boldsymbol{\alpha}^{(d)}\| \ll P^{-d+H\tau+\varepsilon} \quad (2 \leqq d \leqq k),$$

since $\tau_d + \gamma_{d-1} + \cdots + \gamma_2 \leqq \tau_2 = \tau$ $(2 \leqq d \leqq k)$. Here $H\tau + \varepsilon = \Delta\Omega\tau + \varepsilon < \Delta$ if $\varepsilon > 0$ is sufficiently small. As for (15.48), this becomes $\Delta\Omega\tau < 1$, which is true for $\Delta \leqq 1$. Thus, as in the proof of Theorem 15.2, either (i) or (ii) holds. This completes the proof of Theorem 15.3.

16
The invariants g and h

16.1 Invariants $g_{\mathbb{C}}$ and $h_{\mathbb{C}}$

In this chapter we find a relationship between the invariants $g(\mathcal{F})$ and $h(\mathcal{F})$ introduced in Chapter 15. The most useful of our results from the point of view of arithmetic application is

Theorem 16.1. *Let $\mathcal{F} = (\mathcal{F}_1, \ldots, \mathcal{F}_r)$ be a system of r forms of degree $d \geqq 2$ with rational coefficients. Then*

$$h(\mathcal{F}) \leqq \Phi(d)(g(\mathcal{F}) + (d-1)r(r-1)).$$

Here $\Phi(d)$ is a certain function with $\Phi(2) = \Phi(3) = 1$, $\Phi(4) = 3$, $\Phi(5) = 13$, and $\Phi(d) < (\log 2)^{-(d-2)}(d-2)!$ in general.

The exact definition of Φ will be given in Section 16.7. We note the following corollaries, which are needed for the proof of results on Diophantine equations such as Theorem 15.1.

Corollary 1. *The conclusion of Theorem 15.2 remains valid if the hypothesis (15.11) is replaced by:*

$$0 < \Omega < \frac{h(\mathcal{F})}{\Phi(d)2^{d-1}(d-1)r} - \frac{(r-1)}{2^{d-1}}.$$

This result is, of course, vacuous unless $h(\mathcal{F}) > r(r-1)\Phi(d)(d-1)$.

Corollary 2. *The conclusion of Theorem 15.3 remains valid if*

$$h_d = h(\mathcal{F}^{(d)}) > r_d(r_d - 1)\Phi(d)(d-1)$$

whenever $2 \leqq d \leqq k$, $r_d > 0$, and if the hypothesis (15.15) is replaced by

$$0 < \Omega < \sigma^{-1}.$$

Here

$$\sigma = \sum_{\substack{d=2 \\ r_d \neq 0}}^{k} \frac{2^{3d-3}(d-1)r_d\Phi(d)}{h_d - (d-1)r_d(r_d - 1)\Phi(d)}.$$

To get Corollary 2, for instance, we simply deduce from Theorem 16.1 that $\sigma \geqq \tau$ in the notation of (15.14).

In this section we will introduce quantities g_C and h_C which are easier to handle than g and h.

If $\mathscr{F}$ is a form of degree $d > 1$ with complex coefficients, let $h_C = h_C(\mathscr{F})$ be the least number h such that $\mathscr{F}$ may be written in the form

$$\mathscr{F} = \mathscr{A}_1 \mathscr{B}_1 + \cdots + \mathscr{A}_h \mathscr{B}_h, \tag{16.1}$$

where $\mathscr{A}_i$, $\mathscr{B}_i$ are forms of positive degrees with *complex* coefficients. Given an r-tuple $\mathscr{F}$ of forms of degree d, let $h_C(\mathscr{F})$ be the minimum of $h_C(\mathscr{F})$ over all forms of the complex pencil of $\mathscr{F}$. Define the algebraic set $\mathscr{M} = \mathscr{M}(\mathscr{F})$ as in Section 15.3, and put

$$g_C = \operatorname{codim} \mathscr{M}.$$

Lemma 16.1. $g_C \leqq 2^{d-1} h_C$.

Proof. We may suppose that $\mathscr{F} = (\mathscr{F}_1, \ldots, \mathscr{F}_r)$ and that $\mathscr{F}_1$ may be written as in (16.1) with $h = h_C$. Write $e(i) = \deg \mathscr{A}_i$, $f(i) = \deg \mathscr{B}_i$, so that $e(i) + f(i) = d$ $(1 \leqq i \leqq h_C)$. It is easily seen that, writing

$$\mathscr{G}_i = \mathscr{A}_i \mathscr{B}_i,$$

the corresponding multilinear forms are related by

$$\mathscr{G}_i(\mathbf{X}_1 \,|\, \cdots \,|\, \mathbf{X}_d) = \sum \mathscr{A}_i(\mathbf{X}_{j_1} \,|\, \cdots \,|\, \mathbf{X}_{j_{e(i)}}) \mathscr{B}_i(\mathbf{X}_{k_1} \,|\, \cdots \,|\, \mathbf{X}_{k_{f(i)}})$$

where the sum is over all decompositions of $1, \ldots, d$ into disjoint sets $\{j_1 < \cdots < j_{e(i)}\}$ and $\{k_1 < \cdots < k_{f(i)}\}$. After all, the right-hand side is multilinear and symmetric in $\mathbf{X}_1, \ldots, \mathbf{X}_d$, and reduces to

$$\sum (-1)^{e(i)+f(i)} e(i)! \, (d - e(i))! \, \mathscr{G}_i(\mathbf{X}) = (-1)^d d! \, \mathscr{G}_i(\mathbf{X})$$

when $\mathbf{X}_1 = \cdots = \mathbf{X}_d = \mathbf{X}$.

Thus a point $(\mathbf{x}_1, \ldots, \mathbf{x}_{d-1})$ will certainly lie in $\mathscr{M}$ if

$$\mathscr{A}_i(\mathbf{x}_{j_1} \,|\, \cdots \,|\, \mathbf{x}_{j_{e(i)}}) = 0$$

for $1 \leqq i \leqq h_C$ and any $1 \leqq j_1 < \cdots < j_{e(i)} \leqq d - 1$, and if furthermore

$$\mathscr{B}_i(\mathbf{x}_{k_1} \,|\, \cdots \,|\, \mathbf{x}_{k_{f(i)}}) = 0$$

for $1 \leqq i \leqq h_C$ and any $1 \leqq k_1 < \cdots < k_{f(i)} \leqq d - 1$. The number of all these equations is

$$\sum_{i=1}^{h_C} \left(\binom{d-1}{e(i)} + \binom{d-1}{f(i)} \right) \leqq 2^{d-1} h_C.$$

An analogous inequality in the reverse direction is the following proposition, which is very difficult to prove.

Proposition 16.1. *For a single form $\mathscr{F}$ of degree $d > 1$,*

$$h_{\mathbb{C}}(\mathscr{F}) \leqq \Phi(d) g_{\mathbb{C}}(\mathscr{F})$$

with $\Phi(d)$ the function of Theorem 16.1.

The proof will occupy Sections 16.3–16.7.

Corollary. *For a system $\mathscr{F}$ of r forms of degree $d > 1$,*

$$h_{\mathbb{C}}(\mathscr{F}) \leqq \Phi(d)(g_{\mathbb{C}}(\mathscr{F}) + r - 1).$$

Proof. Since $\mathcal{M}(\mathscr{F})$ is the union of $\mathcal{M}(\mathscr{F})$ over all the forms $\mathscr{F}$ of the pencil, and since $\mathcal{M}(\lambda \mathscr{F}) = \mathcal{M}(\mathscr{F})$ for $\lambda \neq 0$, there is some $\mathscr{F}$ in the pencil with

$$\dim \mathcal{M}(\mathscr{F}) \geqq \dim \mathcal{M}(\mathscr{F}) - (r - 1),$$

or

$$g_{\mathbb{C}}(\mathscr{F}) = \operatorname{codim} \mathcal{M}(\mathscr{F}) \leqq g_{\mathbb{C}}(\mathscr{F}) + r - 1.$$

Thus

$$h_{\mathbb{C}}(\mathscr{F}) \leqq h_{\mathbb{C}}(\mathscr{F}) \leqq \Phi(d) g_{\mathbb{C}}(\mathscr{F}) \leqq \Phi(d)(g_{\mathbb{C}}(\mathscr{F}) + r - 1).$$

16.2 The arithmetical case

Now let $\mathscr{F}$ be an r-tuple of forms of degree $d > 1$ with rational coefficients. It is easily seen that

$$h_{\mathbb{C}}(\mathscr{F}) \leqq h(\mathscr{F}), \qquad g_{\mathbb{C}}(\mathscr{F}) \leqq g(\mathscr{F}).$$

The proof of Lemma 16.1 does not seem to have an analogue in the 'arithmetical case', i.e. an upper bound for $g(\mathscr{F})$ in terms of $h(\mathscr{F})$. Fortunately, this is not important for the applications we are concerned with. However, the analogue of Proposition 16.1 holds. That is, for a single form $\mathscr{F}$,

$$h(\mathscr{F}) \leqq \Phi(d)[g(\mathscr{F})]. \tag{16.2}$$

Proposition 16.2. *Suppose that $\mathscr{F}$ is a form of degree $d > 1$ with rational coefficients, and write $\mathcal{M} = \mathcal{M}(\mathscr{F})$. Suppose that for some $P > 1$,*

$$z_P(\mathcal{M}) > C_1 P^{s(d-1) - \gamma - 1}, \tag{16.3}$$

where $C_1 = C_1(d, s)$ is a suitable constant independent of $\mathscr{F}$, and where γ is an integer. Then

$$h(\mathscr{F}) \leqq \Phi(d)\gamma. \tag{16.4}$$

Now if $\gamma = [g(\mathscr{F})]$, then (16.3) is certainly true for some arbitrarily large values of P, so that (16.4) and indeed (16.2) holds.

We shall prove Proposition 16.2 in Section 16.8 by quite modest modifications to the proof of Proposition 16.1.

Once Proposition 16.2 is known, it is easy to deduce Theorem 16.1, as we now show. Put $\gamma = [g(\mathscr{F})] + (d-1)r(r-1)$, and choose $\varepsilon > 0$ with

$$\gamma + 1 > g(\mathscr{F}) + (d-1)r(r-1) + 2\varepsilon. \qquad (16.5)$$

By definition of $g(\mathscr{F})$, there are certain arbitrarily large values of P with

$$z_P(\mathcal{M}(\mathscr{F})) \gg P^{s(d-1)-g-\varepsilon}. \qquad (16.6)$$

Here, and for the remainder of this proof, the implied constants may depend on $\mathscr{F}$.

Suppose $(\mathbf{x}_1, \ldots, \mathbf{x}_{d-1})$ with $|\mathbf{x}_i| \leq P$ lies in $\mathcal{M}(\mathscr{F})$. The matrix (15.6) then has rank less than r. Thus there is a linear combination $\mathscr{F} = a_1 \mathscr{F}_1 + \cdots + a_r \mathscr{F}_r$ with (15.7). The coefficient a_i have to satisfy the system of linear equations

$$\sum_{j=1}^{r} m_{ij}a_j = 0 \qquad (i = 1, \ldots, s). \qquad (16.7)$$

The rank of the matrix (m_{ij}) is $\leq r-1$, and the entries m_{ij} are $\ll P^{d-1}$. Hence there is a non-trivial integer solution $\mathbf{a} = (a_1, \ldots, a_r)$ of (16.7) with

$$|\mathbf{a}| \ll P^{(d-1)(r-1)}.$$

The number of possibilities for such $\mathbf{a}$ is $\ll P^{(d-1)r(r-1)}$. Hence there is a form $\mathscr{F}$ in the rational pencil with

$$z_P(\mathcal{M}(\mathscr{F})) \gg P^{-(d-1)r(r-1)} z_P(\mathcal{M}(\mathscr{F})).$$

In conjunction with (16.5), (16.6) this gives

$$z_P(\mathcal{M}(\mathscr{F})) \gg P^{s(d-1)-\gamma-1+\varepsilon},$$

and hence gives (16.3) when P is large. Thus (16.4) holds, and further

$$h(\mathscr{F}) \leq h(\mathscr{F}) \leq \Phi(d)\gamma.$$

This completes the proof of Theorem 16.1.

In the next section we begin a rather elaborate construction leading to the proof of Proposition 16.1.

16.3 Simple points

Let

$$s = e + t.$$

Let V be an irreducible algebraic variety of dimension e in $\mathbb{C}^s$. Let $\mathscr{I}(V)$ be the ideal of polynomials $f(\mathbf{X}) \in \mathbb{C}(\mathbf{X}) = \mathbb{C}(X_1, \ldots, X_s)$ which vanish on V. We will write $\partial f / \partial X_i$ for the partial derivatives, and $\partial f / \partial x_i$ for the partial derivatives evaluated at a particular point $\mathbf{x}$. Given $\mathbf{x} \in V$, let $\text{Gr}(\mathbf{x})$ be the set of vectors

$$\text{grad} f(\mathbf{x}) = (\partial f / \partial x_1, \ldots, \partial f / \partial x_s)$$

where f runs through $\mathscr{I}(V)$. Then $\text{Gr}(\mathbf{x})$ is a vector space over $\mathbb{C}$. It is well known (see e.g. Lang (1958), Chapter VIII, Section 2) that

$$\dim \text{Gr}(\mathbf{x}) \leqq t. \tag{16.8}$$

Points with $\dim \text{Gr}(\mathbf{x}) = t$ are called *simple* or non-singular; points with $\dim \text{Gr}(\mathbf{x}) < t$ are called singular. Again, it is well known that the singular points form a proper algebraic subset of V.

Let $\mathbf{G} = (G_1, \ldots, G_s)$ be a new vector of variables. For $\mathbf{x} \in V$, let $H(\mathbf{x})$ be the set of linear forms

$$f^{(1)}(\mathbf{G}) = \sum_{i=1}^{s} \frac{\partial f}{\partial x_i} G_i.$$

Then $H(\mathbf{x})$ is a vector space isomorphic to $\text{Gr}(\mathbf{x})$. In view of (16.8), there are t linear forms (which depend on $\mathbf{x}$), say

$$k_1(\mathbf{G}), \ldots, k_t(\mathbf{G}),$$

which generate $H(\mathbf{x})$. Schmidt (1985) generalizes this fact to higher derivatives: see Lemma 16.3 below.

Lemma 16.2. *Suppose that $\mathbf{x}$ is a simple point on V. Suppose that, say, the vectors $(\partial f / \partial x_1, \ldots, \partial f / \partial x_t)$, where f ranges over $\mathscr{I}(V)$, contain t independent ones. Write $\mathbf{x} = (x_1, \ldots, x_t, y_1, \ldots, y_e)$. Then there exist unique formal power series*

$$\chi_i \in \mathbb{C}[[Y_1, \ldots, Y_e]] \qquad (i = 1, \ldots, t)$$

with constant term zero such that

$$f(x_1 + \chi_1(\mathbf{Y}), \ldots, x_t + \chi_t(\mathbf{Y}), y_1 + Y_1, \ldots, y_e + Y_e) = 0$$

for each $f \in \mathscr{I}(V)$.

Moreover, if V is defined over the rationals (that is, definable by polynomial equations with rational coefficients) and if $\mathbf{x}$ has rational components, then the series χ_i have rational coefficients.

Proof. Except possibly for the last assertion, this is well known (Lang (1958), Chapter VIII, Section 4). The uniqueness part of the lemma is obtained by a quite simple argument, which also gives the last assertion (about rational coefficients) without extra effort.

16.4 Higher derivatives

Again, let $V \subseteq \mathbb{C}^s$ be an irreducible variety of dimension e and of codimension t. Let $\mathbf{x} \in V$. Given $f \in \mathcal{I}(V)$ and given $n > 0$, write

$$f^{(n)}(\mathbf{G}) = \frac{1}{n!} \sum_{i_1=1}^{s} \cdots \sum_{i_n=1}^{s} \frac{\partial^n f}{\partial x_{i_1} \cdots \partial x_{i_n}} G_{i_1} \cdots G_{i_n}, \qquad (16.9)$$

so that $f^{(n)}$ is a form of degree n.

Lemma 16.3. *Suppose* $\mathbf{x}$ *is a simple point of V. Then there are forms*

$$k_1^{(p)}(\mathbf{G}), \ldots, k_t^{(p)}(\mathbf{G}) \qquad (p = 1, 2, \ldots)$$

which depend on $\mathbf{x}$, *and where $k_j^{(p)}$ is of degree p, such that for $f \in \mathcal{I}(V)$ and for $n = 1, 2, \ldots$ we have*

$$f^{(n)}(\mathbf{G}) = \sum_{p=1}^{n} \sum_{q=1}^{t} h_q^{(n-p)}(\mathbf{G}) k_q^{(p)}(\mathbf{G}) \qquad (16.10)$$

where $h_q^{(n-p)}$ is a form of degree $n-p$ which depends on f, as well as on $n-p$ and q.

Moreover, when V is defined over the rationals and when $\mathbf{x}$ is rational, then the $k_q^{(p)}$ have rational coefficients. When further f has rational coefficients, then so do the $h_q^{(n-p)}$.

Proof. We make the same conventions as in Lemma 16.2. In particular $\chi_1, \ldots, \chi_t$ will be the formal series of that lemma. We may suppose that $\mathbf{x} = \mathbf{0}$. Let $\mathcal{R}$ be the ring

$$\mathcal{R} = \mathbb{C}[\mathbf{X}][[\mathbf{Y}]] = \mathbb{C}[X_1, \ldots, X_t][[Y_1, \ldots, Y_e]]$$

consisting of formal series in $\mathbf{Y}$ whose coefficients are polynomials in $\mathbf{X}$. For $f = f(\mathbf{X}, \mathbf{Y})$ in $\mathcal{R}$ we have

$$f(\mathbf{X}, \mathbf{Y}) - f(\chi_1(\mathbf{Y}), X_2, \ldots, X_t, \mathbf{Y}) = (X - \chi_1(\mathbf{Y}))h_1(\mathbf{X}, \mathbf{Y})$$

with $h_1 \in \mathcal{R}$. Similarly,

$$f(\chi_1(\mathbf{Y}), X_2, \ldots, X_t, \mathbf{Y}) - f(\chi_1(\mathbf{Y}), \chi_2(\mathbf{Y}), X_3, \ldots, X_t, \mathbf{Y})$$
$$= (X_2 - \chi_2(\mathbf{Y}))h_2(\mathbf{X}, \mathbf{Y})$$

with $h_2 \in \mathcal{R}$. Continuing in this manner we get

$$f(\mathbf{X}, \mathbf{Y}) = (X_1 - \chi_1(\mathbf{Y}))h_1(\mathbf{X}, \mathbf{Y})$$
$$+ \cdots + (X_t - \chi_t(\mathbf{Y}))h_t(\mathbf{X}, \mathbf{Y}) + f(\chi(\mathbf{Y}), \mathbf{Y}).$$

In the case when $f \in \mathcal{I}(V)$, this becomes

$$f(\mathbf{X}, \mathbf{Y}) = \sum_{q=1}^{t} (X_q - \chi_q(\mathbf{Y}))h_q(\mathbf{X}, \mathbf{Y}). \qquad (16.11)$$

Now when $\mathbf{x} = \mathbf{0}$, then $f^{(n)}$ of (16.9) is just the form of degree n in the Taylor expansion

$$f(\mathbf{C}) = f(\mathbf{0}) + f^{(1)}(\mathbf{C}) + \cdots + f^{(n)}(\mathbf{C}) + \cdots$$

Clearly in this way $f^{(n)}$ may be defined for $f \in \mathcal{R}$, and not just for polynomials $f \in \mathbb{C}[X_1, \ldots, X_s] = \mathbb{C}[\mathbf{X}, \mathbf{Y}]$. Further, it is clear that when $f = uv$ with $u, v \in \mathcal{R}$, then

$$f^{(n)} = \sum_{p=0}^{n} u^{(p)} v^{(n-p)}$$

where $u^{(0)} = u(\mathbf{0})$, $v^{(0)} = v(\mathbf{0})$. Applying this remark to (16.11) we get

$$f^{(n)} = \sum_{q=1}^{t} \sum_{p=0}^{n} k_q^{(p)} h_q^{(n-p)},$$

where $k_q(\mathbf{X}, \mathbf{Y}) = X_q - \chi_q(\mathbf{Y})$. Note that k_q $(q = 1, \ldots, t)$ is independent of f. Since $k_q^{(0)} = 0$, formula (16.10) follows.

In the case when V is defined over the rationals and when $\mathbf{x}$ is rational, the series χ_i will have rational coefficients by Lemma 16.2, and hence so will the $k_q^{(p)}$. In the case when f has rational coefficients, we may work in the ring $\mathcal{R}_\mathbb{Q} = \mathbb{Q}[\mathbf{X}][[\mathbf{Y}]]$, and the series h_q will have rational coefficients.

16.5 Operators $\mathcal{D}_\tau$

Given a multilinear form $h(\mathbf{G}_1, \ldots, \mathbf{G}_q)$, and given a set

$$\rho = \{u_1 < \cdots < u_q\}$$

of positive integers, put

$$h(\mathbf{G}_\rho) = h(\mathbf{G}_{u_1}, \ldots, \mathbf{G}_{u_p}),$$

so that $h(\mathbf{G}_\rho)$ is a multilinear form in vectors $\mathbf{G}_{u_1}, \ldots, \mathbf{G}_{u_q}$. Given a form $g(\mathbf{G})$ of degree n, define the multilinear form $g(\mathbf{G}_1 | \cdots | \mathbf{G}_n)$ as in Section 16.4. When $g(\mathbf{G}) = h(\mathbf{G})k(\mathbf{G})$ with forms h, k of respective degrees $n - p$, p, then

$$g(\mathbf{G}_1 | \cdots | \mathbf{G}_n) = \sum_{\rho, \sigma} h(\mathbf{G}_\rho) k(\mathbf{G}_\sigma) \qquad (16.12)$$

where the sum is over the partitions of $\{1, \ldots, n\}$ into subsets ρ, σ with respective cardinalities $n - p$, p (see the proof of Lemma 16.1). In particular, it follows from (16.10) that

$$f^{(n)}(\mathbf{G}_1 | \cdots | \mathbf{G}_n) = \sum_{p=1}^{n} \sum_{\rho, \sigma} \sum_{q=1}^{t} h_q^{(n-p)}(\mathbf{G}_\rho) k_q^{(p)}(\mathbf{G}_\sigma). \qquad (16.13)$$

Now suppose that V is an irreducible variety of codimension t in $\mathbb{C}^{(d-2)s}$; this is what we shall be concerned with in proving Proposition 16.1. We consider polynomials $f(\mathbf{X}) = f(\mathbf{X}_1, \ldots, \mathbf{X}_{d-2})$ where each $\mathbf{X}_i$

has s components. Given a simple point $\mathbf{x} = (\mathbf{x}_1, \ldots, \mathbf{x}_{d-2})$ of V, we may apply everything we have found out in Section 16.4. Forms $f^{(n)}(\mathbf{G})$ as in (16.9) now have $\mathbf{G} = (\mathbf{G}_1, \ldots, \mathbf{G}_{d-2})$; similarly for multilinear forms $f^{(n)}(\mathbf{G}_1 | \cdots | \mathbf{G}_n)$.

Given a subset

$$\tau \subseteq \{1, \ldots, d-2\}$$

we introduce an operator $\mathscr{D}_\tau(\mathbf{G}_\tau)$ as follows. The operator acts on polynomials $f(\mathbf{X})$. When $\tau = \varnothing$, then $\mathscr{D}_\varnothing f = f(\mathbf{x})$, i.e. one substitutes $\mathbf{x}$ for $\mathbf{X}$. When $\tau = \{u_1, \ldots, u_n\}$, then

$$\mathscr{D}_\tau(\mathbf{G}_\tau)f = (-1)^n f^{(n)}(\mathbf{E}_{u_1} | \cdots | \mathbf{E}_{u_n}), \tag{16.14}$$

where

$$\mathbf{E}_l = (0, 0, \ldots, 0, \mathbf{G}_l, 0, \ldots, 0) \qquad (1 \leq l \leq d-2). \tag{16.15}$$

$\overset{\longleftarrow l \longrightarrow}{}$

Thus

$$\mathscr{D}_\tau(\mathbf{G}_\tau)f = \sum_{i_1=1}^{s} \cdots \sum_{i_n=1}^{s} G_{u_1 i_1} \cdots G_{u_n i_n} \frac{\partial^n f}{\partial x_{u_1 i_1} \cdots \partial x_{u_n i_n}}. \tag{16.16}$$

Substituting (16.13) we get (for $f \in \mathscr{I}(V)$ and $\mathbf{x}$ simple on V) that

$$\mathscr{D}_\tau(\mathbf{G}_\tau)f = (-1)^n \sum_{p=1}^{n} \sum_{\sigma,\rho} \sum_{q=1}^{t} h_q^{(n-p)}(\mathbf{E}_\rho) k_q^{(p)}(\mathbf{E}_\sigma),$$

where the sum is over partitions of τ into subsets ρ, σ of cardinalities $|\rho| = n-p$, $|\sigma| = p$. Put

$$h_{q\rho}^{(n-p)}(\mathbf{G}_\rho) = (-1)^n h_q^{(n-p)}(\mathbf{E}_\rho),$$

$$k_{q\sigma}^{(p)}(\mathbf{G}_\sigma) = k_q^{(p)}(\mathbf{E}_\sigma),$$

so that $h_{q\rho}^{(n-p)}$ and $k_{q\sigma}^{(p)}$ are multilinear forms in vectors $\mathbf{G}$ with s components. With this notation we finally get for $f \in \mathscr{I}(V)$ that

$$\mathscr{D}_\tau(\mathbf{G}_\tau)f = \sum_{p=1}^{n} \sum_{\sigma,\rho} \sum_{q=1}^{t} h_{q\rho}^{(n-p)}(\mathbf{G}_\rho) k_{q\sigma}^{(p)}(\mathbf{G}_\sigma). \tag{16.17}$$

We recall that the $k_{q\sigma}^{(p)}$ are independent of $f \in \mathscr{I}(V)$ while the $h_{q\rho}^{(n-p)}$ depend on f. When $\mathbf{x}$ is rational and V is defined over the rationals, then the ks have rational coefficients. If, further $f \in \mathscr{I}(V)$ has rational coefficients, then so do the hs.

16.6   Proof of Proposition 16.1: beginning

The case $d = 2$ is straightforward. Here $\mathcal{M}$ consists of those $\mathbf{x}$ with $\mathscr{F}(\mathbf{x} | \mathbf{Z}) = 0$, and hence $\mathcal{M}$ is a subspace of $\mathbb{C}^s$ of codimension $g_\mathbb{C}$. Since neither g_C nor h_C is affected by a non-singular linear transformation of the variables, we may suppose that $\mathcal{M}$ is the subspace $x_1 = \cdots = x_{g_C} = 0$. Each $\mathbf{X}$ may uniquely be written as $\mathbf{X} = \mathbf{X}_{\mathcal{M}} + \mathbf{X}^\perp$ with $X_{\mathcal{M}} \in \mathcal{M}$, and $\mathbf{X}^\perp$ in the orthogonal complement of $\mathcal{M}$. We have

$$\mathcal{F}(\mathbf{X}) = \tfrac{1}{2}\mathcal{F}(\mathbf{X}_{\mathcal{M}} + \mathbf{X}^\perp \mid \mathbf{X}_{\mathcal{M}} + \mathbf{X}^\perp) = \mathcal{F}(\mathbf{X}^\perp)$$

$$= \sum_{i=1}^{g_C} c_{ij} X_i X_j,$$

with certain coefficients c_{ij}. Since $X_1, \ldots, X_{g_C}$ are linear forms in $\mathbf{X}$, we have $h_C \leqq g_C$.

We now set up the framework of the proof for $d > 2$. Let V be an irreducible component of $\mathcal{M}$, of codimension g_C. Let K be a field of definition of V, containing the coefficients of $\mathcal{F}$. From now on $(\mathbf{x}_1, \ldots, \mathbf{x}_{d-1})$ will be a fixed generic point of V with respect to K. In particular, it will be a simple point of V. Write

$$u = \text{transc. deg. } K(\mathbf{x}_1, \ldots, \mathbf{x}_{d-2})/K, \tag{16.18}$$

$$v = \text{transc. deg. } K(\mathbf{x}_1, \ldots, \mathbf{x}_{d-1})/K(\mathbf{x}_1, \ldots, \mathbf{x}_{d-2}), \tag{16.19}$$

so that $u + v = \dim V$, and set

$$t = s(d-2) - u, \qquad a = s - v. \tag{16.20}$$

Thus

$$a + t = s(d-1) - \dim V = g_C. \tag{16.21}$$

Let S be the subspace consisting of vectors $\mathbf{y}$ such that $\mathcal{F}(\mathbf{x}_1 \mid \cdots \mid \mathbf{x}_{d-2} \mid \mathbf{y} \mid \mathbf{Z}) = 0$ identically in $\mathbf{Z}$, i.e. $\mathbf{y}$ is such that $(\mathbf{x}_1, \ldots, \mathbf{x}_{d-2}, \mathbf{y}) \in \mathcal{M}$. Since $\mathbf{x}_{d-1}$ lies in S and has v components which are algebraically independent over $K(\mathbf{x}_1, \ldots, \mathbf{x}_{d-2})$, it follows that $\dim S \geqq v$. If we had $\dim S > v$, then some $\mathbf{x}'_{d-1}$ in S would have more than v components which are independent over $K(\mathbf{x}_1, \ldots, \mathbf{x}_{d-2})$, contradicting the fact that $\dim \mathcal{M} = \dim V = u + v$. Thus $\dim S = v$ and

$$\text{codim } S = a. \tag{16.22}$$

In what follows write $\mathbf{X} = (\mathbf{X}_1, \ldots, \mathbf{X}_{d-2})$ for vectors of variables and

$$\mathbf{x} = (\mathbf{x}_1, \ldots, \mathbf{x}_{d-2}), \tag{16.23}$$

where $\mathbf{x}_1, \ldots, \mathbf{x}_{d-2}$ are the given vectors. Further introduce the matrix

$$A(\mathbf{X}): \mathcal{F}(\mathbf{X}_1 \mid \cdots \mid \mathbf{X}_{d-2} \mid \mathbf{e}_i \mid \mathbf{e}_j) = b_{ij}(\mathbf{X}) \quad (\text{say}) \qquad (1 \leqq i, j \leqq s).$$

Here $\mathbf{e}_1, \ldots, \mathbf{e}_s$ are the basis vectors. A vector $\mathbf{y} = y_1 \mathbf{e}_1 + \cdots + y_s \mathbf{e}_s$ lies in S precisely if

$$\sum_{i=1}^{s} b_{ij}(\mathbf{x}) y_i = 0 \qquad (1 \leqq j \leqq s).$$

In view of (16.22) the matrix $A(\mathbf{x})$ has

$$\text{rank } A(\mathbf{x}) = a. \qquad (16.24)$$

We may suppose without loss of generality that the submatrix $1 \leq i, j \leq a$ is non-singular.

It would be helpful to display a basis of S explicitly using polynomials in $\mathbf{x}$. To this end, write $\Delta = \Delta(\mathbf{X})$ for the subdeterminant of $A(\mathbf{X})$ with $1 \leq i, j \leq a$. Let

$$\Delta_j^i(\mathbf{X}) \qquad (1 \leq i \leq s - a = v, \ 1 \leq j \leq a)$$

be the subdeterminant formed from the first a rows, and from the columns $1, 2, \ldots, j - 1, a + i, j + 1, \ldots, a$. Put

$$\mathbf{y}^1 = \mathbf{y}^1(\mathbf{X}) = (\Delta_1^1, \ldots, \Delta_a^1, -\Delta, 0, \ldots, 0)$$
$$\mathbf{y}^2 = \mathbf{y}^2(\mathbf{X}) = (\Delta_1^2, \ldots, \Delta_a^2, 0, -\Delta, 0, \ldots, 0),$$
$$\vdots$$
$$\mathbf{y}^v \doteq \mathbf{y}^v(\mathbf{X}) = (\Delta_1^v, \ldots, \Delta_a^v, 0, 0, \ldots, -\Delta).$$

Write $b_{ij} = b_{ij}(\mathbf{X})$, and let $1 \leq i \leq v$. The expression

$$U_{ij}(\mathbf{X}) = \begin{vmatrix} b_{11} & \cdots & b_{1a} & b_{1,a+i} \\ \vdots & & & \\ b_{a1} & \cdots & b_{aa} & b_{a,a+i} \\ b_{j1} & \cdots & b_{ja} & b_{j,a+i} \end{vmatrix}$$

is 0 for $1 \leq j \leq a$, and is an $(a + 1) \times (a + 1)$ subdeterminant of $A(\mathbf{X})$ for $a < j \leq s$.

Expand $U_{ij}(\mathbf{X})$ by row $a + 1$. The minor of b_{jk} is $(-1)^{a-k} \Delta_k^i$ for $k \leq a$. Thus

$$U_{ij}(\mathbf{X}) = \sum_{k=1}^{a} b_{jk}(-1)^{k+a+1}(-1)^{a-k}\Delta_k^i + (-1)^{2a+2}\Delta b_{j,a+i}$$

$$= -\left\{ \sum_{k=1}^{a} b_{jk}\Delta_k^i - \Delta b_{j,a+i} \right\}.$$

Note that, by symmetry of the multilinear form,

$$\mathscr{F}(\mathbf{X}_1 | \cdots | \mathbf{X}_{d-2} | \mathbf{y}^i(\mathbf{X}) | \mathbf{e}_j) = \sum_{k=1}^{s} b_{kj}y_k^i$$

$$= \sum_{k=1}^{s} b_{jk}y_k^i = -U_{ij}(\mathbf{X}).$$

We can sum this up in the following way. Let the $(a + 1) \times (a + 1)$ subdeterminants of $A(\mathbf{X})$ be $D_1(\mathbf{X}), \ldots, D_N(\mathbf{X})$ in some order and $\mathbf{D}(\mathbf{X}) = (D_1(\mathbf{X}), \ldots, D_N(\mathbf{X}))$. Then

$$\mathscr{F}(\mathbf{X}_1 | \cdots | \mathbf{X}_{d-2} | \mathbf{y}^i(\mathbf{X}) | \mathbf{Z}) = \mathscr{B}^i(\mathbf{D}(\mathbf{X}), \mathbf{Z}) \qquad (1 \leq i \leq v) \quad (16.25)$$

where $\mathcal{B}^i$ is a bilinear form, in the vector $\mathbf{D}$ with N components and the vector $\mathbf{Z}$ with s components.

In view of (16.24), we have $\mathbf{D}(\mathbf{x}) = \mathbf{0}$, but $\Delta(\mathbf{x}) \neq 0$ by our remarks above. The vectors

$$\mathbf{y}^1(\mathbf{x}), \ldots, \mathbf{y}^v(\mathbf{x})$$

are linearly independent and lie in S, so they are a basis of S. Now if $\mathbf{y}^1(\mathbf{x}), \ldots, \mathbf{y}^v(\mathbf{x})$ together with, say, $\mathbf{z}^1, \ldots, \mathbf{z}^a$, forms a basis of $\mathbb{C}^s$, each $\mathbf{Z}$ is uniquely

$$\mathbf{Z} = \mathcal{N}_1(\mathbf{Z})\mathbf{y}^1(\mathbf{x}) + \cdots + \mathcal{N}_v(\mathbf{Z})\mathbf{y}^v(\mathbf{x}) + \mathcal{L}_1(\mathbf{Z})\mathbf{z}^1 + \cdots + \mathcal{L}_a(\mathbf{Z})\mathbf{z}^a.$$

$$(16.26)$$

with linear forms $\mathcal{N}_1, \ldots, \mathcal{N}_v, \mathcal{L}_1, \ldots, \mathcal{L}_a$.

16.7   Completion of the proof of Proposition 16.1

We have to bound $h_{\mathbb{C}}$ above in terms of a and t. What we shall actually do is find, and enumerate, polynomials $f_j(\mathbf{X}_1, \ldots, \mathbf{X}_d)$ $(j = 1, 2, \ldots, m)$ such that $\mathcal{F}(\mathbf{X}_1 \mid \cdots \mid \mathbf{X}_d)$ lies in the ideal generated by $f_1, \ldots, f_m$. (One has $h_{\mathbb{C}} \leqq m$ here, on setting $\mathbf{X}_1 = \cdots = \mathbf{X}_d$.) The critical formulae are the representation (16.17) for the operator $\mathcal{D}_\tau(\mathbf{G}_\tau)$ of (16.16), the expression (16.25) giving $\mathcal{F}(\mathbf{X}_1 \mid \cdots \mid \mathbf{X}_{d-2} \mid \mathbf{y}^i(\mathbf{X}) \mid \mathbf{Z})$ as a bilinear form, and the decomposition (16.26) of vectors in $\mathbb{C}^s$. (It is instructive to compare what follows with the simpler Section 7 of Schmidt (1982*e*).)

Write $E = \{1, 2, \ldots, d-2\}$. Given $\sigma \subseteq E$, we put

$$\mathcal{F}_\sigma(\mathbf{G}_\sigma, \mathbf{Y}, \mathbf{Z}) = \mathcal{F}(\mathbf{w}_1 \mid \cdots \mid \mathbf{w}_{d-2} \mid \mathbf{Y} \mid \mathbf{Z})$$

with

$$\mathbf{w}_i = \begin{cases} \mathbf{G}_i & \text{for } i \in \sigma \\ \mathbf{x}_i & \text{for } i \notin \sigma. \end{cases}$$

We are going to apply $\mathcal{D}_\tau(\mathbf{G}_\tau)$ with $\tau \subseteq E$ to the identity (16.25). When $\tau = \varnothing$ we get nothing interesting: both sides become zero. Suppose now that τ consists of a single element u. In this case, applying $\mathcal{D}_\tau(\mathbf{G}_\tau)$ to (16.25) we obtain

$$\mathcal{F}(\mathbf{x}_1 \mid \cdots \mid \mathbf{G}_u \mid \cdots \mid \mathbf{x}_{d-2} \mid \mathbf{y}^i(\mathbf{x}) \mid \mathbf{Z}) + \mathcal{F}(\mathbf{x}_1 \mid \cdots \mid \mathbf{x}_{d-2} \mid \mathcal{D}_\tau(\mathbf{G}_\tau)\mathbf{y}^i \mid \mathbf{Z})$$
$$= \mathcal{B}^i(\mathcal{D}_\tau(\mathbf{G}_\tau)\mathbf{D}, \mathbf{Z})$$

where $\mathcal{D}_\tau$ applied to a vector acts componentwise. The last relation may be rewritten as

$$\mathcal{F}_\tau(\mathbf{G}_\tau, \mathbf{y}^i(\mathbf{x}), \mathbf{Z}) + \mathcal{F}_\varnothing(\mathbf{G}_\varnothing, \mathcal{D}_\tau(\mathbf{G}_\tau)\mathbf{y}^i, \mathbf{Z}) = \mathcal{B}^i(\mathcal{D}_\tau(\mathbf{G}_\tau)\mathbf{D}, \mathbf{Z}).$$

More generally, we have for arbitrary $\tau \subseteq E$ the relation

$$\sum_{\sigma \subseteq \tau} \mathcal{F}_\sigma(\mathbf{G}_\sigma, \mathcal{D}_{\tau \setminus \sigma}(\mathbf{G}_{\tau \setminus \sigma})\mathbf{y}^i, \mathbf{Z}) = \mathcal{B}^i(\mathcal{D}_\tau(\mathbf{G}_\tau)\mathbf{D}, \mathbf{Z}). \qquad (16.27)$$

Rather than (16.27) it is easier to prove the formula

$$\sum_{\sigma \subseteq \tau} \mathscr{F}'_\sigma(\mathbf{G}_\sigma, \mathscr{D}'_{\tau \backslash \sigma}(\mathbf{G}_{\tau \backslash \sigma})\mathbf{y}^i, \mathbf{Z}) = \mathscr{B}^i(\mathscr{D}'_\tau(\mathbf{G}_\tau)\mathbf{D}, \mathbf{Z}), \qquad (16.28)$$

where $\mathscr{F}'_\lambda$, $\mathscr{D}'_\lambda$ are defined exactly like $\mathscr{F}_\lambda$, $\mathscr{D}_\lambda$ but with $\mathbf{X}$ in place of $\mathbf{x}$. (Thus in (16.28) both sides are polynomials in $\mathbf{G}_u$ ($u \in \tau$), $\mathbf{X}$ and $\mathbf{Z}$.) Note the convenient property

$$\mathscr{D}'_\sigma(\mathbf{G}_\sigma)\mathscr{D}'_v(\mathbf{G}_v) = \mathscr{D}'_{\sigma \cup v}(\mathbf{G}_{\sigma \cup v})$$

for disjoint sets σ, v in S.

We prove (16.28) by induction on $|\tau|$; the proof for $|\tau| = 0$ or 1 has essentially been done above. In the induction step we suppose for simplicity that $1 \in \tau$ and write $\tau = \lambda \cup \{1\}$ with $|\lambda| = |\tau| - 1$. Apply $\mathscr{D}'_{\{1\}}(\mathbf{G}_{\{1\}})$ to both sides of

$$\sum_{v \subseteq \lambda} \mathscr{F}'_v(\mathbf{G}_v, \mathscr{D}'_{\lambda \backslash v}(\mathbf{G}_{\lambda \backslash v})\mathbf{y}^i, \mathbf{Z}) = \mathscr{B}^i(\mathscr{D}'_\lambda(\mathbf{G}_\lambda)\mathbf{D}, \mathbf{Z}).$$

This gives

$$\sum_{v \subseteq \lambda} \mathscr{F}(\mathbf{G}_1 \,|\, \mathbf{w}_2 \,|\, \cdots \,|\, \mathbf{w}_{d-2} \,|\, \mathscr{D}'_{\lambda \backslash v}(\mathbf{G}_{\lambda \backslash v})\mathbf{y}^i, \mathbf{Z})$$

$$+ \sum_{v \subseteq \lambda} \mathscr{F}(\mathbf{X}_1 \,|\, \mathbf{w}_2 \,|\, \cdots \,|\, \mathbf{w}_{d-2} \,|\, \mathscr{D}'_\rho(\mathbf{G}_\rho)\mathbf{y}^i, \mathbf{Z}) = \mathscr{B}^i(\mathscr{D}'_\tau(\mathbf{G}_\tau)\mathbf{D}, \mathbf{Z}),$$

$$(16.29)$$

where $\rho = \rho(v) = \{1\} \cup (\lambda \backslash v)$ and where $\mathbf{w}_j = \mathbf{G}_j$ for $j \in v$, $\mathbf{w}_j = \mathbf{X}_j$ for $j \notin v$. The first sum on the left-hand side of (16.29) corresponds to the terms of the left-hand side of (16.28) with σ of the form $v \cup \{1\}$, $v \subseteq \lambda$; from the second sum we get the remaining terms $\sigma = v$ with $v \subseteq \lambda$. Hence (16.29) is equivalent to (16.28). This completes the proof of the induction step. (16.28) is true in general for $1 \leq i \leq v$ and we get (16.27) on inserting the value $\mathbf{X} = \mathbf{x}$.

We observe that

$$\mathscr{F}_\varnothing(\mathbf{G}_\varnothing, \mathbf{Y}, \mathbf{y}^i(\mathbf{x})) = \mathscr{F}(\mathbf{x}_1 \,|\, \cdots \,|\, \mathbf{x}_{d-2} \,|\, \mathbf{Y} \,|\, y^i(\mathbf{x})) = 0 \qquad (1 \leq j \leq v).$$

Hence substituting $\mathbf{Z} = \mathbf{y}^i(\mathbf{x})$ into (16.27) we obtain

$$\sum_{\varnothing \neq \sigma \subseteq \tau} \mathscr{F}_\sigma(\mathbf{G}_\sigma, \mathscr{D}_{\tau \backslash \sigma}(\mathbf{G}_{\tau \backslash \sigma})\mathbf{y}^i, y^i(\mathbf{x})) = \mathscr{L}_{ij}(\mathscr{D}_\tau(\mathbf{G}_\tau)\mathbf{D}) \qquad (16.30)$$

with certain linear forms $\mathscr{L}_{ij}$. This is an identity of multilinear forms in vectors $\mathbf{G}_u$ with $\mathbf{u} \in \tau$.

For $\sigma \subseteq E$, we define a linear transformation $A_\sigma(\mathbf{G}_\sigma) : \mathbb{C}^s \to \mathbb{C}^s$ as follows: $A_\varnothing(\mathbf{G}_\varnothing)$ is the identity map, and for $\sigma \neq \varnothing$ we stipulate that

$$A_\sigma(\mathbf{G}_\sigma)\mathbf{y}^i(\mathbf{x}) = \mathscr{D}_\sigma(\mathbf{G}_\sigma)\mathbf{y}^i, \qquad (16.31)$$

and that $A_\sigma(\mathbf{G}_\sigma)$ vanishes on the linear span of $\mathbf{z}^1, \ldots, \mathbf{z}^a$.

Note that $A_\sigma(\mathbf{G}_\sigma)\mathbf{Y}$ is linear in the $\mathbf{G}_i$ $(i \in \sigma)$ as well as in $\mathbf{Y}$.
The forms $\mathscr{L}_1, \ldots, \mathscr{L}_a$ of (16.26) reappear in the next lemma.

Lemma 16.4. $\mathscr{F}_\tau(\mathbf{G}_\tau, \mathbf{Y}, \mathbf{Z})$ *(as a polynomial in* $\mathbf{G}_\tau$, $\mathbf{Y}$, $\mathbf{Z}$*) lies in the ideal generated by the forms*

(A) $\qquad\qquad\qquad \mathscr{L}_i(\mathbf{Z}) \qquad (1 \leqq i \leqq a),$

by

(B) $\qquad\quad \mathscr{L}_i(A_{\sigma_1}(\mathbf{G}_{\sigma_1}) \cdots A_{\sigma_p}(\mathbf{G}_{\sigma_p})\mathbf{Y}) \qquad (1 \leqq i \leqq a)$

where $\sigma_1, \ldots, \sigma_p$ *are disjoint subsets of* τ *whose union has cardinality* $< |\tau|$, *and by*

(C) $\qquad\qquad\qquad k_{q\sigma}^{(p)}(\mathbf{G}_\sigma)$

where $1 \leqq p \leqq |\tau|$, $1 \leqq q \leqq t$, $\sigma \subseteq \tau$ *with* $|\sigma| = p$ *and where the forms* $k_{q\sigma}^{(p)}$ *come from* (16.17).

Proof. We proceed by induction on $|\tau|$, beginning with the case $\tau = \varnothing$. In this case

$$\mathscr{F}_\varnothing(\mathbf{G}_\varnothing, \mathbf{Y}, \mathbf{Z}) = \mathscr{F}(\mathbf{x}_1 | \cdots | \mathbf{x}_{d-2} | \mathbf{Y} | \mathbf{Z}). \qquad (16.32)$$

Writing $\mathbf{Y}$, $\mathbf{Z}$ in the form (16.26), we see that (16.32) is a bilinear form in $(\mathscr{L}_1(\mathbf{Y}), \ldots, \mathscr{L}_a(\mathbf{Y}))$ and $(\mathscr{L}_1(\mathbf{Z}), \ldots, \mathscr{L}_a(\mathbf{Z}))$, hence lies in the ideal generated by (A).

Next, let us consider the case when $|\tau| = 1$. Here (16.30) reduces to

$$\mathscr{F}_\tau(\mathbf{G}_\tau, \mathbf{y}^i(\mathbf{x}), \mathbf{y}^j(\mathbf{x})) = \mathscr{L}_{ij}(\mathscr{D}_\tau(\mathbf{G}_\tau)\mathbf{D}) \qquad (1 \leqq i, j \leqq v).$$

Again writing $\mathbf{Y}$, $\mathbf{Z}$ in the form (16.26), we find that $\mathscr{F}_\tau(\mathbf{G}_\tau, \mathbf{Y}, \mathbf{Z})$ lies in the ideal generated by $\mathscr{L}_i(\mathbf{Z})$ $(1 \leqq i \leqq a)$, by

$$\mathscr{L}_i(\mathbf{Y}) = \mathscr{L}_i(A_\varnothing(\mathbf{G}_\varnothing)\mathbf{Y}) \qquad (1 \leqq i \leqq a)$$

of (B), and by $\mathscr{L}_{ij}(\mathscr{D}_\tau(\mathbf{G}_\tau)\mathbf{D})$. Now each component D_l of $\mathbf{D}$ vanishes on the variety V whose generic point was $\mathbf{x}$, i.e. each component lies in $\mathscr{I}(V)$. So (16.17) may be applied, and $\mathscr{D}_\tau(\mathbf{G}_\tau)D_l$ lies in the ideal generated by the forms $k_{q\sigma}^{(p)}(\mathbf{G}_\sigma)$ with $p = 1 = |\tau|$, with $1 \leqq q \leqq t$ and with $\sigma = \tau$.

Suppose now that $|\tau| > 1$ and that the lemma has been shown for the proper subsets of τ. We may rewrite (16.30) as

$$\mathscr{F}_\tau(\mathbf{G}_\tau, \mathbf{y}^i(\mathbf{x}), \mathbf{y}^j(\mathbf{x})) = \mathscr{L}_{ij}(\mathscr{D}_\tau(\mathbf{G}_\tau)\mathbf{D})$$
$$- \sum_{\varnothing \neq \rho \subsetneqq \tau} \mathscr{F}_\rho(\mathbf{G}_\rho, \mathscr{D}_{\tau\backslash\rho}(\mathbf{G}_{\tau\backslash\rho})\mathbf{y}^i, \mathbf{y}^j(\mathbf{x})).$$

Therefore $\mathscr{F}_\tau(\mathbf{G}_\tau, \mathbf{Y}, \mathbf{Z})$ lies in the ideal generated by $\mathscr{L}_i(\mathbf{Y})$, $\mathscr{L}_i(\mathbf{Z})$ $(i = 1, \ldots, a)$, by the forms

$$\mathscr{L}_{ij}(\mathscr{D}_\tau(\mathbf{G}_\tau)\mathbf{D}) \qquad (1 \leqq i, j \leqq v), \qquad (16.33)$$

plus forms

$$\mathcal{F}_\rho\left(\mathbf{G}_\rho, \sum_{i=1}^v \mathcal{N}_i(\mathbf{Y})\mathcal{D}_{\tau\backslash\rho}(\mathbf{G}_{\tau\backslash\rho})\mathbf{y}^i, \sum_{j=1}^v \mathcal{N}_i(\mathbf{Z})\mathbf{y}^j(\mathbf{x})\right) \tag{16.34}$$

with $\varnothing \neq \rho \subsetneqq \tau$. We write $\hat{\mathbf{Z}} = \sum_{j=1}^v \mathcal{N}_j(\mathbf{Z})\mathbf{y}^j(\mathbf{x})$ and note that $\hat{\mathbf{Z}}$ lies in S. We may rewrite (16.34) as

$$\mathcal{F}_\rho(\mathbf{G}_\rho, A_{\tau\backslash\rho}(\mathbf{G}_{\tau\backslash\rho})\mathbf{Y}, \hat{\mathbf{Z}}). \tag{16.35}$$

Again by (16.17), each $\mathcal{D}_\tau(\mathbf{G}_\tau)D_l$, and hence each form (16.33), lies in the ideal generated by (C). By induction, $\mathcal{F}_\rho(\mathbf{G}_\rho, \mathbf{Y}, \mathbf{Z})$ with $\varnothing \neq \rho \subsetneqq \tau$ lies in the ideal generated by (A), by (B) with disjoint subsets $\sigma_1, \ldots, \sigma_p$ whose union is less than ρ, and by (C) with $1 \leq p \leq |\rho|$ and $\sigma \subseteq \rho$. Since $\mathcal{L}_i(\hat{\mathbf{Z}}) = 0$, the form (16.35) lies in the ideal generated by (C) and by

$$\mathcal{L}_i(A_{\sigma_1}(\mathbf{G}_{\sigma_1}) \cdots A_{\sigma_p}(\mathbf{G}_{\sigma_p})A_{\tau\backslash\rho}(\mathbf{G}_{\tau\backslash\rho})\mathbf{Y}).$$

Since $\sigma_1, \ldots, \sigma_p, \tau\backslash\rho$ are disjoint subsets of τ whose union is less than τ, this is of the type (B), and Lemma 16.4 follows.

The lemma will now be applied with $\tau = E = \{1, \ldots, d-2\}$. The number of forms is as follows. The number of forms (A) is a. The number of forms (B) is

$$a(1 + \theta_{d-2})$$

where θ_m is the number of disjoint non-empty subsets $\sigma_1, \ldots, \sigma_p$ of $\{1, \ldots, m\}$ whose union has cardinality less than m. Here the ordering of $\sigma_1, \ldots, \sigma_p$ matters, but each σ_l itself is an unordered set. The number of forms (C) is

$$t\sum_{p=1}^{d-2}\binom{d-2}{p} = t(2^{d-2} - 1).$$

In summary, the multilinear form $\mathcal{F}(\mathbf{X}_1 | \cdots | \mathbf{X}_{d-2} | \mathbf{Y} | \mathbf{Z})$ lies in the ideal generated by the

$$a + a(1 + \theta_{d-2}) + t(2^{d-2} - 1) \tag{16.36}$$

forms (A), (B) and (C).

Substituting $\mathbf{X}_1 = \cdots = \mathbf{X}_{d-2} = \mathbf{Y} = \mathbf{Z} = \mathbf{X}$, we see that $\mathcal{F}(\mathbf{X})$ lies in an ideal generated by forms of degrees between 1 and $d-1$, the number of these forms given by (16.36). Since $\mathcal{L}_i(\mathbf{Y})$ from (B) and $\mathcal{L}_i(\mathbf{Z})$ from (A) both become $\mathcal{L}_i(\mathbf{X})$, we may in fact save the summand a in (16.36), and infer that

$$h_\mathbb{C}(\mathcal{F}) \leq a(1 + \theta_{d-2}) + t(2^{d-2} - 1).$$

By (16.21),

$$h_\mathbb{C}(\mathcal{F}) \leq g_\mathbb{C}\Phi(d)$$

where $\Phi(d)$ is defined by

$$\Phi(d) = \max\,(1 + \theta_{d-2},\, 2^{d-2} - 1).$$

(In fact, $1 + \theta_{d-2} \geqq 2^{d-2}$, so

$$\Phi(d) = 1 + \theta_{d-2}.)$$

All that we lack is the set of assertions made about Φ in stating Theorem 16.1; these follow immediately from

Lemma 16.5. *The quantity* $\eta_m = 1 + \theta_m$ *has* $\eta_1 = 1$, $\eta_2 = 3$, $\eta_3 = 13$, *and in general* $\eta_m < (\log 2)^{-m} m!$

Proof. Setting $q = p + 1$ in the definition, we see that θ_m is the number of partitions of $1, \ldots, m$ into non-empty subsets $\sigma_1, \ldots, \sigma_q$ where $q \geqq 2$. Hence η_m is the numbers of partitions into non-empty subsets $\sigma_1, \ldots, \sigma_q$ where $q \geqq 1$.

η_1 'counts' only $\{1\}$, so that $\eta_1 = 1$.

η_2 counts $\{1, 2\}$, $\{1\} \cup \{2\}$, $\{2\} \cup \{1\}$, so that $\eta_2 = 3$.

Similarly, $\eta_3 = 13$.

In general,

$$\eta_m = \sum_{q=1}^{m} \sum_{\substack{u_1 + \cdots + u_q = m \\ u_i > 0}} \frac{m!}{u_1! \cdots u_q!}.$$

Setting $u_1 + \cdots + u_{q-1} = u$ and $q - 1 = p$, we obtain

$$\eta_m = 1 + \sum_{u=1}^{m-1} \sum_{p=1}^{u} \sum_{u_1 + \cdots + u_p = u} \frac{u!}{u_1! \cdots u_p!} \binom{m}{u}$$

$$= 1 + \sum_{u=1}^{m-1} \binom{m}{u} \eta_u = \sum_{u=0}^{m-1} \binom{m}{u} \eta_u$$

if we put $\eta_0 = 1$. The quantities $\zeta_m = \eta_m (\log 2)^m / m!$ have

$$\zeta_m = \sum_{u=0}^{m-1} \frac{(\log 2)^{m-u}}{(m-u)!} \zeta_u.$$

Hence when $\zeta_0, \zeta_1, \ldots, \zeta_{m-1}$ are $\leqq 1$, then $\zeta_m < e^{\log 2} - 1 = 1$. Therefore each of $\zeta_1, \zeta_2, \ldots$ is < 1, and the lemma follows.

16.8 Proof of Proposition 16.2

Let us look at the case $d = 2$ first. When $\mathcal{F}$ has rational coefficients, then $\mathcal{M}$ is a subspace defined over the rationals. The number of integer points $\mathbf{x}$ counted by $z_P(\mathcal{M})$ is

$$\ll P^{\dim \mathcal{M}}$$

with a constant in $\ll$ which depends only on s. Hence if C_1 in (16.3) is sufficiently large, we have

$$s(d-1) - \gamma - 1 < \dim \mathcal{M},$$

so that $\operatorname{codim} \mathcal{M} \leqq \gamma$. Since for $d = 2$ we have $h(\mathcal{F}) \leqq \operatorname{codim} \mathcal{M}$, the estimate (16.4) follows.

Before dealing with the case $d > 2$, we need some general facts from algebraic geometry. Suppose that V is an algebraic set in $\mathbb{C}^S$. We will say that V belongs to the class $\mathscr{C}(l)$ if it is the set of zeros of polynomials $f_1, \ldots, f_l$, each of total degree $\leqq l$.

Lemma 16.6. *Suppose $V \in \mathscr{C}(l)$. There is a $C_2 = C_2(l, S)$ such that, decomposing V as a union of varieties $V_1, \ldots, V_m$ as in Lemma 15.1, we have $m \leqq C_2$, and each V_i lies in $\mathscr{C}(C_2)$.*

Proof. See Seidenberg (1974), Section 65.

Lemma 16.7. *Suppose $V \in \mathscr{C}(l)$ contains integer points in a given bounded domain $\mathscr{D}$, and write $z_{\mathscr{D}}(V)$ for the number of these integer points. There is a subset $V' \subseteq V$ such that*

   (A) *V' is an irreducible algebraic variety,*
   (B) *V' is defined over the rationals,*
   (C) *there is an integer point in $V' \cap \mathscr{D}$ which is a simple point of V',*
   (D) *$z_{\mathscr{D}}(V') \geqq C_3 z_{\mathscr{D}}(V)$ where $C_3 = C_3(l, S) > 0$.*

Proof. We will construct a sequence

$$V = V^0 \supset V^1 \supset V^2 \supset \cdots \tag{16.37}$$

where V^i is an algebraic set which (a) belongs to $\mathscr{C}(l_i)$ with $l_i = l_i(l, S)$; and (b) has $z_{\mathscr{D}}(V^i) \geqq m_i z_{\mathscr{D}}(V)$ with $m_i = m_i(l, S) > 0$.

Suppose V^i has been constructed. What we do to get V^{i+1} splits into three cases.

Case I. Suppose V^i is not an irreducible variety. Then let V^{i+1} be the irreducible component of V^i for which $z_{\mathscr{D}}(V^{i+1})$ is largest possible. By Lemma 16.6 and by the inductive hypothesis, V^{i+1} will have properties (a) and (b).

Case II. Suppose V^i is an irreducible variety that is not defined over the rationals. Let $f_1, \ldots, f_n$ be n polynomials of degree $\leqq l_i$ defining V^i. The total number of coefficients of $f_1, \ldots, f_n$ is bounded in terms of S and l_i, and hence so is the dimension D of the $\mathbb{Q}$-vector space spanned by these coefficients. If $\beta_1, \ldots, \beta_D$ is a basis of this vector space, we may write

$$f_m = \sum_j \beta_j f_{mj} \qquad (1 \leqq m \leqq n)$$

where the polynomials f_{mj} have rational coefficients. Let V^{i+1} be the algebraic set defined by $f_{mj} = 0$ $(1 \leqq m \leqq n, \; 1 \leqq j \leqq D)$. Then V^{i+1} is defined over the rationals and hence is a proper subset of V^i, so that $\dim V^{i+1} < \dim V^i$ (Lang (1958), Chapter II, Section 3, Corollary 1). Further $V^{i+1} \in \mathscr{C}(l_iD)$ and $z_{\mathscr{D}}(V^{i+1}) = z_{\mathscr{D}}(V^i)$ so that (a) and (b) hold.

Case III. Suppose V^i is an irreducible variety which is defined over $\mathbb{Q}$, but all the integer points of $V^i \cap \mathscr{D}$ are singular points of V^i. In this case let V^{i+1} be the set of singular points of V^i. Again, (a) and (b) hold.

The chain (16.37) must end after a bounded number of steps, since Case I cannot occur twice in a row, and since in Cases II and III the dimension is reduced. The last set in the chain has the properties (A), (B), (C) and (D).

Proof of Proposition 16.2. We suppose that $d > 2$. For given $\mathbf{x}_1, \ldots, \mathbf{x}_{d-2}$, the $\mathbf{x}_{d-1}$ with $(\mathbf{x}_1, \ldots, \mathbf{x}_{d-2}, \mathbf{x}_{d-1}) \in \mathcal{M}$ form a subspace $S(\mathbf{x}_1, \ldots, \mathbf{x}_{d-2})$, with a certain codimension a. Given $\mathbf{x}_1, \ldots, \mathbf{x}_{d-2}$, the number of integer points $\mathbf{x}_{d-1} \in P\mathscr{C}$ with $(\mathbf{x}_1, \ldots, \mathbf{x}_{d-1}) \in \mathcal{M}$ is then $\leqq C_4(s)P^{s-a}$. Let $V_a \subseteq \mathbb{C}^{s(d-2)}$ be the algebraic set consisting of $(\mathbf{x}_1, \ldots, \mathbf{x}_{d-2})$ for which $\operatorname{codim} S(\mathbf{x}_1, \ldots, \mathbf{x}_{d-2}) \leqq a$. Then

$$z_P(\mathcal{M}) \leqq C_4(s) \sum_{a=0}^{s} z_P(V_a) P^{s-a}.$$

By the hypothesis (16.3), there must be an a in $0 \leqq a \leqq s$ with

$$z_P(V_a) > C_5(s) C_1 P^{s(d-2) - \gamma - 1 + a}.$$

Since $z_P(V_a) \leqq C_6(s, d) P^{s(d-2)}$, it follows that for sufficiently large C_1 we must have $a - \gamma - 1 < 0$; i.e.

$$0 \leqq a \leqq \gamma.$$

Now $V_a \subseteq \mathbb{C}^{s(d-2)}$ is in some class $\mathscr{C}(l)$ with $l = l(s, d)$. Define $V' \subseteq V_a$ according to Lemma 16.7. Thus V' is an irreducible variety, and is defined over the rationals. We have $z_P(V') \geqq C_3 z_P(V_a)$, hence

$$z_P(V') \geqq C_7(s, d) C_1 P^{s(d-2) - \gamma - 1 + a}.$$

Since $V' \in \mathscr{C}(C_8)$ where $C_8 = C_8(s, d)$, it follows for a sufficiently large value of C_1 that $\dim V' \geqq s(d-2) + a - \gamma$, or

$$a + t \leqq \gamma, \tag{16.38}$$

where $t = \operatorname{codim} V'$.

Further by part (C) of Lemma 16.7, there is an integer point $\mathbf{x} = (\mathbf{x}_1, \ldots, \mathbf{x}_{d-2}) \in V'$ which is simple on V'. The whole construction for the proof of Proposition 16.1 can be carried over, but this time our point

x has integer coefficients. All the polynomials occurring have rational coefficients. Whereas in Section 16.7 we used the fact that each component D_l of **D** lay in $\mathscr{I}(V)$, we now use the fact that $D_l \in \mathscr{I}(V_a) \subseteq \mathscr{I}(V')$. Thus (16.17) holds for $f = D_l$, where the forms k are defined in terms of V' and **x**. The inequality (16.38) takes the place of (16.21), and we may indeed conclude that $h(\mathscr{F}) \leqq \gamma\Phi(d)$.

17
Exponential sums: polynomials on a finite group

17.1 Introduction

Let $\mathbb{F}_q$ be the finite field with $q = p^l$ elements where p is a prime. Suppose $\mathcal{G}(\mathbf{X}) = \mathcal{G}(X_1, \ldots, X_s)$ is a polynomial with coefficients in $\mathbb{F}_q$. We are interested in sums

$$S = \sum_{\mathbf{x} \in \mathbb{F}_q^s} e(p^{-1}\tau(\mathcal{G}(\mathbf{x}))),$$

where τ denotes the trace from $\mathbb{F}_q$ to the prime field $\mathbb{F}_p$;

$$\tau(x) = x + x^p + \cdots + x^{p^{l-1}}.$$

(Schmidt (1976) gives helpful background in Chapter I on finite fields and on τ in particular.)

Suppose that $d \geqq 2$, and that

$$\mathcal{G} = \mathcal{F}^{(d)} + \cdots + \mathcal{F}^{(1)} + \mathcal{F}^{(0)} \tag{17.1}$$

where $\mathcal{F}^{(j)}$ is a form of degree j, or 0, and $\mathcal{F}^{(d)} \neq 0$. We write

$$h = h(\mathcal{G}) = h(\mathcal{F}^{(d)})$$

for the smallest number h such that $\mathcal{F}^{(d)}$ may be written as

$$\mathcal{F}^{(d)} = \mathcal{A}_1 \mathcal{B}_1 + \cdots + \mathcal{A}_h \mathcal{B}_h \tag{17.2}$$

with forms $\mathcal{A}_1, \mathcal{B}_1, \ldots, \mathcal{A}_h, \mathcal{B}_h$ of positive degrees and with coefficients in $\mathbb{F}_q$. The quantity h is invariant under linear substitutions of the variables with coefficients in $\mathbb{F}_q$. We define $\bar{h} = \bar{h}(\mathcal{G})$ in the same way as h but with $\mathcal{A}_1, \mathcal{B}_1, \ldots, \mathcal{A}_h, \mathcal{B}_h$ permitted to have coefficients in the algebraic closure $\bar{\mathbb{F}}_q$ of $\mathbb{F}_q$.

In this chapter we are going to give an estimate of S which depends on $h = h(\mathcal{G})$.

Theorem 17.1. *For $p > d$ we have*

$$|S| \ll q^{s-\kappa} \tag{17.3}$$

with

$$\kappa = 2^{1-d} |h/\Phi(d)|. \tag{17.4}$$

Here $|\bar{\alpha}|$, denotes the smallest integer $\geqq \alpha$, and the function $\Phi(d)$ is the one appearing in Theorem 16.1. The constant implied by '$\ll$' depends only on s and d.

Theorem 17.1, and the other results in Chapter 17, are due to Schmidt (1984b). The method of proof is very close to that of Chapters 15 and 16. (The case $d = 3$ of Theorem 17.1 is due to Davenport and Lewis (1962), who point out that the case $d = 2$ is fairly obvious. These authors draw on Davenport (1963).)

The common zeros of $\mathscr{A}_1, \mathscr{B}_1, \ldots, \mathscr{A}_h, \mathscr{B}_h$ in (17.2) are singular points of $\mathscr{F}^{(d)}$, and hence $s - 2h \leqq \dim V^*$, where V^* is the locus of singular points of $\mathscr{F}^{(d)}$ (in some 'universal domain'). Theorem 17.1 is obviously most effective when $\dim V^*$ is small. In the case when $\mathscr{F}^{(d)}$ is non-singular, $h \geqq s/2$, but Deligne (1974) has given the much sharper estimate $|S| \ll q^{s/2}$ for such forms. (No paper previous to Schmidt (1984) gives useful bounds for 'arbitrary' forms of degree $d > 3$.)

Theorem 17.1 yields information about Betti numbers. For according to the theorem, the sums

$$S_l = \sum_{\mathbf{x} \in \mathbb{F}_{q^l}^s} e(p^{-1}\tau_l(\mathscr{G}(\mathbf{x}))),$$

where τ_l is the trace from $\mathbb{F}_{q^l}$ to $\mathbb{F}_p$, have

$$|S_l| \ll q^{l(s-\bar{\kappa})}.$$

Here $\bar{\kappa}$ is defined just like κ, but with $\bar{h}$ in place of h. It follows (see Katz (1980), Chapter 2) that the Betti numbers B_r with $r > 2s - 2\bar{\kappa}$ vanish.

The sums of Theorem 17.1 are 'complete' sums. Our second result concerns 'incomplete' sums of the type

$$S_{\mathscr{B}} = \sum_{\mathbf{x} \in \mathscr{B}} e(p^{-1}\tau(\mathscr{G}(\mathbf{x}))),$$

where $\mathscr{B}$ is a 'box' contained in $\mathbb{F}_q^s$. That is,

$$\mathscr{B} = \mathscr{F}_1 \times \cdots \times \mathscr{F}_s, \tag{17.5}$$

$\mathscr{F}_i$ being a set of elements $\alpha_i + c_{i1}\alpha_{i1} + \cdots + c_{il}\alpha_{il}$ in $\mathbb{F}_q$, where α_i is a fixed element of $\mathbb{F}_q$, $\alpha_{i1}, \ldots, \alpha_{il}$ is a fixed basis of $\mathbb{F}_q$ over $\mathbb{F}_p$, and the c_{ij} ($1 \leqq i \leqq s$, $1 \leqq j \leqq l$) run through the integers in $1 \leqq c_{ij} \leqq P_{ij}$ with fixed natural numbers $P_{ij} \leqq p$. In the case when

$$P_{ij} = p \qquad (1 \leqq i \leqq s, \ 1 \leqq j \leqq l)$$

the box $\mathscr{B} = \mathbb{F}_q^s$, and $S_{\mathscr{B}}$ is the complete sum S. In general, when

$$1 \leqq P_{ij} \leqq P \leqq p \qquad (1 \leqq i \leqq s, \ 1 \leqq j \leqq l)$$

we will say that $\mathscr{B}$ is *of size* $\leqq P$. On the other hand we will say that $\mathscr{B}$ is of size $\geq P$ when $1 \leq P \leq P_{ij} \leq p$.

Theorem 17.2. *Let* $P = p^{\theta}$ *where* $d^{-1} < \theta \leq 1$, *and put*

$$\kappa = \kappa(\theta) = (d - \theta^{-1})(d - 1)^{-1} 2^{1-d} \overline{|h/\Phi(d)|}. \tag{17.6}$$

Then if $\mathscr{B}$ *is a box of size* $\leq P$, *we have*

$$|S_{\mathscr{B}}| \ll P^{sl-\kappa l + \varepsilon} \tag{17.7}$$

with the constant in $\ll$ *depending only on* $s, d, l, \theta, \varepsilon$.

Except for the ε in the exponent, and the dependence of the constant on ε and l (which renders the assertion useless for small p), Theorem 17.2 contains Theorem 17.1 as the special case $\theta = 1$. Before Schmidt's work, non-trivial estimates for incomplete sums $S_{\mathscr{B}}$ were available only for $\theta > \frac{1}{2}$; the same applies to arithmetic applications of the type given in Theorem 17.3 below. See Serre (1977), Théorème A.5; Chalk and Williams (1965), Smith (1970), Baker (1983). (One must except the results about *diagonal* congruences given in Chapter 12.)

The restriction $\theta > d^{-1}$ is a natural one. For example,

$$\operatorname{Re} e_p(x_1^d + \cdots + x_s^d) > \frac{1}{2}$$

for $\mathbf{x}$ in a box $\mathscr{B}$: $1 \leq x_i \leq p^{1/d}/(6s)$, so no non-trivial estimate for $S_{\mathscr{B}}$ is possible for $\mathscr{G}(\mathbf{x}) = x_1^d + \cdots + x_s^d$.

Now let $\mathscr{G} = (\mathscr{G}_1, \ldots, \mathscr{G}_r)$ be an *r*-tuple of polynomials. The *pencil* generated by $\mathscr{G}$ is the set of polynomials

$$\mathscr{G} = \mathbf{a}\mathscr{G} = a_1\mathscr{G}_1 + \cdots + a_r\mathscr{G}_r$$

with $\mathbf{a} = (a_1, \ldots, a_r) \in \mathbb{F}_q^r$, but $\mathbf{a} \neq \mathbf{0}$. We will suppose that each polynomial of the pencil is of degree ≥ 2. Then we define

$$h(\mathscr{G}) = \min h(\mathscr{G}),$$

with the minimum to be taken over polynomials $\mathscr{G}$ of the pencil. In the special case when each polynomial of the pencil is of degree d, and using the notation (17.1) for each $\mathscr{G}_i$, we have $h(\mathscr{G}) = h(\mathscr{F}^{(d)})$ with $\mathscr{F}^{(d)} = (\mathscr{F}_1^{(d)}, \ldots, \mathscr{F}_r^{(d)})$. More generally, when each polynomial is of degree $\leq d$, one has $s - 2h \leq \dim V^*$, where V^* is the manifold of singular points of $\mathscr{F}^{(d)}$, i.e. points where the matrix $\partial \mathscr{F}_i^{(d)}/\partial x_j$ $(1 \leq i \leq r, 1 \leq j \leq s)$ has rank $< r$. Thus $h \geq s/2$ when $\mathscr{F}^{(d)}$ is non-singular.

Give a box $\mathscr{B} \subseteq \mathbb{F}_q^s$, denote its cardinality by $|\mathscr{B}|$, and write $N_{\mathscr{B}} = N_{\mathscr{B}}(\mathscr{G})$ for the number of common zeros of $\mathscr{G}$ in $\mathscr{B}$.

Theorem 17.3. *Let* $\mathscr{G}$ *be an r-tuple of polynomials such that every polynomial in its pencil has a degree between* d_0 *and* d_1, *where* $2 \leq d_0 \leq d_1$ *are given. Suppose that* $p > d_1$, *and* $P = p^{\theta}$ *with* $d_0^{-1} < \theta \leq 1$. *Write* $h = h(\mathscr{G})$ *and let* $\kappa = \kappa(\theta)$ *be defined by* (17.6). *Then given a box* $\mathscr{B}$ *of* $\mathbb{F}_q^s$

of size $\leqq P$, we have, for some $d = d_0, \ldots, d_1$,

$$N_{\mathscr{B}} = q^{-r} |\mathscr{B}| + O(P^{sl - \kappa l + \varepsilon}) \tag{17.8}$$

with an implied constant depending only on s, d, l, θ, ε.

In particular, $N_{\mathscr{B}} > 1$ when $\mathscr{B}$ is a box of size $\geqq p^{\theta}$, provided that

$$h > r(d - 1)(d\theta - 1)^{-1} 2^{d-1} \Phi(d), \tag{17.9}$$

and that $p > C_1(s, d, r, l, \theta)$.

Our next theorem was mentioned in Chapter 1.

Theorem 17.4. *Given d, r and $\theta > \frac{1}{2}$, there is a $C_2 = C_2(d, r, \theta)$ as follows. Let $\mathscr{F}$ be a system of r forms of degree $\leqq d$ with integer coefficients in $s > C_2$ variables. Then for every prime p, the system of congruences*

$$\mathscr{F}(\mathbf{x}) \equiv 0 \pmod{p} \tag{17.10}$$

has a solution $\mathbf{x} \neq \mathbf{0}$ with

$$|\mathbf{x}| \ll p^{\theta}, \tag{17.11}$$

with a constant in $\ll$ which depends only on d, r, θ.

Two comments are in order. In the first place, we are going to prove this theorem in Chapter 18 with the prime p replaced by an arbitrary modulus m. This is very complicated, so Theorem 17.4 acts as a 'warm up'. In the second place, the interesting case of Theorem 17.4 is when the forms of $\mathscr{F}$ are of even degree. We know much more for forms of odd degree (see Chapter 14).

We also prove a theorem which will be needed in Chapter 18, but fits well into the present chapter. To state it, we define $h(\mathscr{F})$ for a form $\mathscr{F}$ of degree $d \geqq 2$ with coefficients in a commutative ring A. As one would expect, it is the smallest integer h such that $\mathscr{F}$ may be written as (17.2), with $\mathscr{A}_i$, $\mathscr{B}_i$ being forms of positive degree with coefficients in A. In particular, where $A = A_m = \mathbb{Z}/m\mathbb{Z}$, and when $\mathscr{F}$ is a form in $A_m[\mathbf{X}]$, the invariant $h(\mathscr{F})$ is well defined. Now let a be a divisor of m. There is a natural map $A_m \mapsto A_a$, and a natural map of polynomial rings $A_m[\mathbf{X}] \to A_a[\mathbf{X}]$. Write $\mathscr{F}_a(\mathbf{X})$ for the image of $\mathscr{F}(\mathbf{X})$ under this map; write $h_a(\mathscr{F}) = h(\mathscr{F}_a)$. We then define $h_a(\mathscr{G})$ for polynomials $\mathscr{G}$ in $A_m[\mathbf{X}]$ of degree $\leqq d$ by using their homogeneous part $\mathscr{F}^{(d)}$.

Again when $\mathscr{G} \in A_m[\mathbf{X}]$, we consider sums

$$S_{\mathscr{B}} = \sum_{\mathbf{x} \in \mathscr{B}} e(m^{-1} \mathscr{G}(\mathbf{x}))$$

where $\mathscr{B}$ is a box in A_m^s. That is, $\mathscr{B}$ is a product set (17.5), where $\mathscr{F}_i$ is a set of elements $\alpha_i + c_i\beta_i$ in A_m, where α_i is fixed, β_i is a fixed unit of A_m, and c_i $(i = 1, \ldots, s)$ runs through $1, 2, \ldots, P_i$. The P_i are fixed integers in $1 \le P_i \le m$; we say that the box is of size $\le P$ if $1 \le P_i \le P \le m$.

Theorem 17.5. *Let $\mathcal{G}$ be a polynomial of degree $\le d$ with coefficients in A_m where m is square-free. Let $\mathcal{B}$ be a box of size $\le P = m^\theta$ where $d^{-1} < \theta \le 1$. Suppose that*

$$|S_{\mathcal{B}}| \ge P^{s-J}$$

where $J \ge 1$. When $\Gamma > 1$ is an integer and when $m \ge C_3(s, d, \theta, \Gamma)$, there is a factorization $m = ab$ such that $b \le P^{1/\Gamma}$, and

$$h_a(\mathcal{F}) \le (d - \theta^{-1})^{-1} d^2 2^d \Phi(d) J\Gamma. \tag{17.12}$$

17.2   Weyl's inequality for groups

Let G, H be additive groups. Just as in Section 15.4, for any mapping $\mathcal{F}: G \to H$, we define

$$\mathcal{F}_d(g_1, \ldots, g_d) = \sum_{\varepsilon_1=0}^{1} \cdots \sum_{\varepsilon_d=0}^{1} (-1)^{\varepsilon_1 + \cdots + \varepsilon_d} \mathcal{F}(\varepsilon_1 g_1 + \cdots + \varepsilon_d g_d),$$

so that $\mathcal{F}_d$ is a symmetric function from $G \times \cdots \times G$ into H. Given a subset A of G, and $g \in G$, we write $A - g$ for the set of differences $a - g$ with $a \in A$. The difference set A^D is defined to be the union of the sets $A - g$ with $g \in A$. Again following Section 15.4, let

$$A(g_1, \ldots, g_t) = \bigcap_{\varepsilon_1=0}^{1} \cdots \bigcap_{\varepsilon_t=0}^{1} (A - \varepsilon_1 g_1 - \cdots - \varepsilon_t g_t).$$

Lemma 17.1. *Let $\mathcal{F}$ be a map from G into H where G and H are additive groups. Then $\mathcal{F}_d(g_1, \ldots, g_d)$ is 'multilinear', i.e. it is a homomorphism in each argument g_i, if and only if $\mathcal{F}_{d+1}(g_1, \ldots, g_{d+1})$ is identically zero.*

Proof. We have the identity

$$\mathcal{F}_{d+1}(g_1, g_2, g_3, \ldots, g_{d+1})$$
$$= \mathcal{F}_d(g_1, g_3, \ldots, g_{d+1}) + \mathcal{F}_d(g_2, g_3, \ldots, g_{d+1})$$
$$- \mathcal{F}_d(g_1 + g_2, g_3, \ldots, g_{d+1})$$

established just as in Section 15.4. Now $\mathcal{F}_d$ is linear in its first argument if and only if the right-hand side vanishes identically. Given the symmetry of $\mathcal{F}_d$, the result follows.

We shall define a *polynomial of degree* $\le d$ from G into H to be a map $\mathcal{F}$ with the two equivalent properties of Lemma 17.1. If $f(x_1, \ldots, x_s)$ is a linear combination of monomials $x_{i_1} \cdots x_{i_l}$ with $0 \le l \le k$ with coefficients in A_m, then f is a polynomial of degree $\leq k$ from A_m^s into A_m. (Argue as in Lemma 15.3.) If further $\mathcal{F}(x_1, \ldots, x_s) = m^{-1} f(x_1, \ldots, x_s)$, then $\mathcal{F}$ may be interpreted as a polynomial of degree $\leq k$ from A_m^s into $\mathbb{Q}/\mathbb{Z}$. But not every polynomial from A_m^s into $\mathbb{Q}/\mathbb{Z}$ is of this form. For instance, $\mathcal{F}(x) = x^2/4$ may be interpreted as a polynomial from $\mathbb{Z}/2\mathbb{Z}$ into $\mathbb{Q}/\mathbb{Z}$. There is no need to go into this here; see Schmidt (1982a, b) for further discussion.

For maps from G into $\mathbb{R}/\mathbb{Z}$ we have the following version of Weyl's inequality (and for polynomial maps, it is an effective tool).

Lemma 17.2. *Let $\mathcal{F}$ be a map from G into H where H is a subgroup of $\mathbb{R}/\mathbb{Z}$. Let A be a finite subset of G, and put*

$$S = S_A = \sum_{g \in A} e(\mathcal{F}(g)). \tag{17.13}$$

Then for each $d \geq 2$,

$$|S|^{2^{d-1}} \leq |A^D|^{2^{d-1}-d} \sum_{g_1 \in A^D} \cdots \sum_{g_{d-1} \in A^D} \left| \sum_{g_d \in A(g_1, \ldots, g_{d-1})} e(\mathcal{F}_d(g_1, \ldots, g_d)) \right|.$$

Proof. This is practically a verbatim repetition of the proof of Lemma 15.2.

Now let $\mathcal{G}: G \to H$ be a polynomial of degree $\leq d$ where $d \geq 2$. With $\mathcal{G}$ we associate the subset $\mathcal{M}$ of $G^{d-1} = G \times \cdots \times G$ consisting of $(g_1, \ldots, g_{d-1})$ for which

$$\mathcal{G}_d(g_1, \ldots, g_{d-1}, g) = 0$$

identically in g. (Thus $\mathcal{M}$ reduces to $\mathcal{M}(\mathcal{G})$ in the case $G = \mathbb{C}^s$, $H = \mathbb{C}$, $\mathcal{G}$ a form of degree d: see Section 15.3.) If, say, $e_1, \ldots, e_T$ is a set of generators of G, then $(g_1, \ldots, g_{d-1})$ lies in $\mathcal{M}$ precisely when

$$\mathcal{G}_i(g_1, \ldots, g_{d-1}, e_i) = 0 \qquad (1 \leq i \leq T).$$

Lemma 17.3. *Suppose G is finite, $\mathcal{G}$ is a polynomial of degree $\leq d$ from G into $H \subseteq \mathbb{R}/\mathbb{Z}$ where $d \geq 2$, and S is given by (17.13) with $\mathcal{F} = \mathcal{G}$ and $A = G$, so that S is a 'complete sum'. Then*

$$|S|^{2^{d-1}} \leq |G|^{2^{d-1}-d+1} |\mathcal{M}|.$$

Proof. Take $A = G$ in Lemma 17.2. In the inner sum we have $A(g_1, \ldots, g_{d-1}) = G$. Write

$$e(\mathcal{G}_d(g_1, \ldots, g_{d-1}, g)) = \chi(g).$$

Then χ is a character on G, that is $\chi: G \to T = \{e(\theta): \theta \in \mathbb{R}\}$,

$$\chi(g_1 + g_2) = \chi(g_1)\chi(g_2) \qquad (g_1, g_2 \in G).$$

The usual argument for Dirichlet characters now applies to our inner sum

$$T_\chi = \sum_{g \in G} \chi(g).$$

Namely,

$$T_\chi = \sum_{g \in G} \chi(g + a) = \chi(a) T_\chi$$

for any $a \in G$, so $T_\chi = 0$ if χ is not the identity map. In other words the inner sum is equal to $|G|$ when $(g_1, \ldots, g_{d-1}) \in \mathcal{M}$, and equal to 0 otherwise; the lemma follows.

Lemma 17.4. *Make the assumptions of Lemma 17.3, in the special case when $G = \mathbb{F}_q^s$. Then for each $J > 0$ we have either*

$$|S| \leq q^{s-J}$$

or

$$|\mathcal{M}| > q^{s(d-1) - 2^{d-1}J}.$$

Proof. This is an immediate consequence of the preceding lemma.

Lemma 17.5. *Let $G = \mathbb{F}_q^s$ and $H = (p^{-1}\mathbb{Z}/\mathbb{Z})$, i.e. the rationals $p^{-1}a$ with $a \in \mathbb{Z}$, taken modulo 1. Let $\mathcal{F}(\mathbf{X}) = \mathcal{F}(X_1, \ldots, X_s)$ be a polynomial in the traditional sense of the type (17.1) with coefficients in $\mathbb{F}_q$. Then*

$$\mathcal{P}(\mathbf{X}) = p^{-1}\tau(\mathcal{F}(\mathbf{X}))$$

is a polynomial of degree $\leq d$ from G to H in the sense of this section. Moreover, $\mathcal{M}(\mathcal{P}) = \mathcal{M}(\mathcal{F}^{(d)})$, i.e. $\mathcal{M} = \mathcal{M}(\mathcal{P})$ consists of $(\mathbf{x}_1, \ldots, \mathbf{x}_{d-1})$ with $\mathbf{x}_i \in \mathbb{F}_q^s$ having

$$\mathcal{F}_d^{(d)}(\mathbf{x}_1, \ldots, \mathbf{x}_{d-1}, \mathbf{X}) = 0$$

identically in $\mathbf{X}$.

Proof. First of all, the map $\mathbf{x} \mapsto \mathcal{F}(\mathbf{x})$ is a polynomial of degree $\leq d$ in the sense of this section from G into $H_1 = \mathbb{F}_q$. We have

$$\mathcal{F}_d(\mathbf{x}_1, \ldots, \mathbf{x}_d) = \mathcal{F}_d^{(d)}(\mathbf{x}_1, \ldots, \mathbf{x}_d).$$

Next, the map $x \mapsto p^{-1}\tau(x)$ is a homomorphism from H_1 into H. Thus $\mathcal{P}_d(\mathbf{x}_1, \ldots, \mathbf{x}_d) = p^{-1}\tau(\mathcal{F}_d^{(d)}(\mathbf{x}_1, \ldots, \mathbf{x}_d))$ and this is multilinear. So $\mathcal{P}$ is a polynomial of degree $\leq d$.

Now if $x \in \mathbb{F}_q$ is such that $p^{-1}\tau(\alpha x) = 0$ (the zero element of H) for each $\alpha \in \mathbb{F}_q$, then necessarily $x = 0$. Further $(\mathbf{x}_1, \ldots, \mathbf{x}_{d-1})$ lies in $\mathcal{M} = \mathcal{M}(\mathcal{P})$ precisely when

$$p^{-1}\tau(\mathcal{F}_d^{(d)}(\mathbf{x}_1, \ldots, \mathbf{x}_{d-1}, \mathbf{x})) = 0$$

for each $\mathbf{x} \in \mathbb{F}_q^s$. Replacing $\mathbf{x}$ by $\alpha \mathbf{x}$ and noting that $\mathcal{F}_d^{(d)}$ is linear in each argument, we may conclude that

$$\mathscr{F}_d^{(d)}(\mathbf{x}_1, \ldots, \mathbf{x}_{d-1}, \mathbf{x}) = 0$$

for every $\mathbf{x} \in \mathbb{F}_q^s$, and hence

$$\mathscr{F}_d^{(d)}(\mathbf{x}_1, \ldots, \mathbf{x}_{d-1}, \mathbf{X}) = 0,$$

identically in $\mathbf{X}$.

17.3 $\mathcal{M}$ and incomplete sums

In this section we carry over Lemma 17.4 to incomplete sums as far as possible. The function $\|\theta\|$ (distance from the nearest integer) can be thought of as a function on $\mathbb{R}/\mathbb{Z}$.

Let $P \geqq 1$. Let $e_1, \ldots, e_B$ be members of the group G. We assume in the present section that

> the P^B elements $c_1 e_1 + \cdots + c_B e_B$ with $1 \leqq c_1, \ldots, c_B \leqq P$ (17.14)
> are distinct.

Given natural numbers $Q_1, \ldots, Q_s$ all $\leqq P$ we can define a *box* in G of size $\leqq P$ to consist of all

$$e_0 + c_1 e_1 + \cdots + c_B e_B$$

where e_0 is an element of G and each c_i runs through the integers $1, 2, \ldots, Q_i$. The $Q_1 Q_2 \cdots Q_B$ elements so obtained are distinct. We call $e_1, \ldots, e_B$ a *basis* of $\mathscr{B}$ (though the basis may not be uniquely determined by $\mathscr{B}$).

Lemma 17.6. *Let $\mathscr{P}$ be a polynomial of degree $\leqq d$ (where $d \geqq 2$) from G into $H \subseteq \mathbb{R}/\mathbb{Z}$. Let $\mathscr{B} \subseteq G$ be a box of size $\leqq P$ with basis $e_1, \ldots, e_B$. Define the sum $S_{\mathscr{B}}$ as in (17.13) with $\mathscr{P}$ instead of $\mathscr{F}$. Then*

$$|S_{\mathscr{B}}|^{2^{d-1}} \ll P^{(2^{d-1}-d)B} \sum \prod_{i=1}^{B} \min (P, \|\mathscr{P}_d(g_1, \ldots, g_{d-1}, e_i)\|^{-1}),$$

where the sum is over $(d-1)$-tuples of elements $g_1, \ldots, g_{d-1}$ of $\mathscr{B}^D$. The constant in $\ll$ depends only on B, d.

Proof. We may assume that $P > 4$. Suppose first that $Q_1, \ldots, Q_B$ in the definition of $\mathscr{B}$ have $Q_i \leqq P/4$. An intersection $\mathscr{B} \cap (\mathscr{B} + g)$ can only be non-empty if

$$g = c_1 e_1 + \cdots + c_B e_B$$

with integers c_i in $(-P/4, P/4)$. Moreover, a point of $\mathscr{B} \cap (\mathscr{B} + g)$ is of the form

$$e_0 + x_1 e_1 + \cdots + x_B e_B = e_0 + (y_1 + c_1) e_1 + \cdots + (y_B + c_B) e_B$$

with $1 \leq x_i, y_i \leq Q_i \leq P/4$. We see from (17.14) that $x_i - y_i - c_i$, which lies in $(-3P/4, 3P/4)$, must be zero for $i = 1, \ldots, B$. Thus $\mathscr{B} \cap (\mathscr{B} + g)$ is the box

$$\{e_0 + x_1 e_1 + \cdots + x_B e_B: 1 \leq x_i \leq \min(Q_i, Q_i - c_i)\}.$$

Indeed, each non-empty set $\mathscr{B}(g_1, \ldots, g_{d-1})$ appearing in Lemma 17.2 (with $A = \mathscr{B}$) is a box. The inner sum in Lemma 17.2 can be written

$$\sum_{c_1=1}^{R_1} \cdots \sum_{c_B=1}^{R_B} e(\mathscr{L}(e_0 + c_1 e_1 + \cdots + c_B e_B))$$

where $\mathscr{L}$ is a homomorphism and $R_1, \ldots, R_B$ are integers $\leq P$. ($\mathscr{L}, R_1, \ldots, R_B$ depend on $g_1, \ldots, g_{d-1}$.) This has modulus

$$\prod_{i=1}^{B} \left| \sum_{e_i=1}^{R_i} e(c_i \mathscr{L}(e_i)) \right| \leq \prod_{i=1}^{B} \min(R_i, \|\mathscr{L}(e_i)\|^{-1})$$

$$\leq \prod_{i=1}^{B} \min(P, \|\mathscr{L}(e_i)\|^{-1}).$$

The lemma follows for $\mathscr{B}$ on noting the form of $\mathscr{L}$ and observing that $|\mathscr{B}^D| \ll P^B$. The restriction $Q_i \leq P/4$ may be removed on decomposing an arbitrary box $\mathscr{B}$ of size $\leq P$ into $O(1)$ smaller boxes.

Lemma 17.7. *Make the same assumptions as in the preceding lemma. Suppose further that*

$$|S_\mathscr{B}| \geq P^{B-J}$$

where $J > 0$. Then the number N of $(d-1)$-tuples of elements $g_1, \ldots, g_{d-1}$ in $\mathscr{B}^D$ with

$$\|\mathscr{P}_d(g_1, \ldots, g_{d-1}, e_i)\| < P^{-1} \qquad (i = 1, \ldots, B)$$

satisfies

$$N \gg P^{B(d-1)-2^{d-1}J-\varepsilon}$$

where the implied constant depends only on B, d, ε.

Proof. This is a slight variant of the proof of Lemma 15.9; it is necessary to decompose $\mathscr{B}^D$ into $O(1)$ sets $\mathscr{A}$ with $\mathscr{A}^D \subset \mathscr{B}^D$.

Note that $\mathscr{B}^D$ is contained in the set $\mathscr{C}(P)$ of elements $c_1 e_1 + \cdots + c_B e_B$ with $|c_i| \leq P$ $(i = 1, \ldots, B)$. More generally, when $1 \leq R \leq P$, write $\mathscr{C}(R)$ for the set of elements $c_1 e_1 + \cdots + c_B e_B$ with $|c_i| \leq R$ $(i = 1, \ldots, B)$. These elements are not necessarily distinct, but by (17.14) no element is repeated more than 3^B times.

Lemma 17.8. *Make the same assumptions as in the preceding lemma. Let*

$0 < \sigma \leqq 1$. *Then the number* $N(\sigma)$ *of* $(d-1)$-*tuples* $g_1, \ldots, g_{d-1}$ *in* $\mathscr{C}(P^\sigma)$ *with*

$$\|\mathscr{P}_d(g_1, \ldots, g_{d-1}, e_i)\| < P^{-d+(d-1)\sigma} \qquad (i = 1, \ldots, B) \quad (17.15)$$

satisfies

$$N(\sigma) \gg P^{B(d-1)\sigma - 2^{d-1}J - \varepsilon}.$$

The implied constant depends only on B, d, σ, ε.

Proof. This is essentially the case $d = k$ of Lemma 15.11. By the remark preceding Lemma 17.8 we lose a factor of at most 3^B at each stage of the proof.

Given $e_1, \ldots, e_B$ and a polynomial $\mathscr{P}$ as above, write $\mathscr{M}(R)$ for the subset of $\mathscr{M}$ consisting of $(g_1, \ldots, g_{d-1})$ in $\mathscr{M}$ with each $g_i \in \mathscr{C}(R)$.

Lemma 17.9. *Suppose* $d \geqq 2$ *and* $\mathscr{P}$ *is a polynomial of degree* $\leqq d$ *from* G *into* $H = (m^{-1}\mathbb{Z}/\mathbb{Z})$, *i.e. the rationals with denominator* m, *taken modulo* 1. *Suppose that* $d^{-1} < \theta \leqq 1$,

$$0 < \sigma < (d - \theta^{-1})/(d-1). \quad (17.16)$$

Suppose $\mathscr{B} \subseteq G$ *is a box of size* $\leqq P = m^\theta$, *and with a basis* $e_1, \ldots, e_B$ *which generates* G *and satisfies* (17.14). *Then given* $J > 0$, *we have either*

$$|S_\mathscr{B}| \leqq P^{B-J},$$

or $R = P^\sigma$ *has*

$$|\mathscr{M}(R)| \gg R^{B(d-1) - 2^{d-1}(J/\sigma) - \varepsilon}. \quad (17.17)$$

The constant in $\gg$ *here depends only on* B, d, σ, ε.

Proof. We may assume $m > 1$. By (17.16), and since $P = m^\theta$, the right-hand side of (17.15) is less than m^{-1}. Since the values of $\mathscr{P}_d$ lie in $(m^{-1}\mathbb{Z}/\mathbb{Z})$, the relation (17.15) then leads to $\mathscr{P}_d(g_1, \ldots, g_{d-1}, e_i) = 0$ $(i = 1, \ldots, B)$ and indeed to $(g_1, \ldots, g_{d-1}) \in \mathscr{M}(R)$. This proves Lemma 17.9.

17.4   Invariants h, $\bar{h}$, g

Let $\mathbb{F}$ be an arbitrary field and $\mathscr{F}$ a form in s variables of degree $d \geqq 2$ with coefficients in $\mathbb{F}$. Since $\mathbb{F}$ is a commutative ring, we have already defined $h(\mathscr{F})$ as the least number h such that $\mathscr{F}$ may be written as

$$\mathscr{F} = \mathscr{A}_1\mathscr{B}_1 + \cdots + \mathscr{A}_h\mathscr{B}_h \quad (17.18)$$

with forms $\mathscr{A}_i$, $\mathscr{B}_i$ of positive degree and with coefficients in $\mathbb{F}$. We define $\bar{h}(\mathscr{F})$ to be the least integer h such that $\mathscr{F}$ may be written as (17.18) with forms $\mathscr{A}_i$, $\mathscr{B}_i$ having coefficients in the algebraic closure of $\mathbb{F}$. We have $\bar{h} \leqq h$.

We define $\mathcal{M}$ as the subset of $\mathbb{F}^{s(d-1)}$ consisting of $(d-1)$-tuples $\mathbf{x}_1, \ldots, \mathbf{x}_{d-1}$ with $\mathscr{F}_d(\mathbf{x}_1, \ldots, \mathbf{x}_{d-1}, \mathbf{X}) = 0$, identically in $\mathbf{X}$. Now let Ω be some universal domain over $\mathbb{F}$, i.e. a field containing $\mathbb{F}$ that is algebraically closed and of infinite transcendence degree over $\mathbb{F}$. We define $\bar{\mathcal{M}}$ as the set of $(d-1)$-tuples $\mathbf{x}_1, \ldots, \mathbf{x}_{d-1}$ with components in Ω for which $\mathscr{F}_d(\mathbf{x}_1, \ldots, \mathbf{x}_{d-1}, \mathbf{X}) = 0$. Then $\bar{\mathcal{M}}$ is an algebraic set in $\Omega^{s(d-1)}$; we denote its codimension by $\bar{g} = \bar{g}(\mathscr{F})$. Just as in Lemma 16.1 we have

$$\bar{g} \leqq 2^{d-1}\bar{h}. \tag{17.19}$$

Proposition 17.1. *When $\mathbb{F}$ is of characteristic $> d$, we have*

$$\bar{h} \leqq \Phi(d)\bar{g}.$$

The case $\mathbb{F} = \mathbb{C}$ is Proposition 16.1. Practically no changes are necessary in the general case. We can avoid the formula (16.9), with $n!$ in the denominator, by using the expansion

$$f(\mathbf{C}) = f(\mathbf{0}) + f^{(1)}(\mathbf{C}) + \cdots + f^{(n)}(\mathbf{C}) + \cdots$$

to define $f^{(n)}(\mathbf{C})$. Writing $\mathscr{P}(\mathbf{X}) = \mathscr{F}_d(\mathbf{X}, \ldots, \mathbf{X})$ one finds (corresponding to the conclusions drawn after Lemma 16.4) that $\bar{h}(\mathscr{P}) \leqq \Phi(d)\bar{g}(\mathscr{F})$. When the characteristic exceeds d, then the relation $\mathscr{P}(\mathbf{X}) = (-1)^d d!\, \mathscr{F}(\mathbf{X})$ shows that also $\bar{h}(\mathscr{F}) \leqq \Phi(d)\bar{g}(\mathscr{F})$.

Now let $\mathscr{D}(Z)$ be a subset of $\mathbb{F}^s$ such that the projection on any coordinate axis contains at most Z elements. Thus $\mathscr{D}(Z) \subseteq \mathscr{D}_1 \times \cdots \times \mathscr{D}_s$ where $\mathscr{D}_i \subseteq \mathbb{F}$ has cardinality $|\mathscr{D}_i| \leqq Z$. Let $\mathcal{M}(\mathscr{D}(Z))$ consist of $(d-1)$-tuples $(\mathbf{x}_1, \ldots, \mathbf{x}_{d-1}) \in \mathcal{M}$ with each $\mathbf{x}_j \in \mathscr{D}(Z)$.

Proposition 17.2. *Suppose that $\mathscr{F}$ is a form of degree $d > 1$ with coefficients in a perfect field $\mathbb{F}$ of characteristic $> d$. Suppose that for some $\mathscr{D}(Z)$ where $Z > 1$ we have*

$$|\mathcal{M}(\mathscr{D}(Z))| > C_4 Z^{s(d-1)-\gamma-1}, \tag{17.20}$$

where $C_4 = C_4(s, d)$ is a suitable constant, and where γ is an integer. Then

$$h(\mathscr{F}) \leqq \Phi(d)\gamma.$$

This corresponds to Proposition 16.2, which contains essentially the case $\mathbb{F} = \mathbb{Q}$. As in Chapter 16, it is Proposition 17.2 rather than Proposition 17.1 that is applicable to exponential sums. For background material on perfect fields, see e.g. van der Waerden (1949), Chapter V.

For the proof of Proposition 17.2, which is very close to that of Proposition 16.2, we need the following lemma. We first generalize the definition of the class of algebraic sets $\mathscr{C}(l)$ in the obvious way. Given a universal domain Ω and a natural number S, let $\mathscr{C}(l)$ be the class of algebraic sets of Ω^S which can be defined by a set of equations $f_1 = \cdots = f_i = 0$, where each f_i is a polynomial of total degree $\leq l$. The analogue of Lemma 16.6 in Ω^S rather than $\mathbb{C}^S$ is true (Seidenberg (1974)).

Lemma 17.10.   *Let $\mathcal{D}(Z)$ be a subset of Ω^S as above. Now if V is an algebraic set of dimension e belonging to $\mathcal{C}(l)$ then*

$$|V \cap \mathcal{D}(Z)| \leq C_5(S, l)Z^e.$$

Proof.   By the remark preceding Lemma 17.10, we may suppose that V is irreducible.

Let $\mathbb{F}$ be a field of definition of V, and without loss of generality let $(\zeta, \eta) = (\zeta_1, \ldots, \zeta_e, \eta_1, \ldots, \eta_t)$ with $t = S - e$ be a generic point of V over $\mathbb{F}$, such that ζ has transcendence degree e over $\mathbb{F}$. Each η_j is algebraic over $\mathbb{F}(\zeta)$, and since V lies in $\mathcal{C}(l)$, there are non-zero polynomials $g_j(\mathbf{X}, Y_j)$ $(h = 1, \ldots, t)$ with coefficients in $\mathbb{F}$, with $g_j(\zeta, \eta_j) = 0$ and of degree $\leq l_1$ where $l_1 = l_1(S, l)$. The number of

$$\mathbf{x} = (x_1, \ldots, x_e) \in \mathcal{D}_1 \times \cdots \times \mathcal{D}_e$$

is $\leq Z^e$, and given such $\mathbf{x}$ with

$$g_j(\mathbf{x}, Y_j) \neq 0 \qquad (j = 1, \ldots, t),$$

the number of $\mathbf{y}$ with $(\mathbf{x}, \mathbf{y}) \in V$ is $\leq l_1^t$. So there are at most $l_1^t Z^e$ such points.

When $e = 0$, then $l_1^t Z^e = l_1^S$ is a bound for the number of points of V. When $e > 0$, then we also have to consider points $(\mathbf{x}, \mathbf{y})$ on V with $g_1(\mathbf{x}, Y_1) \cdots g_t(\mathbf{x}, Y_t) = 0$. These form a proper algebraic subset W of V, and since V was irreducible, $\dim W < \dim V = e$. Moreover, W lies in $\mathcal{C}(l_2)$ with $l_2 = l_2(S, l)$. So if we assume inductively that the lemma is true for dimension less than e, we may infer that

$$|W \cap \mathcal{D}(Z)| \leq C_5(S, l_2)Z^{e-1},$$

and therefore

$$|V \cap \mathcal{D}(Z)| \leq l_1^t Z^e + C_5(S, l_2)Z^{e-1} \leq C_5(S, l)Z^e.$$

This completes the proof of Lemma 17.10.

With this lemma available, the whole of the proof of Proposition 16.2 can be carried over to prove Proposition 17.2. We just have to replace $\mathbb{Q}$ by $\mathbb{F}$, and $\mathbb{C}$ by a universal domain Ω over $\mathbb{F}$. The following remark explains where we need the hypothesis that $\mathbb{F}$ is a perfect field. When V is an algebraic subset of Ω^S, we could define 'V is defined over $\mathbb{F}$' by either

(i)   the ideal $\mathcal{I}(V)$ of polynomials $f \in \Omega(\mathbf{X})$ which vanish on V has a basis in $\mathbb{F}[\mathbf{X}]$, or

(ii) V is the set of zeros of an ideal $\mathcal{R}$ which has a basis in $\mathbb{F}[\mathbf{X}]$.

Both of these concepts occur in the proof. For instance, it is property (i) that we need in Lemma 16.2, more precisely for the last assertion of that lemma. On the other hand, in Lemma 16.7, it is seen that 'defined over $\mathbb{F}$' refers to property (ii).

However, for a perfect field $\mathbb{F}$, every algebraic extension is separable, and hence (i), (ii) are the same by the equivalence of C6, C7 in Lang (1958), Chapter 3, Section 5.

The proof of Proposition 17.2 therefore proceeds as for Proposition 16.2; Lemma 17.10 is needed to get the inequality (16.38) in the general case.

17.5   Proof of Theorems 17.1 and 17.2

Let κ be given by (17.4), and set $\gamma = \overline{|h(\mathscr{F})/\Phi(d)|} - 1 = 2^{d-1}\kappa - 1$. Put

$$J = \kappa - ((\log C_4)/(2^{d-1}\log q))$$

where $C_4 = C_4(s, d)$ is the constant in Proposition 17.2. Now if S is a complete sum of the type considered in Theorem 17.1, then by Lemma 17.4 we have either

$$|S| \leqq q^{s-J} = C_4^{2^{1-d}} q^{s-\kappa}, \tag{17.21}$$

or

$$|\mathcal{M}| > q^{s(d-1)-2^{d-1}J} = C_4 q^{s(d-1)-2^{d-1}\kappa}$$
$$= C_4 q^{s(d-1)-\gamma-1}. \tag{17.22}$$

When $p > d$, we may apply Proposition 17.2 with $Z = q$ and $\mathscr{D}(Z) = \mathbb{F}_q^s$. We see that (17.22), which is now the same as (17.20), leads to $h(\mathscr{F}) \leqq \Phi(d)\gamma$, contradicting our choice of γ. Thus (17.21) must hold, and Theorem 17.1 is proved.

We now turn to Theorem 17.2. Again, set $\gamma = \overline{|h(\mathscr{F})/\Phi(d)|} - 1$, but this time let κ be given by (17.6), so that

$$2^{d-1}\kappa(d-1)(d-\theta^{-1})^{-1} = \gamma + 1.$$

Let $\varepsilon > 0$ be given; we may assume ε is so small that

$$2^{d-1}(d-1)(d-\theta^{-1})^{-1}(\kappa - (\varepsilon/l)) + \varepsilon^2/l < \gamma + 1.$$

Pick σ with (17.16), that is, with $0 < \sigma < (d - \theta^{-1})/(d - 1)$, and so close to the right-hand endpoint of this interval that

$$2^{d-1}\sigma^{-1}(\kappa - (\varepsilon/l)) + \varepsilon^2/l < \gamma + 1. \tag{17.23}$$

We can choose $\sigma = \sigma(s, d, l, \theta, \varepsilon)$. Finally, let

$$J = \kappa l - \varepsilon.$$

The map $\mathbf{x} \to \mathscr{P}(\mathbf{x}) = p^{-1}\tau(\mathscr{G}(\mathbf{x}))$ occurring in the sum $S_\mathscr{B}$ is a polynomial of degree $\leqq d$ from $G = \mathbb{F}_q^s$ into $H = (p^{-1}\mathbb{Z}/\mathbb{Z})$. We now apply

Lemma 17.9 with $m = p$, $P = p^\theta$, $B = sl$ and ε^2 instead of ε. The sum of Theorem 17.2 then either has

$$|S_{\mathcal{B}}| \le P^{B-J} = P^{sl-\kappa l+\varepsilon}, \tag{17.24}$$

or $R = P^\sigma$ satisfies (17.17). The constant in $\gg$ in (17.17) depends on $B = sl$, d, σ, ε; hence only on s, d, l, θ, ε.

The projection of $\mathcal{M}(R)$ on each of the s coordinate axes of $\mathbb{F}_q^s$ contains at most $Z = (2R+1)^l$ elements. Hence $\mathcal{M}(R) = \mathcal{M}(\mathcal{D}(Z))$, and (17.17) becomes

$$\mathcal{M}(\mathcal{D}(Z)) \gg Z^{s(d-1)-2^{d-1}((\kappa/\sigma)-(\varepsilon/(l\sigma))-\varepsilon^2/l}.$$

Thus in view of (17.23) we have

$$\mathcal{M}(\mathcal{D}(Z)) > C_4 Z^{s(d-1)-\gamma-1}$$

provided p, and hence P, R and Z is large. Here $C_4 = C_4(s, d)$ is the constant of Proposition 17.2. When $p > d$, this proposition yields $h(\mathcal{F}) \le \Phi(d)\gamma$, which contradicts our choice of γ. Thus (17.24) must hold, and Theorem 17.2 is proved.

17.6   Proof of Theorem 17.3

It is not difficult to see that

$$\sum_{\mathbf{a}\in\mathbb{F}_q^r} e(p^{-1}\tau(\mathbf{ay})) = \begin{cases} q^r & \text{when } \mathbf{y} = 0 \\ 0 & \text{when } \mathbf{y} \in \mathbb{F}_q^r, \ \mathbf{y} \ne 0. \end{cases}$$

Thus

$$N_{\mathcal{B}} = q^{-r}|\mathcal{B}| + q^{-r} \sum_{\mathbf{a}\in\mathbb{F}_q^r, \mathbf{a}\ne 0} \sum_{\mathbf{x}\in\mathcal{B}} e(p^{-1}(\tau(\mathbf{a}\,\mathcal{G}(\mathbf{x}))). \tag{17.25}$$

Each polynomial $\mathbf{a}\mathcal{G}$ occurring here is of degree $\ge d_0$. Since $d_0^{-1} < \theta \le 1$, given a box $\mathcal{B}$ of size $\le P = p^\theta$ we may apply Theorem 17.2. Thus the inner sum on the right-hand side of (17.25) is $\ll P^{sl-\kappa l+\varepsilon}$. This gives (17.8).

Since h is an integer, (17.9) implies that h divided by the right-hand side of (17.9) is $\ge 1+2\zeta$ with $\zeta = \zeta(d, r, \theta)$. As a consequence, $\theta\kappa \ge r(1+2\zeta)$. We choose $\varepsilon = \varepsilon(d, r, \theta)$ with

$$r(1+2\zeta)(1-\varepsilon) - \varepsilon \ge r + \zeta.$$

Now suppose that $\mathcal{B}$ is a box with $P_{ij} = P = [p^\theta]$ ($1\le i\le s$, $1\le j\le l$). Then $|\mathcal{B}| = P^{sl}$, and (17.8) yields

$$N_{\mathcal{B}}/|\mathcal{B}| = q^{-r} + O(P^{-\kappa l+\varepsilon}).$$

When $p > C_6(d, r, \theta)$, then

$$P \ge p^{\theta(1-\varepsilon)} = q^{(\theta/l)(1-\varepsilon)}.$$

Since

$$(\kappa l - \varepsilon)(\theta/l)(1 - \varepsilon) > \theta\kappa(1 - \varepsilon) - \varepsilon \geqq r(1 + 2\zeta)(1 - \varepsilon) - \varepsilon$$
$$\geqq r + \zeta,$$

we have

$$N_{\mathscr{B}}/|\mathscr{B}| = q^{-r}(1 + O(q^{-\zeta})).$$

The constant implied by O here depends on s, d, r, l, θ, and hence we certainly have $N_{\mathscr{B}} > 1$ when $p > C_1(s, d, r, l, \theta)$.

17.7   Proof of Theorem 17.4

Let $\mathscr{F} = (\mathscr{F}^{(k)}, \ldots, \mathscr{F}^{(2)}, \mathscr{F}^{(1)})$ be a system of forms over $\mathbb{Z}$, with the subsystem $\mathscr{F}^{(d)}$ consisting of $r_d \geqq 0$ forms of degree d. Such a system will be called of type $\mathbf{r} = (r_k, \ldots, r_2, r_1)$. We have to prove that given $\mathbf{r}$ and $\theta > \frac{1}{2}$, there is a $C_7 = C_7(\mathbf{r}, \theta)$ such that for a system of forms of type $\mathbf{r}$ in $s > C_7$ variables, the congruences (17.10) have a non-trivial solution $\mathbf{x}$ with (17.11).

Given $\mathbf{r} = (r_k, \ldots, r_1)$ and $\mathbf{r}' = (r'_l, \ldots, r'_1)$ with non-negative integer components and with $r_k \neq 0$, we write $\mathbf{r} > \mathbf{r}'$ if either $k > l$, or if $k = l$ and there is a t in $1 \leqq t \leqq k$ having $r_t > r'_t$ and $r_i = r'_i$ for $t < i \leqq k$. Then $>$ establishes a well ordering among the vectors $\mathbf{r}$, and we may prove Theorem 17.4 by induction.

We begin with the observation that the theorem is true for $\mathbf{r} = (r_1)$, and that the truth of the theorem for $\mathbf{r} = (r_k, \ldots, r_2, 0)$ implies its truth for $(r_k, \ldots, r_2, r_1)$. This is an easy application of 'Siegel's lemma' (Cassels (1957), Chapter VI, Lemma 3). By this lemma, given linear forms $\mathscr{L}_i$ $(i = 1, \ldots, r_1)$ in s variables, and with coefficients of absolute value $< p$, they have a common integer zero $\mathbf{y}$ with $0 < |\mathbf{y}| < (sp)^{r_1/(s-r_1)}$. Thus when $s > \varepsilon^{-1}(1 + \varepsilon)r_1$, there is a non-trivial zero with $|\mathbf{y}| \leqq (sp)^{\varepsilon}$; when $s > l(\varepsilon^{-1}(1 + \varepsilon)r_1 + 1)$ there are l linearly independent such zeros $\mathbf{y}_1, \ldots, \mathbf{y}_l$. We now set $x = z_1\mathbf{y}_1 + \cdots + z_l\mathbf{y}_l$. With each form $\mathscr{F}_i$ of $\mathscr{F}$ we associate a new form

$$\mathscr{F}_i^*(\mathbf{Z}) = \mathscr{F}_i(Z_1\mathbf{y}_1 + \cdots + Z_l\mathbf{y}_l).$$

Since the r_1 linear forms $\mathscr{L}_i^*(\mathbf{Z})$ vanish identically, it remains to solve $\mathscr{F}_i^*(\mathbf{z}) \equiv 0 \pmod{p}$ for a system $\mathscr{F}^*$ of type $\mathbf{r} = (r_2, \ldots, r_k, 0)$. When $l > C_7(\mathbf{r}, (\theta/2) + \frac{1}{4})$ there is such a $\mathbf{z}$ with $|\mathbf{z}| \ll p^{(\theta/2)+(1/4)}$. Thus with $\varepsilon = (\theta/2) - \frac{1}{4}$, the vector $\mathbf{x} = z_1\mathbf{y}_1 + \cdots + z_l\mathbf{y}_l$ will have both (17.10) and (17.11).

It will thus suffice to prove the theorem for $\mathbf{r} = (r_k, \ldots, r_2, 0)$, assuming its truth for each $\mathbf{r}' < \mathbf{r}$. Write $r = r_k + \cdots + r_2$, and let N be the number of solutions of (17.10) with (17.11). We have $N = N_{\mathscr{B}}$, where $\mathscr{B}$ is the box $|\mathbf{x}| \leq p^{\theta}$. As in the preceding proof,

$$N = p^{-r}\left(|\mathcal{B}| + \sum_{\substack{\mathbf{a} \,(\mathrm{mod}\,p) \\ \mathbf{a} \neq 0}} \sum_{\mathbf{x} \in \mathcal{B}} e(p^{-1}\mathbf{a}\mathcal{F}(\mathbf{x}))\right).$$

Suppose for the moment that for each $\mathbf{a} \neq 0$, the inner sum over $\mathbf{x} \in \mathcal{B}$ has absolute value $\leq |\mathcal{B}| \, p^{-r-1}$. Then since the number of possibilities for $\mathbf{a}$ is less than p^r, we obtain

$$N \geq p^{-r} |\mathcal{B}| \, (1 - p^{-1}).$$

In particular, when s is large, we may infer that $N > 1$, and there is a non-trivial solution in the box $\mathcal{B}$.

We may therefore suppose that one of the inner sums is $> |\mathcal{B}| \, p^{-r-1}$ in absolute value. The box $\mathcal{B}$ is of size $\leq P$ with $P = 2[p^{\theta}] + 1$, and thus there is an $\mathbf{a} = (\mathbf{a}^{(r_k)}, \ldots, \mathbf{a}^{(r_2)})$ with a sum

$$|S_{\mathcal{B}}| > |\mathcal{B}| \, p^{-r-1} \geq P^{s-(r+1)/\theta}. \tag{17.26}$$

Here $\mathbf{a}^{(r_j)}$ has r_j coordinates and

$$S_{\mathcal{B}} = \sum_{\mathbf{x} \in \mathcal{B}} e(p^{-1}\mathcal{G}(\mathbf{x}))$$

with

$$\mathcal{G}(\mathbf{x}) = \mathbf{a}\mathcal{F}(\mathbf{x}) = \sum_{d=2}^{k} \mathbf{a}^{(r_d)}\mathcal{F}^{(r_d)}(\mathbf{x})$$

There is a unique d in $2 \leq d \leq k$ such that $\mathbf{a}^{(r_d)} \neq 0$, while $\mathbf{a}^{(r_t)} = 0$ for $d < t \leq k$. Let

$$\mathcal{F} = \mathbf{a}^{(r_d)}\mathcal{F}^{(r_d)}(\mathbf{x}) = a_1\mathcal{F}_1^{(r_d)} + \cdots + a_{r_d}\mathcal{F}_{r_d}^{(r_d)}$$

(say). *Either* (i) $\mathcal{G}$ is of degree d with $h(\mathcal{G}) = h(\mathcal{F})$, or (ii) $\mathcal{F} = 0$. In case (i), Theorem 17.2 tells us that $h(\mathcal{G})$ is small, say $h(\mathcal{G}) \leq C_8(d, r, \theta)$; in other words $h(\mathcal{F}) \leq C_8(d, r, \theta)$. Suppose without loss of generality $a_1 \in \mathbb{F}_p$ is not zero. Then in our given system $\mathcal{F}$, we may replace $\mathcal{F}_1^{(r_d)}$ by the form $\mathcal{F}$. This does not change the type $\mathbf{r}$ of the system. Moreover, we may replace $\mathcal{F}$ by a system of at most $C_8(d, r, \theta)$ forms of degree less than d, i.e. less than the degree of $\mathcal{F}$. The situation is similar in case (ii); here we replace $\mathcal{F}_1^{(r_d)}$ by 0.

Hence we may replace $\mathcal{F}$ by a system $\mathcal{F}'$ of type $\mathbf{r}' < \mathbf{r}$. Since each component of $\mathbf{r}'$ is bounded in terms of $\mathbf{r}$ and θ, it is easy to complete our inductive proof of Theorem 17.4.

17.8 The invariants h_a

Let $\mathcal{F}$ be a form of degree $d > 1$ with coefficients in $A_m = \mathbb{Z}/m\mathbb{Z}$, where m is square-free. The map $\mathbf{x} \mapsto m^{-1}\mathcal{F}(\mathbf{x})$ is a polynomial map of degree

$\leqq d$ from $G = A_m^s$ into $H = (m^{-1}\mathbb{Z}/\mathbb{Z})$. Thus $\mathcal{M}$ and $\mathcal{M}(R)$ may be defined as in Section 17.2; we take $B = s$ and

$$e_i = (0, \ldots, 0, \underset{\underleftarrow{\qquad i \qquad}}{1}, 0, \ldots, 0).$$

Proposition 17.3. *Suppose that $\Gamma > 1$ is an integer, and that*

$$|\mathcal{M}(R)| \geqq R^{s(d-1)-\gamma} \tag{17.27}$$

where $R \geqq C_9(s, d, \Gamma)$ and where $\gamma > 0$ is an integer. Then there is a factorization $m = ab$ with

$$h_a(\mathscr{F}) \leqq d\Phi(d)\gamma\Gamma \tag{17.28}$$

and $b \leqq R^{1/\Gamma}$.

Proof. For each divisor n of m, we write $\mathcal{M}_n$ for the set of $(d-1)$-tuples $\mathbf{x}_1, \ldots, \mathbf{x}_{d-1}$ with $\mathbf{x}_i \in A_n^s$ such that

$$(\mathscr{F}_n)_d(\mathbf{x}_1, \ldots, \mathbf{x}_{d-1}, \mathbf{X}) = 0,$$

i.e.

$$\mathscr{F}_d(\mathbf{x}_1, \ldots, \mathbf{x}_{d-1}, \mathbf{X}) \equiv 0 \pmod{n}.$$

Since m is square-free, and by the Chinese remainder theorem,

$$|\mathcal{M}_{nl}| = |\mathcal{M}_n| \, |\mathcal{M}_l| \tag{17.29}$$

when nl divides m.

Let $C_{10} = \max(C_4, d)$, with $C_4 = C_4(s, d)$ as in Proposition 17.2. We may suppose that $R \geqq 3^{sd} C_{10}(s, d)$. We now divide the prime divisors p of m into four classes. The first class consists of primes $\geqq 3R$. The second class contains primes p in $C_{10} < p < 3R$ with

$$|\mathcal{M}_p| \geqq p^{s(d-1)-2\gamma\Gamma}.$$

The third class consists of prime divisors in $C_{10} < p < 3R$ with

$$|\mathcal{M}_p| < p^{s(d-1)-2\gamma\Gamma}, \tag{17.30}$$

and the fourth class of primes $\leqq C_{10}$.

We write $m = ab$, where a is the product of primes of the first and second class, and b is the product of primes of the third and fourth class.

Fix a prime p of the first or second class. The set

$$\mathcal{M}_p(R)$$

$$= \{(\mathbf{x}_1, \ldots, \mathbf{x}_{d-1}) \in A_p^s : \mathscr{F}_d(\mathbf{x}_1, \ldots, \mathbf{x}_{d-1}, \mathbf{X}) \equiv 0 \,(\mathrm{mod}\,p), \max_i |\mathbf{x}_i| \leqq R\}$$

is a set $\mathcal{M}(\mathscr{D}(Z))$ as in Proposition 17.2 with $\mathbb{F} = \mathbb{F}_p$, and $Z = 2R + 1 \leqq$

3R. For primes of the first class, $p \geq Z$ and $Z > 3^{sd}C_{10}(s, d)$ so that (17.27) implies

$$|\mathcal{M}_p(R)| \geq R^{s(d-1)-\gamma} \geq (Z/3)^{s(d-1)-\gamma} \geq C_4(s, d)Z^{s(d-1)-\gamma-1}.$$

Proposition 17.2 yields $h_p(\mathcal{F}) \leq \Phi(d)\gamma$. For primes of the second class, we apply Proposition 17.2 with $Z = p$, $\mathcal{D}(Z) = A_p$; then

$$|\mathcal{M}(\mathcal{D}(Z))| = |\mathcal{M}_p| \geq p^{s(d-1)-2\gamma\Gamma} > C_4(s, d)Z^{s(d-1)-2\gamma\Gamma-1},$$

and we find that

$$h_p(\mathcal{F}) \leq 2\gamma\Gamma\Phi(d).$$

Thus when p is a prime factor of a, we may write

$$\mathcal{F} = \sum_{i=1}^{[d/2]} (\mathcal{A}_1^{(i)}\mathcal{B}_1^{(i)} + \cdots + \mathcal{A}_k^{(i)}\mathcal{B}_k^{(i)}) \pmod{p}$$

where $k = 2\gamma\Gamma\Phi(d)$ and where $\deg \mathcal{A}_j^{(i)} = i$, and $\deg \mathcal{B}_j^{(i)} = d - i$. An application of the Chinese remainder theorem yields

$$h_a(\mathcal{F}) \leq [d/2]k \leq d\Phi(d)\gamma\Gamma,$$

as asserted in (17.28).

Suppose now that l is a divisor of m in $1 < l \leq 3R$ with

$$|\mathcal{M}_l| < l^{s(d-1)-2\gamma\Gamma}. \tag{17.31}$$

Then considering the possible images in A_l of members of $\mathcal{M}(R)$, we readily see that

$$|\mathcal{M}(R)| \leq |\overline{3R/l}|^{s(d-1)} |\mathcal{M}_l| \leq (6R)^{s(d-1)}l^{-2\gamma\Gamma},$$

which in conjunction with (17.27) gives

$$l \leq 6^{s(d-1)}R^{1/(2\Gamma)} \leq R^{2/(3\Gamma)} < R^{1/2}, \tag{17.32}$$

when $R \geq C_9(s, d, \Gamma)$. Hence, in particular, the primes of the third class are $\leq R^{2/(3\Gamma)}$. We claim that the product of primes of the third class is $\leq R^{2/(3\Gamma)}$. For suppose we know that a product of some of these primes is $\leq R^{2/(3\Gamma)}$, say that $p_1 \cdots p_t \leq R^{2/(3\Gamma)}$, and let p_{t+1} be further prime of the third class. Then $p_1 \cdots p_{t+1} = l$ say, has $l < R^{1/2} \cdot R^{1/2} < 3R$. Moreover, repeated application of (17.29), (17.30) yields (17.31). Thus indeed (17.32) holds.

Finally, the primes of the fourth class have a product $\leq C_{11}(s, d) \leq R^{1/(3\Gamma)}$, so that altogether $b \leq R^{1/\Gamma}$. This completes the proof of Proposition 17.3.

Proof of Theorem 17.5. Set

$$\sigma = \tfrac{4}{5}(d - \theta^{-1})/(d - 1).$$

We are going to apply Lemma 17.9 with $G = A_m^s$ and with $B = s$. Then either $|S_{\mathcal{B}}| \leqq P^{s-J}$, or $R = P^{\sigma}$ has

$$|\mathcal{M}(R)| \gg R^{s(d-1)-2^{d-1}(J/\sigma)-\varepsilon}.$$

Thus if we set $\gamma = [2^{d-1}(J/\sigma) + \frac{3}{2}]$, then

$$|\mathcal{M}(R)| \geqq R^{s(d-1)-\gamma}$$

provided only that R is large, i.e. provided only that $m \geqq C_{12}(s, d, \sigma)$.

Again, if R is large, i.e. if $m \geqq C_{13}(s, d, \sigma, \Gamma)$, Proposition 17.3 gives us a factorization $m = ab$ with $b \leqq R^{1/\Gamma} = m^{\theta\sigma/\Gamma} \leqq m^{\theta/\Gamma}$, and with

$$h_a(\mathcal{F}^{(d)}) \leqq d\Phi(d)\gamma\Gamma.$$

In view of

$$\gamma \leqq 2^{d-1} \cdot \tfrac{5}{4}(d-1)(d-\theta^{-1})^{-1}J + \tfrac{3}{2}$$
$$\leqq 2^d(d-1)(d-\theta^{-1})^{-1}J,$$

this gives (17.12), and Theorem 17.5 is proved.

18
Small solutions of congruences to general modulus

18.1 Introduction

Our goal in the present chapter is a proof of the following theorem of Schmidt (to appear).

Theorem 18.1. *Let k, r be natural numbers. Let $\mathcal{F}_1(x_1, \ldots, x_s), \ldots,$ $\mathcal{F}_r(x_1, \ldots, x_s)$ be forms over $\mathbb{Z}$ with degrees between 1 and k, with $s > C_1(k, r, \varepsilon)$. Then the system of congruences*

$$\mathcal{F}_i(\mathbf{x}) \equiv 0 \pmod{m} \qquad (i = 1, \ldots, r) \tag{18.1}$$

has a solution $\mathbf{x} = (x_1, \ldots, x_s)$ with

$$0 < |\mathbf{x}| \ll m^{(1/2)+\varepsilon}. \tag{18.2}$$

The implied constant depends only on k, r and ε.

As pointed out in Chapter 1, the exponent in (18.2) is essentially best possible. (Conceivably, ε could be replaced by 0.) The theorem clearly remains true with the forms $\mathcal{F}_i$ replaced by polynomials with constant term zero. Another apparently more general formulation is that the $\mathcal{F}_1, \ldots, \mathcal{F}_r$ are any polynomials of degree $\leq k$ and if $\mathbf{x}_0$ is a solution of (18.1), then there is another solution $\mathbf{x}$ with $|\mathbf{x} - \mathbf{x}_0| \ll m^{(1/2)+\varepsilon}$.

The case when m is a prime was proved in Chapter 17. In the present chapter we do not need any new work on exponential sums. Nevertheless some rather formidable complications are involved in bringing to bear what we know about such sums. Fortunately it is not hard to restrict all the complicated work to square-free moduli m; this is similar to the situation in Chapter 12.

The present proof does *not* provide solutions of (18.1), (18.2) with g.c.d. $(m, x_1, \ldots, x_s) = 1$. It would be interesting to have such a stronger theorem; it appears to be quite difficult even in a simple case, such as a system of quadratic congruences.

To prove the theorem we have to set up more general systems of congruences. So let $2 \leq l \leq k$, and let

$$\mathcal{F} = (\mathcal{F}^{(k)}, \mathcal{F}^{(k-1)}, \ldots, \mathcal{F}^{(l)})$$

be a system of $r = r_k + r_{k-1} + \cdots + r_l$ forms with integer coefficients, where the subsystem $\mathscr{F}^{(d)}$ consists of $r_d \geqq 0$ forms of degree d. When $r_d > 0$ write

$$\mathscr{F}^{(d)} = (\mathscr{F}_1^{(d)}, \ldots, \mathscr{F}_{r_d}^{(d)}).$$

Further let $\mathbf{m} = (\mathbf{m}^{(k)}, \mathbf{m}^{(k-1)}, \ldots, \mathbf{m}^{(l)})$, where $\mathbf{m}^{(d)}$ for $l \leqq d \leqq k$ is an r_d-tuple of positive integers, say $\mathbf{m}^{(d)} = (m_1^{(d)}, \ldots, m_{r_d}^{(d)})$. Consider the system of congruences

$$\mathscr{F}_i^{(d)}(\mathbf{x}) \equiv 0 \quad (\mathrm{mod}\, m_i^{(d)}) \qquad (1 \leqq i \leqq r_d, \, l \leqq d \leqq k). \tag{18.3}$$

(These conditions should be interpreted as empty for d with $r_d = 0$.) Given $P > 1$, write N_P for the number of solutions of this system with $\mathbf{x}$ in the cube $\mathscr{C}_P$ given by $1 \leqq x_j \leqq P$ ($j = 1, \ldots, s$). Heuristically one should expect that

$$N_P \sim P^s / M,$$

where M is the product of the components of $\mathbf{m}$.

For each d with $r_d > 0$ write $m^{(d)}$ for the least common multiple of $m_1^{(d)}, \ldots, m_{r_d}^{(d)}$, and let m be a common multiple of the numbers $m^{(d)}$ so obtained. Given a prime factor p of $m^{(d)}$, let $\mathscr{F}_p^{(d)}$ consist of the reductions modulo p of those forms $\mathscr{F}_i^{(d)}$ of $\mathscr{F}^{(d)}$ for which $p \mid m_i^{(d)}$. Thus $\mathscr{F}_p^{(d)}$ has a positive number of components, but at most r_d components. (In general, the p in $\mathscr{F}_p$ or $\mathscr{F}_p$ will always refer to reduction modulo p, and a confusion with components $\mathscr{F}_i$ of $\mathscr{F}$ should not arise.)

We recall that for a form $\mathscr{F}$ with coefficients in $\mathbb{F}_p$, $h(\mathscr{F})$ is the least integer h such that

$$\mathscr{F} = \mathscr{A}_1 \mathscr{B}_1 + \cdots + \mathscr{A}_h \mathscr{B}_h$$

with forms $\mathscr{A}_i$, $\mathscr{B}_i$ of positive degrees over $\mathbb{F}_p$. For a system $\mathscr{F} = (\mathscr{F}_1, \ldots, \mathscr{F}_t)$ of forms of equal degree, we define $h(\mathscr{F})$ just as in Chapter 17 by considering the pencil of $\mathscr{F}$:

$$h(\mathscr{F}) = \min_{\mathbf{a} \in \mathbb{F}_p^t, \mathbf{a} \neq \mathbf{0}} h(\mathbf{a}\mathscr{F}).$$

In particular, $h(\mathscr{F}_p^{(d)})$ (with $\mathscr{F}_p^{(d)}$ as above) is defined for each d with $r_d > 0$ and each prime p dividing $m^{(d)}$. We now set

$$h(\mathscr{F}, \mathbf{m}) = \min_{l \leqq d \leqq k; r_d > 0} \min_{p \mid m^{(d)}} h(\mathscr{F}_p^{(d)}). \tag{18.4}$$

Proposition 18.1. *With the above conventions for $\mathscr{F}$, M, m and so on, suppose that $P = m^{(1/l) + \varepsilon}$ with $\varepsilon > 0$ and $h(\mathscr{F}, \mathbf{m}) \geqq C_2(k, l; r_k, \ldots, r_l; \varepsilon)$. Then if m is square-free with each prime divisor $\geqq C_3(k, l; r_k, \ldots, r_l; \varepsilon, s)$, we have*

$$|N_P - P^s / M| < P^s / (2M). \tag{18.5}$$

The proposition will be proved in Sections 18.2 and 18.3. We then set up a rather elaborate induction to deal with systems (18.3) with small $h(\mathscr{F}, \mathbf{m})$. Theorem 18.1, for m not necessarily square-free, then follows readily.

18.2 Estimation of exponential sums

We will employ vectors $(\mathbf{a}^{(k)}, \ldots, \mathbf{a}^{(l)})$ where $\mathbf{a}^{(d)}$ has r_d components; say $\mathbf{a}^{(d)} = (a_1^{(d)}, \ldots, a_{r_d}^{(d)})$ when $r_d > 0$. Given $\mathbf{a}^{(d)}$, put

$$\mathbf{A}^{(d)} = (a_1^{(d)}/m_1^{(d)}, \ldots, a_{r_d}^{(d)}/m_{r_d}^{(d)}).$$

The notation

$$\sum_{\mathbf{a}^{(d)}}$$

will stand for the r_d-fold sum where $a_i^{(d)}$ ranges from 1 to $m_i^{(d)}$ $(1 \leq i \leq r_d)$. With these conventions, we have

$$\sum_{\mathbf{a}^{(k)}} \cdots \sum_{\mathbf{a}^{(l)}} e(\mathbf{A}^{(k)}\mathscr{F}^{(k)}(\mathbf{x}) + \cdots + \mathbf{A}^{(l)}\mathscr{F}^{(l)}(\mathbf{x})) = \begin{cases} M & \text{when (18.3) holds} \\ 0 & \text{otherwise.} \end{cases}$$

(Summations $\sum_{\mathbf{a}^{(d)}}$ with $r_d = 0$ are omitted on the left-hand side.) It follows that

$$N_P = M^{-1} \sum_{\mathbf{a}^{(k)}} \cdots \sum_{\mathbf{a}^{(l)}} S(\mathbf{a}^{(k)}, \ldots, \mathbf{a}^{(l)}) \qquad (18.6)$$

with

$$S(\mathbf{a}^{(k)}, \ldots, \mathbf{a}^{(l)}) = \sum_{\mathbf{x} \in \mathscr{C}_P} e(\mathbf{A}^{(k)}\mathscr{F}^{(k)}(\mathbf{x}) + \cdots + \mathbf{A}^{(l)}\mathscr{F}^{(l)}(\mathbf{x})).$$

Given $\mathbf{a}^{(h)}, \ldots, \mathbf{a}^{(l)}$, write Δ_k for the least common denominator of $\mathbf{A}^{(k)}$ in case $r_k > 0$, and $\Delta_k = 1$ if $r_k = 0$. Let Δ_{k-1} be the positive integer such that $\Delta_k \Delta_{k-1}$ is the least common denominator of the point $(\mathbf{A}^{(k)}, \mathbf{A}^{(k-1)})$ when $r_{k-1} > 0$, and set $\Delta_{k-1} = 1$ when $r_{k-1} = 0$. In general, choose $\Delta_k, \Delta_{k-1}, \ldots, \Delta_l$ such that $\Delta_k \Delta_{k-1} \cdots \Delta_d$ for $l \leq d \leq k$ is the least common denominator of $(\mathbf{A}^{(k)}, \ldots, \mathbf{A}^{(d)})$. Finally set $\Delta = \Delta_k \Delta_{k-1} \cdots \Delta_l$. In the sum in (18.6) there is precisely one summand with $\Delta = 1$, namely the summand with $a_i^{(d)} = m_i^{(d)}$ throughout. This summand has $S(\mathbf{a}^{(k)}, \ldots, \mathbf{a}^{(l)}) = |\mathscr{C}_P| = P^s + O(P^{s-1})$, and hence contributes

$$P^s M^{-1} + O(P^{s-1}M^{-1}) \qquad (18.7)$$

to (18.6).

The summands with $\Delta > 1$ will be estimated by appealing to the following lemma. We shall need boxes of the type $a_j \leq x < b_j$ with integers a_j, b_j. Such a box will be said to be of side $\leq R$ if $b_j - a_j \leq R$ $(j = 1, \ldots, s)$.

Lemma 18.1. *Let $\mathcal{G} = \mathcal{F}^{(d)} + \mathcal{F}^{(d-1)} + \cdots + \mathcal{F}^{(0)}$ be a polynomial of degree d, with $\mathcal{F}^{(j)}$ homogeneous of degree j and with coefficients in the ring $\mathbb{Z}/n\mathbb{Z}$ where n is square-free. Let $\mathcal{B}$ be a box of side $\leq R$ where $R \geq n^\theta$, $d^{-1} < \theta \leq 1$. Let $S_\mathcal{B}$ be the sum*

$$S_\mathcal{B} = \sum_{\mathbf{x} \in \mathcal{B}} e(n^{-1}\mathcal{G}(\mathbf{x})). \tag{18.8}$$

Suppose that $J \geq 1$, and that $\Gamma > 1$ is an integer. Then either

$$|S_\mathcal{B}| \ll R^s n^{-\theta J}, \tag{18.9}$$

or there is a factorization $n = ab$ with $b \leq R^{1/\Gamma}$ and with

$$h(\mathcal{F}_p^{(d)}) \leq (d - \theta^{-1})^{-1} d^2 2^d \Phi(d) J\Gamma \tag{18.10}$$

for every prime factor p of a. Here $\mathcal{F}_p^{(d)}$ is the reduction of $\mathcal{F}^{(d)}$ modulo p, and $\Phi(d)$ is the function of Chapters 16 and 17. The constant in $\ll$ depends only on s, d, θ, Γ.

Proof. This is an easy consequence of Theorem 17.5. We will assume that n has no factorization as indicated, and we will derive (18.9). The box $\mathcal{B}$ is the disjoint union of boxes $\mathcal{B}^*$ of side $\leq n^\theta = R_1$, say; the number of such boxes $\mathcal{B}^*$ required is $\ll (R/R_1)^s$. We now apply Theorem 17.5 to $S_{\mathcal{B}^*}$. There is no factorization $n = ab$ with $b \leq n^{1/\Gamma}$ and (18.10) for all $p \mid a$. A fortiori, there is no factorization with $b \leq n^{1/\Gamma}$ and (17.12). Thus

$$|S_{\mathcal{B}^*}| < R_1^{s-J} = R_1^s n^{-\theta J}$$

provided that $n \geq C_4(s, d, \theta, \Gamma)$, and hence

$$|S_{\mathcal{B}^*}| \ll R_1^s n^{-\theta J}$$

in general. (Because $h(\mathcal{F}_p^{(d)}) \leq s$ trivially, the non-existence of a factorization n as above is possible only if the right-hand side of (18.10) is $< s$; hence $J < s$.)

Taking the sum over the boxes $\mathcal{B}^*$ making up $\mathcal{B}$, we get

$$S_\mathcal{B} \ll (R/R_1)^s R_1^s n^{-\theta J} = R^s n^{-\theta J}.$$

18.3 Proof of Proposition 18.1

We now return to the situation described before Lemma 18.1.

Lemma 18.2. *Make the hypotheses of Proposition 18.1 and let $\mathbf{a}^{(k)}, \ldots, \mathbf{a}^{(l)}$ with $\Delta > 1$ be given. Then*

$$|S(\mathbf{a}^{(k)}, \ldots, \mathbf{a}^{(l)})| \ll P^s \Delta^{-h\varphi}$$

where $h = h(\mathcal{F}, \mathbf{m})$, where $\varphi = \varphi(k, l, \varepsilon)$, and where the constant in $\ll$ depends only on k, l, ε and s.

Proof. It is no more difficult to prove the lemma for an arbitrary box $\mathscr{B}$ of side $\leq P$ where $P \geq m^{(1/l)+\varepsilon}$. We may suppose that $0 < \varepsilon < 1$. Set

$$\Lambda = 2^k/\varepsilon, \qquad \Gamma = \Lambda + 1. \tag{18.11}$$

Let

$$V_d = \Delta_k \Delta_{k-1} \cdots \Delta_{d+1} \qquad \text{when } l \leq d < k, \text{ and } V_k = 1;$$
$$W_d = \Delta_d \Delta_{d-1} \cdots \Delta_l \qquad \text{when } l \leq d \leq k, \text{ and } W_{l-1} = 1.$$

Then

$$\Delta = V_d W_d \qquad (l \leq d \leq k).$$

There is some j, $l \leq j \leq k$ for which the inequality

$$\Delta_j^\Lambda \leq W_{j-1} \tag{18.12}$$

is false. For if (18.12) held for $l \leq j \leq k$, we would find that $\Delta_l^\Lambda \leq W_{l-1} = 1$, whence $\Delta_l = 1$. Next, $\Delta_{l+1} \leq W_l = \Delta_l = 1$, so that $\Delta_{l+1} = 1$, etc. We would obtain $\Delta = 1$, contrary to hypothesis. From now on, let d be the largest number in $l \leq d \leq k$ for which (18.12) fails for $j = d$. Thus

$$\Delta_d^\Lambda > W_{d-1}, \tag{18.13}$$

but (18.12) is true for $d < j \leq k$. In the case when $d < k$ it follows that $\Delta_{d+1}^\Lambda \leq W_d$, next that

$$\Delta_{d+2}^\Lambda \leq W_{d+1} = \Delta_{d+1} W_d < W_d^2,$$

next that

$$\Delta_{d+3}^\Lambda \leq W_{d+2} = \Delta_{d+2} W_{d+1} \leq W_d^{2^2},$$

and so on; finally, that $\Delta_k^\Lambda \leq W_d^{2^{k-d-1}}$. Therefore $V_d^\Lambda \leq W_d^{2^{k-d}}$, and by our choice of Λ this yields $V_d \leq W_d^{\varepsilon/2}$. We may infer on the one hand that

$$W_d \geq \Delta^{1/2}, \tag{18.14}$$

and on the other hand that

$$P \geq m^{(1/l)+\varepsilon} \geq W_d^{(1/l)+\varepsilon} \geq W_d^{(1/d)+\varepsilon} \geq V_d W_d^{(1/d)+(\varepsilon/2)}. \tag{18.15}$$

We have $\mathbf{A}^{(d)} = (V_d \Delta_d)^{-1} \mathbf{b}$ where $\mathbf{b}$ is an integer point with g.c.d. $(\Delta_d, b_1, \ldots, b_{r_d}) = 1$. Thus

$$\mathbf{A}^{(d)} \mathscr{F}^{(d)} = (V_d \Delta_d)^{-1} \mathscr{F} \qquad \text{with } \mathscr{F} = \mathbf{b} \mathscr{F}^{(d)}. \tag{18.16}$$

Every prime factor p of Δ_d also divides $m^{(d)}$, and hence $h(\mathscr{F}_p^{(d)}) \geq h$ by the definition (18.4) of $h = h(\mathscr{F}, \mathbf{m})$. Now if i is a subscript with $p \nmid m_i^{(d)}$, then p is not in the denominator of $A_i^{(d)} = a_i^{(d)}/m_i^{(d)}$, and we must have $p \mid b_i$. On the other hand since p does occur in the denominator of $\mathbf{A}^{(d)}$, there must be some subscripts i with $p \mid m_i^{(d)}$ and $p \nmid b_i$. The components of $\mathscr{F}_p^{(d)}$ stem from forms $\mathscr{F}_i^{(d)}$ with $p \mid m_i^{(d)}$, and $\mathscr{F}_p$ (i.e. the reduction of

$\mathscr{F}$ modulo p) therefore belongs to the pencil of $\mathscr{F}_p^{(d)}$. We draw the conclusion that

$$h(\mathscr{F}_p) \geq h. \tag{18.17}$$

Points $\mathbf{x}$ in the given box $\mathscr{B}$ will be written as $\mathbf{x} = V_d \mathbf{y} + \mathbf{z}$ with $0 \leq z_j < V_d$. For given $\mathbf{z}$, for $\mathbf{x}$ to be in the box $\mathscr{B}$ of side $\leq P$, the point $\mathbf{y}$ will be in a box $\mathscr{B}(\mathbf{z})$ of side $\leq 2P/V_d$. (Note that $V_d \leq P$ by (18.15).) For given $\mathbf{z}$ we have modulo 1

$$\mathbf{A}^{(k)}\mathscr{F}^{(k)}(\mathbf{x}) + \cdots + \mathbf{A}^{(l)}\mathscr{F}^{(l)}(\mathbf{x})$$
$$= \mathbf{A}^{(d)}\mathscr{F}^{(d)}(V_d \mathbf{y}) + W_d^{-1}\mathscr{F}^{(d-1)}(\mathbf{y}) + \cdots + W_d^{-1}\mathscr{F}^{(1)}(\mathbf{y}) + \Delta^{-1}\mathscr{F}^{(0)}$$
$$\tag{18.18}$$

where $\mathscr{F}^{(d-1)}, \ldots, \mathscr{F}^{(0)}$ are forms of respective degrees $d-1, \ldots, 0$ with integer coefficients. In the first place, on expanding the terms of the left-hand side of (18.18) according to (14.47), we collect terms $\sum_{j=l}^{k} \mathbf{A}^{(j)}\mathscr{F}^{(j)}(\mathbf{z})$ as $\Delta^{-1}\mathscr{F}^{(0)}$. For $j > d$, terms depending on $V_d \mathbf{y}$ make a zero contribution (modulo 1). For $l \leq j \leq d$, we separate the term $\mathbf{A}^{(d)}\mathscr{F}^{(d)}(V_d \mathbf{y})$; all terms not yet accounted for are of degree $< d$ and have a denominator which is a divisor of $\Delta V_d^{-1} = W_d$. This leads to the expression (18.18). (There is an analogous argument in the proof of Lemma 8.2.)

In view of (18.16) we have

$$\mathbf{A}^{(d)}\mathscr{F}^{(d)}(V_d \mathbf{y}) = \Delta_d^{-1}V_d^{d-1}\mathscr{F}(\mathbf{y}) = W_d^{-1}\mathscr{F}^{(d)}(\mathbf{y}) \tag{18.19}$$

with $\mathscr{F}^{(d)} = W_{d-1}V_d^{d-1}\mathscr{F}$. Since m is square-free, Δ_d is coprime to W_{d-1} and to V_d, and hence (18.17) yields

$$h(\mathscr{F}_p^{(d)}) \geq h \tag{18.20}$$

for every prime factor p of Δ_d.

We are going to apply Lemma 18.1 to the polynomial

$$\mathscr{G} = \mathscr{F}^{(d)} + \mathscr{F}^{(d-1)} + \cdots + \mathscr{F}^{(1)}$$

with the forms $\mathscr{F}^{(j)}$ coming from (18.18), (18.19), and to a box $\mathscr{B}(\mathbf{z})$. Such a box has side $\leq 2P/V_d = R$, say. Setting $n = W_d$, $\theta = (1/d) + (\varepsilon/2)$, we have $R \geq n^{\theta}$ by (18.15). Let Γ be given by (18.11), and put

$$J = (d - \theta^{-1})d^{-2}2^{-1-d}\Phi(d)^{-1}\Gamma^{-1}h \tag{18.21}$$

so that $J \geq 1$ when $h \geq C_2$. One alternative of Lemma 18.1 gives a factorization $W_d = ab$. By (18.10), (18.20) and our choice of J, a prime factor p of Δ_d cannot divide a, so that $\Delta_d \mid b$, and hence $\Delta_d \leq n^{1/\Gamma}$, or $\Delta_d^{\Gamma} \leq W_d = \Delta_d W_{d-1}$, that is, $\Delta_d^{\wedge} \leq W_{d-1}$, contradicting (18.13). Thus the other alternative must hold. This means, in view of (18.18) and (18.19), that the part $S(\mathbf{z})$ of the sum $S(\mathbf{a}^{(k)}, \ldots, \mathbf{a}^{(l)})$ with $\mathbf{x} = V_d \mathbf{y} + \mathbf{z}$ and given $\mathbf{z}$

has

$$S(\mathbf{z}) \ll R^s n^{-\theta J} \ll (P/V_d)^s W_d^{-\theta J}.$$

Taking the sum over $\mathbf{z}$ we obtain

$$S(\mathbf{a}^{(k)}, \ldots, \mathbf{a}^{(l)}) \ll P^s W_d^{-\theta J} \ll P^s \Delta^{-(\theta/2)J} \ll P^s \Delta^{-\varphi h},$$

by virtue of (18.14) and (18.21). This proves Lemma 18.2.

Now that Lemma 18.2 has been established, the proof of Proposition 18.1 is easily completed. Given Δ, there are not more than Δ^r vectors $(\mathbf{a}^{(k)}, \ldots, \mathbf{a}^{(l)})$. Thus all the sums $S(\mathbf{a}^{(k)}, \ldots, \mathbf{a}^{(l)})$ with given Δ contribute together not more than $\ll P^s \Delta^{r-\varphi h}$, that is $\ll P^s \Delta^{-2}$ when $h \geq C_2$. Since the least prime factor of m is $\geq C_3$, the sum over all the $S(\mathbf{a}^{(k)}, \ldots, \mathbf{a}^{(l)})$ with $\Delta > 1$ is $\ll P^s C_3^{-1}$. In view of (18.6), and the contribution (18.7) from $\Delta = 1$, we obtain

$$N_P = P^s M^{-1}(1 + O(P^{-1} + C_3^{-1})).$$

Thus (18.5) is certainly true when C_3, and hence P, is large.

18.4   An inductive argument

Throughout this section, λ will be a fixed positive number. With a given $\mathbf{r} = (r_k, \ldots, r_1)$ we will associate a set of vectors $\mathbf{u}$, as follows. Given r_d put·

$$t_d = [2^d r_d \lambda^{-1}] + 1. \tag{18.22}$$

The symbol $\mathbf{u}^{(d)}$ will denote the zero vector when $r_d = 0$, but when $r_d > 0$ it will denote vectors $\mathbf{u}^{(d)} = (u_1^{(d)}, \ldots, u_{r_d}^{(d)})$ with integer components in $1 \leq u_i^{(d)} \leq t_d$. Given d and r_d, let $<$ be an ordering of the vectors $\mathbf{u}^{(d)}$ such that $\mathbf{u}'^{(d)} \neq \mathbf{u}^{(d)}$ and $\mathbf{u}_i'^{(d)} \leq \mathbf{u}_i^{(d)}$ $(i = 1, \ldots, r_d)$ implies $\mathbf{u}'^{(d)} < \mathbf{u}^{(d)}$. We now consider vectors of the form

$$(r_k, \mathbf{u}^{(k)}, \ldots, r_2, \mathbf{u}^{(2)}, r_1, \mathbf{u}^{(1)}) = (\mathbf{r}, \mathbf{u}), \tag{18.23}$$

say, where $r_k > 0$, where each $r_d \geq 0$ and each $\mathbf{u}^{(d)}$ is of the type described. We order the $(\mathbf{r}, \mathbf{u})$ by the convention that $(\mathbf{r}', \mathbf{u}') < (\mathbf{r}, \mathbf{u})$ if either

$$\mathbf{r}' = (r_l', \ldots, r_1'), \qquad \mathbf{r} = (r_k, \ldots, r_1)$$

with $l < k$, or if $l = k$ and there is a d in $1 \leq d \leq k$ such that $r_j' = r_j$, $\mathbf{u}'^{(j)} = \mathbf{u}^{(j)}$ for $d < j \leq k$, and either $r_d' < r_d$, or $r_d' = r_d$ and $\mathbf{u}'^{(d)} < \mathbf{u}^{(d)}$. The vectors $(\mathbf{r}, \mathbf{u})$ are then well ordered.

In proving Theorem 18.1 we will initially suppose m to be square-free. Given $\mathcal{F} = (\mathcal{F}^{(k)}, \ldots, \mathcal{F}^{(1)})$, where $\mathcal{F}^{(d)}$ consists of $r_d \geq 0$ forms of degree d, and given divisors $m_i^{(d)}$ of m $(1 \leq i \leq r_d$ with $1 \leq d \leq k$ and

$r_d > 0$), it will suffice to show that the system

$$\mathscr{F}_i^{(d)}(\mathbf{x}) \equiv 0 \quad (\mathrm{mod}\ m_i^{(d)}) \qquad (1 \leq i \leq r_d,\ 1 \leq d \leq k) \qquad (18.24)$$

has a solution $\mathbf{x}$ satisfying (18.2), with the constant in $\ll$ depending only on $\mathbf{r} = (r_k, \ldots, r_1)$ and ε, provided that the number s of variables exceeds some $C_4 = C_4(\mathbf{r}, \varepsilon)$.

Given $\lambda > 0$ and given $(\mathbf{r}, \mathbf{u})$, we now formulate the following

Assertion $(r, u)_\lambda$. *Let* $\mathscr{F} = (\mathscr{F}^{(k)}, \ldots, \mathscr{F}^{(1)})$ *be a system of forms as above, let m be square-free, and let $m_i^{(d)}$ be divisors of m with*

$$m_i^{(d)} \leq m^{u_i^{(d)}/t_d} \qquad (1 \leq i \leq r_d,\ 1 \leq d \leq k). \qquad (18.25)$$

Then if $\mu > \lambda$ and $s > C_5(\mathbf{r}, \mathbf{u}, \lambda, \mu)$, the system (18.24) has a solution $\mathbf{x}$, $0 < |\mathbf{x}| \ll m^{(1/2)+\mu}$, and with the constant in $\ll$ depending only on $(\mathbf{r}, \mathbf{u})$, λ and μ.

The truth of the assertion for every $\lambda > 0$ and every $(\mathbf{r}, \mathbf{u})$ will give the truth of Theorem 18.1 for square-free m, since for given ε we may set $\mu = \varepsilon$, $\lambda = \varepsilon/2$, and since we may take $u_i^{(d)} = t_d$ throughout, in which case (18.25) is automatically satisfied. The assertion itself will be proved for given $\lambda > 0$ by induction with respect to $<$. (Obviously we have to hold λ fixed, as the ordering, and indeed the ordered set, depends on λ.) The following lemma gets the procedure started.

Lemma 18.3. *The assertion is true for every $\varepsilon > 0$ and every vector $(r_1, \mathbf{u}^{(1)})$. Moreover, when it is true for some particular*

$$(r_k, \mathbf{u}^{(k)}, \ldots, r_2, \mathbf{u}^{(2)}, 0, 0) = (\mathbf{r}, \mathbf{u}) \qquad (18.26)$$

say, then it is true for every vector of the type

$$(r_k, \mathbf{u}^{(k)}, \ldots, r_2, \mathbf{u}^{(2)}, r_1, \mathbf{u}^{(1)}).$$

Proof. We will only prove the second claim. Let $\mathscr{F} = (\mathscr{F}^{(k)}, \ldots, \mathscr{F}^{(l)})$ be given. Let $\mu > \lambda$ and set $\sigma = (\mu - \lambda)/2$. We may suppose the coefficients of our forms to have absolute values less than m.

Let $l = [C_5(\mathbf{r}, \mathbf{u}, \lambda, \lambda + \sigma)] + 1$ with $(\mathbf{r}, \mathbf{u})$ as in (18.26). (Such an l is defined in view of the assertion $(\mathbf{r}, \mathbf{u})_\lambda$.) Exactly as in the proof of Theorem 17.4, Siegel's lemma provides l linearly independent vectors $\mathbf{y}_1, \ldots, \mathbf{y}_l$ with

$$|\mathbf{y}_i| < (sm)^\sigma \qquad (i = 1, \ldots, l)$$
$$\mathscr{F}_j^{(1)}(Z_1 \mathbf{y}_1 + \cdots + Z_l \mathbf{y}_l) = 0 \qquad (j = 1, \ldots, r_1) \qquad (18.27)$$

identically in $\mathbf{Z} = (Z_1, \ldots, Z_l)$, provided only that

$$s > l(\sigma^{-1}(1 + \sigma)r_1 + 1).$$

With each form $\mathscr{F}$ of $\boldsymbol{\mathscr{F}}$ we associate a new form

$$\mathscr{F}^*(\mathbf{Z}) = \mathscr{F}(Z_1\mathbf{y}_1 + \cdots + Z_l\mathbf{y}_l)$$

in $\mathbf{Z}$. In view of (18.27) we need only deal with the congruences

$$\mathscr{F}_i^{(d)*}(\mathbf{z}) \equiv 0 \pmod{m_i^{(d)}} \tag{18.28}$$

where $1 \leq i \leq r_d$ and $2 \leq d \leq k$. By our choice of l there is a non-trivial solution $\mathbf{z}$ of (18.28) with $|\mathbf{z}| \ll m^{(1/2)+\lambda+\sigma}$. But then $\mathbf{x} = z_1\mathbf{y}_1 + \cdots + z_l\mathbf{y}_l$ solves all the congruences (18.24), and

$$0 < |\mathbf{x}| \ll |\mathbf{y}|\,|\mathbf{z}| \ll m^{(1/2)+\lambda+2\sigma} = m^{(1/2)+\mu}.$$

Here the implied constant depends on s. However, much as in Chapter 14, we only need prove assertion $(\mathbf{r}, \mathbf{u})_\lambda$ for $s = [C_5(\mathbf{r}, \mathbf{u}, \lambda, \mu)] + 1$, and it will then follow for larger values of s. In this way we can remove the dependence on s, and Lemma 18.3 is proved. (The same remark about dependence on s will be used again later on without explicit mention.)

Proof of Theorem 18.1. It will be enough to prove the assertion for $(\mathbf{r}, \mathbf{u})$ of the type (18.26), assuming its truth for each $(\mathbf{r}', \mathbf{u}') < (\mathbf{r}, \mathbf{u})$. So let $(\mathscr{F}^{(k)}, \ldots, \mathscr{F}^{(2)})$ be given, let m be square-free and let $m_i^{(d)}$ be divisors of m with (18.25).

In what follows, let

$$C_2 = C_2(k, 2; r_k, \ldots, r_2; \lambda/2)$$

be as in Proposition 18.1. As before, we write $m^{(d)}$ for the least common multiple of $m_1^{(d)}, \ldots, m_{r_d}^{(d)}$ when $r_d > 0$. For each prime factor p of $m^{(d)}$, we define $\mathscr{F}_p^{(d)}$ and $h(\mathscr{F}_p^{(d)})$ as in Section 18.1. When $r_d > 0$, we factor $m^{(d)} = a^{(d)}b^{(d)}$ such that $a^{(d)}$ is the product of exactly those prime factors p of $m^{(d)}$ for which

$$h(\mathscr{F}_p^{(d)}) \geq C_2. \tag{18.29}$$

We now distinguish two cases.

(I) *There is an e in $2 \leq e \leq k$ with $r_e > 0$ and $b^{(e)} \geq m^{r_e/t_e}$.* Let such an e be fixed. For each prime factor p of $b^{(e)}$ we know that some form

$$\mathscr{F}_p = a_{p1}\mathscr{F}_1^{(e)} + \cdots + a_{pr_e}\mathscr{F}_{r_e}^{(e)}$$

has $h(\mathscr{F}_p) < C_2$, where $a_{pi} \equiv 0 \pmod{p}$ for subscripts i with $p \nmid m_i^{(e)}$, but where there is at least one i with $p \mid m_i^{(e)}$ and $a_{pi} \not\equiv 0 \pmod{p}$. There is an i_0 in $1 \leq i_0 \leq r_e$ and a divisor b' of $b^{(e)}$ with $b' \geq (b^{(e)})^{1/r_e} \geq m^{1/t_e}$ such that for each prime factor p of b' we have $p \mid m_{i_0}^{(e)}$ and $a_{pi_0} \not\equiv 0 \pmod{p}$. Say $i_0 = 1$.

The congruences

$$\mathscr{F}_i^{(e)}(\mathbf{x}) \equiv 0 \pmod{m_i^{(e)}} \qquad (1 \leq i \leq r_e)$$

then we have the same solution set as the congruences

$$\mathscr{F}_i(\mathbf{x}) \equiv 0 \pmod{m_i^{(e)}} \qquad (2 \leq i \leq r_e), \qquad (18.30a)$$

$$\mathscr{F}_1(\mathbf{x}) \equiv 0 \pmod{m_1^{(e)}/b'}, \qquad (18.30b)$$

$$\mathscr{F}_p(\mathbf{x}) \equiv 0 \pmod{p} \qquad \text{for each prime factor } p \text{ of } b'. \qquad (18.30c)$$

Now since $h(\mathscr{F}_p) < C_2$, each $\mathscr{F}_p$ can be written as

$$\mathscr{F}_p = \sum_{i=1}^{[e/2]} \sum_{j=1}^{C_2-1} \mathscr{A}_{pj}^{(i)} \mathscr{B}_{pj}^{(i)}$$

where $\mathscr{A}_{pj}^{(i)}$, $\mathscr{B}_{pj}^{(i)}$ are forms with degrees i and $e - i$ respectively. By the Chinese remainder theorem, there are forms $\mathscr{A}_j^{(i)}$ such that (coefficient-wise) $\mathscr{A}_j^{(i)} \equiv \mathscr{A}_{pj}^{(i)} \pmod{p}$ for each prime factor p of b'. The $\mathscr{A}_j^{(i)}$ have degrees less than e. The condition (18.30c) is certainly satisfied if $\mathscr{A}_j^{(i)}(x) \equiv 0 \pmod{b'}$ for $1 \leq i \leq [e/2]$, $1 \leq j \leq C_2 - 1$. The original system (18.24) of congruences is therefore satisfied whenever the following new system is satisfied:

$$\mathscr{F}_i^{(d)}(\mathbf{x}) \equiv 0 \pmod{m_i^{(d)}} \qquad \begin{array}{l} (1 \leq i \leq r_d, \ 2 \leq d \leq k \\ \text{but excluding } i = 1, \ d = e), \end{array} \qquad (18.31a)$$

$$\mathscr{F}_1^{(e)}(\mathbf{x}) \equiv 0 \pmod{m_1'^{(e)}} \qquad \text{where } m_1'^{(e)} = m_1^{(e)}/b', \qquad (18.31b)$$

$$\mathscr{A}_j^{(i)}(\mathbf{x}) \equiv 0 \pmod{b'} \qquad (1 \leq i \leq [e/2], \ 1 \leq j \leq C_2). \qquad (18.31c)$$

Note that

$$m_1'^{(e)} = m_1^{(e)}/b' \leq m^{(u_1^{(e)}-1)/t_e}. \qquad (18.32)$$

Now (18.31) is of the type

$$(\mathbf{r}', \mathbf{u}') = (r_k, \mathbf{u}^{(k)}, \ldots, r_{e+1}, \mathbf{u}^{(e+1)}, r_e', \mathbf{u}'^{(e)}, \ldots, r_2', \mathbf{u}'^{(2)}, r_1', \mathbf{u}'^{(1)})$$

since the congruences of degree $d > e$ have not been changed. But when $u_1^{(e)} > 1$, we may take $r_e' = r_e$ and

$$\mathbf{u}'^{(e)} = (u_1^{(e)} - 1, u_2^{(e)}, \ldots, u_{r_e}^{(e)}) \qquad \text{by (18.32),}$$

while when $u_1^{(e)} = 1$, we may take $r_e' = r_e - 1$. Since the components of $\mathbf{r}'$ are bounded in terms of $\mathbf{r}$ and λ, our inductive assumption yields for $s > C_5(\mathbf{r}, \mathbf{u}, \lambda, \mu)$ a solution $\mathbf{x}$ of (18.31) with

$$0 < |\mathbf{x}| \ll m^{(1/2)+\mu}.$$

This leaves us with the case

(II) $b^{(d)} < m^{r_d/t_d}$ *whenever* $2 \leq d \leq k$ *and* $r_d > 0$. Let g be the product of the primes $\leq C_3$ where

$$C_3 = C_3(k, 2; r_k, \ldots, r_2; \lambda/2, s)$$

is the quantity of the proposition. Consider the congruences

$$\mathscr{F}_i^{(d)}(\mathbf{y}) \equiv 0 \pmod{m_i^{*(d)}} \qquad (1 \leq i \leq r_d,\ 2 \leq d \leq k) \qquad (18.33)$$

where $m_i^{*(d)} = m_i^{(d)}/(m_i^{(d)}, gb^{(d)})$. Then $m^{*(d)}$ (defined in the obvious way) is coprime to $b^{(d)}$, and by (18.29) we have $h(\mathscr{F}_p^{(d)}) \geq C_2$ for each prime factor p of $m^{*(d)}$. The least common multiple m^* of the numbers $m^{*(d)}$ is square-free and its prime factors exceed C_3. Thus by Proposition 18.1, (18.33) has a solution $\mathbf{y} \neq \mathbf{0}$ with $|\mathbf{y}| \ll (m^*)^{(1/2)+\lambda/2} \leq m^{(1/2)+\lambda/2}$. We now set

$$\mathbf{x} = gb^{(2)}b^{(3)} \cdots b^{(k)}\mathbf{y}.$$

Then $\mathbf{x}$ is a non-trivial solution of the original congruences, and

$$|\mathbf{x}| \ll m^{(r_2/t_2)+\cdots+(r_k/t_k)} m^{(1/2)+\lambda/2} \ll m^{(1/2)+\mu}$$

by our choice of $t_2, \ldots, t_k$ in (18.22).

This finishes the proof of Theorem 18.1 for m square-free. It is easy to extend the result to arbitrary m for systems of forms of the type $\mathbf{r} = (r_k, \ldots, r_2, 0)$. We write $m = m_1^2 m_2$ with m_2 square-free and set $\mathbf{x} = m_1\mathbf{y}$ where $\mathbf{y}$ is a small solution of the congruences modulo m_2. Then

$$|\mathbf{x}| = m_1 |\mathbf{y}| \ll m_1 m_2^{(1/2)+\varepsilon} \ll m^{(1/2)+\varepsilon},$$

and

$$\mathscr{F}^{(d)}(\mathbf{x}) = m_1^d \mathscr{F}^{(d)}(\mathbf{y}) \equiv 0 \pmod{m} \qquad (2 \leq d \leq k).$$

Finally, an application of Siegel's lemma such as in the proof of Lemma 18.3 leads from this to systems of forms of arbitrary type $(r_k, \ldots, r_2, r_1)$.

References

Baker, R. C. (1977). Fractional parts of several polynomials. *Quart. J. Math.* Oxford (2) **28**, 453–71.

—— (1978). Fractional parts of several polynomials II. *Mathematika* **25**, 76–93.

—— (1980*a*). Fractional parts of several polynomials III. *Quart. J. Math.* Oxford (2), **31**, 19–36.

—— (1980*b*). Small solutions of quadratic and quartic congruences. *Mathematika* **27**, 30–45.

—— (1981*a*). On the distribution modulo one of the sequence $\alpha n^3 + \beta n^2 + \gamma n$. *Acta Arith.* **39**, 399–405.

—— (1981*b*). Small fractional parts of the sequence αn^k. *Michigan Math. J.* **28**, 224–8.

—— (1981*c*). Metric number theory and the large sieve. *J. London Math. Soc.* (2) **24**, 34–40.

—— (1981*d*). On the fractional parts of αn^2 and βn. *Glasgow Math. J.* **22**, 181–3.

—— (1982*a*). Cubic Diophantine inequalities. *Mathematika* **29**, 83–92.

—— (1982*b*). Weyl sums and Diophantine approximation. *J. London Math. Soc.* (2), **25**, 25–34.

—— (1983). Small solutions of congruences. *Mathematika* **20**, 164–88.

—— and Gajraj, J. (1976*a*). On the fractional parts of certain additive forms. *Math. Proc. Camb. Phil. Soc.* **79**, 463–7.

—— and —— (1976*b*). Some nonlinear Diophantine approximations. *Acta Arith.* **31**, 325–41.

—— and Harman, G. (1981). Small fractional parts of quadratic and additive forms. *Math. Proc. Camb. Phil. Soc.* **90**, 5–12.

—— and —— (1982*a*). Small fractional parts of quadratic forms. *Proc. Edinburgh Math. Soc.* **25**, 269–77.

—— and —— (1982*b*). Diophantine approximation by prime numbers. *J. London Math. Soc.* (2), **25**, 201–15.

—— and —— (1984*a*). Diophantine inequalities with mixed powers. *J. Number Theory* **18**, 69–85.

—— and —— (1984*b*). Small fractional parts of polynomials. *Colloq. Math. Soc. Janos Bolyai*, **34**. *Topics in Classical Number Theory* (Budapest, 1981), 69–110. Elsevier, North Holland.

—— and Kolesnik, G. (1985). On the distribution of p^α modulo one. *J. Reine Angew. Math.* **356**, 174–93.

—— and Schmidt, W. M. (1980). Diophantine problems in variables restricted to the values 0 and 1. *J. Number Theory* **12**, 460–86.

Birch, B. J. (1957). Homogeneous forms of odd degree in a large number of variables. *Mathematika* **4**, 102–5.

—— (1962). Forms in many variables. *Proc. Royal Soc.* A, **265**, 245–63.

—— (1970). Small zeros of diagonal forms of odd degree in many variables. *Proc. Lond. Math. Soc.* (3) **21,** 12–18.

—— and Davenport, H. (1958*a*). On a theorem of Davenport and Heilbronn. *Acta Math.* **100,** 259–79.

—— and —— (1958*b*). Indefinite quadratic forms in many variables. *Mathematika* **5,** 8–12.

Borevich, Z. I. and Shafarevich, I. R. (1966). *Number theory.* Academic Press, New York and London.

Brauer, R. (1945). A note on systems of homogeneous algebraic equations. *Bull. Amer. Math. Soc.* **51,** 749–55.

Cassels, J. W. S. (1957). *An introduction to Diophantine approximation.* Cambridge Tracts in Math. and Math. Physics, 45. Cambridge.

—— (1959). *An introduction to the geometry of numbers.* Springer, Berlin.

Chalk, J. H. H. and Williams, K. S. (1965). The distribution of solutions of congruences. *Mathematika* **12,** 176–80.

Chen, J.-R. (1977). On Professor Hua's estimate of exponential sums. *Sci. Sinica* **20,** 711–19.

Chong, K. K. and Liu, M. C. (1976). The fractional parts of a sum of polynomials. *Monatsh. Math.* **81,** 195–202.

Cochrane, T. (To appear). The distribution of solutions of equations over finite fields.

Cook, R. J. (1972*a*). On the fractional parts of a set of points. *Mathematika* **19,** 63–8.

—— (1972*b*). The fractional parts of an additive form. *Proc. Camb. Phil. Soc.* **72,** 209–12.

—— (1973). On the fractional parts of a set of points. II. *Pacific J. Math.* **45,** 81–5.

—— (1975). On the fractional parts of a set of points. III. *J. London Math. Soc.* (2) **9,** 490–4.

—— (1976*a*). On the fractional parts of a polynomial. *Canad. J. Math.* **28,** 168–73.

—— (1976*b*). Diophantine inequalities with mixed powers (mod 1). *Proc. Amer. Math. Soc.* **57,** 29–34.

—— (1977). On the fractional parts of a set of points. IV. *Indian J. Math.* **19,** 7–23.

—— (1978). On the fractional parts of cubic forms. *Acta Arith.* **34,** 91–102.

—— (1980). Small fractional parts of quadratic forms in many variables. *Mathematika* **27,** 25–29.

—— (1983). Indefinite quadratic polynomials. *Glasgow Math. J.* **24,** 133–8.

—— (1984). Small fractional parts of quadratic and cubic polynomials in many variables. *Colloq. Math. Soc. Janos Bolyai* **34.** *Topics in Classical Number Theory.* (*Budapest, 1981*), 281–308. Elsevier, North Holland.

Corput, J. G. van der (1939). Une inégalité relative aux sommes de Weyl. *Proc. Kon. Akad. v. Wetensch. Amsterdam* **42,** 461–7.

—— and Pisot, Ch. (1939*a*). Sur la discrépance modulo un, I. *Proc. Kon. Ned. Akad. v. Wetensch. Amsterdam* **42,** 476–85.

—— and —— (1939*b*). Sur la discrépance modulo un, II. *Proc. Kon. Ned. Akad. v. Wetensch. Amsterdam.* **42,** 554–65.

Danicic, I. (1957). Contributions to number theory. Ph.D thesis, University of London.

—— (1958). An extension of a theorem of Heilbronn. *Mathematika* **5,** 30–7.

—— (1959). On the fractional parts of θx^2 and ϕx^2. *J. Lond. Math. Soc.* **34,** 353–7.

—— (1967). The distribution (mod 1) of pairs of quadratic forms with integer variables. *J. Lond. Math. Soc.* **42,** 618–23.

Davenport, H. (1939). On sums of positive integral kth powers. *Proc. Roy. Soc.* A, **170,** 293–9.

—— (1956). Indefinite quadratic forms in many variables. *Mathematika* **3,** 81–101.

—— (1958). Indefinite quadratic forms in many variables II. *Proc. Lond. Math. Soc.* (3) **8,** 109–26.

—— (1959). Cubic forms in 32 variables. *Phil. Trans. Royal Soc.* A, **251,** 193–232.

—— (1962a). *Analytic methods for Diophantine equations and Diophantine inequalities.* Ann Arbor Publishers, Ann Arbor.

—— (1962b). Cubic forms in 29 variables. *Proc. Royal Soc.* A, **266,** 287–98.

—— (1963). Cubic forms in sixteen variables. *Proc. Royal Soc.* A, **272,** 285–303.

—— (1967). On a theorem of Heilbronn. *Quart. J. Math.* Oxford (2) **18,** 339–44.

—— (1977). *The collected works of Harold Davenport,* vol. III, ed. B. J. Birch, H. Halberstam and C. A. Rogers. Academic Press, London.

—— (1980). *Multiplicative Number Theory.* 2nd edn. Revised by H. L. Montgomery. Springer Verlag, New York–Heidelberg–Berlin.

—— and Erdös, P. (1939). On sums of positive integral kth powers. *Ann. Math.* **40,** 533–6.

—— and Heilbronn, H. (1946). On indefinite quadratic forms in five variables. *J. Lond. Math. Soc.* **21,** 185–93.

—— and Lewis, D. J. (1962). Exponential sums in many variables. *Amer. J. Math.* **84,** 649–65.

—— and —— (1963). Homogeneous additive equations. *Proc. Royal Soc.* A, **274,** 443–60.

—— and Ridout, D. (1959). Indefinite quadratic forms. *Proc. Lond. Math. Soc.* (3) **9,** 544–55.

Deligne, P. (1974). La conjecture de Weil I. *Publ. Math. IHES.* **43,** 273–307.

Dirichlet, G. P. L. (1842). Verallgemeinerung eines Satzes aus der Lehre von den Kettenbruchen nebst einigen Anwendungen auf die Theorie der Zahlen. *S-B. Preuss. Akad. Wiss.* 93–5.

Erdös, P. and Turán, P. (1948a). On a problem in the theory of uniform distribution I. *Indag. Math.* **10,** 370–8.

—— and —— (1948b). On a problem in the theory of uniform distribution II. *Indag. Math.* **10,** 406–13.

Estermann, T. (1936). Proof that every large integer is the sum of seventeen biquadrates. *Proc. Lond. Math. Soc.* (2) **41,** 126–42.

Gajraj, J. (1976a). An analytical approach to some Diophantine inequalities. Ph.D. thesis, University of London.

—— (1976b). Simultaneous approximation to certain polynomials. *J. Lond. Math. Soc.* (2) **14,** 527–34.

Gelfond, A. O. and Linnik, Yu. V. (1966). *Elementary methods in analytic number theory.* George Allen and Unwin Ltd., London.

Hall, R. R. and Tenenbaum, G. (1982). On the average and normal orders of Hooley's Δ-function. *J. Lond. Math. Soc.* **25,** 392–406.

Hardy, G. H. and Littlewood, J. E. (1914). Some problems of Diophantine approximation I. On the fractional part of $n^k\theta$. *Acta Math.* **37,** 155–91.

—— and —— (1922). Some problems of 'Partitio Numerorum' IV. The singular series in Waring's problem. *Math. Zeit.* **12,** 161–88.

—— and Wright, E. M. (1979). *An introduction to the theory of numbers.* 5th edn. Oxford University Press, Oxford.

Harman, G. (1981). Trigonometric sums over primes I. *Mathematika* **28**, 249–54.
—— (1982). Diophantine approximation and prime numbers. Ph.D. thesis, University of London.
—— (1983*a*). Trigonometric sums over primes II. *Glasgow Math. J.* **24**, 23–37.
—— (1983*b*). On the distribution of αp modulo one. *J. Lond. Math. Soc.* (2) **27**, 9–18.
—— (1984). Diophantine approximation with square free integers. *Math. Proc. Camb. Phil. Soc.* **95**, 381–8.
Hartshorne, R. (1977). *Algebraic geometry*. Springer-Verlag, New York–Heidelberg–Berlin.
Heath-Brown, D. R. (1983). Cubic forms in 10 variables. *Proc. Lond. Math. Soc.* (3) **47**, 225–57.
—— (1984). Diophantine approximation with squarefree integers. *Math. Zeit.* **187**, 335–44.
—— (to appear). Small solutions of quadratic congruences.
Heilbronn, H. (1948). On the distribution of the sequence θn^2 (mod 1). *Quart. J. Math.* Oxford (2) **19**, 249–56.
Hooley, C. (1979). On a new technique and its applications to the theory of numbers. *Proc. Lond. Math. Soc.* (3) **38**, 115–51.
Hua, L.-K. (1940). Sur une somme exponentielle. *C.R. Acad. Sci. Paris* **210**, 520–3.
—— (1949). An improvement of Vinogradov's mean value theorem and several applications. *Quart. J. Math.* Oxford (2) **20**, 48–61.
—— (1957). On exponential sums. *Sci. Rec.* **1**, 1–4.
—— (1965). *Additive theory of prime numbers*. Amer. Math. Soc., Providence.
Jameson, G. J. O. (1970). *A first course on complex functions*. Chapman and Hall, London.
Katz, N. (1980). *Sommes exponentielles*. Astérisque 79.
Kuipers, L. and Niederreiter, H. (1974). *Uniform distribution of sequences*. Wiley, New York.
Lang, S. (1958). *Introduction to algebraic geometry*. Interscience, New York.
Lewis, D. J. (1957). Cubic forms over algebraic number fields. *Mathematika* **4**, 97–101.
Linnik, Y. V. (1943). An elementary solution of Waring's problem by Schnirelman's method. *Mat. Sb.* **12**, 225–30.
Liu, M.-C. (1970*a*). On a theorem of Heilbronn concerning the fractional part of θn^2. *Canad. J. Math.* **22**, 784–8.
—— (1970*b*). On the fractional parts of θn^k and ϕn^k. *Quart. J. Math.* Oxford (2) **21**, 481–6.
—— (1974). Simultaneous approximation of two additive forms. *Proc. Camb. Phil. Soc.* **75**, 77–82.
—— (1975). Simultaneous approximation of additive forms. *Trans. Amer. Math. Soc.* **206**, 361–73.
Loxton, J. H. and Vaughan, R. C. (to appear). Bounds for exponential sums.
Montgomery, H. L. (1978). The analytic principle of the large sieve. *Bull. Amer. Math. Soc.* **84**, 547–67.
Pitman, J. (1968). Cubic inequalities. *J. London Math. Soc.* **43**, 119–26.
—— (1971*a*). Bounds for solutions of diagonal inequalities. *Acta Arith.* **18**, 179–90.
—— (1971*b*). Bounds for solutions of diagonal equations. *Acta Arith.* **19**, 223–47.
—— and Ridout, D. (1967). Diagonal cubic equations and inequalities. *Proc. Roy. Soc. Lond.* A, **297**, 476–502.
Ridout, D. (1958). Indefinite quadratic forms. *Mathematika* **5**, 122–4.

Schinzel, A., Schlickewei, H.-P., and Schmidt, W. M. (1980). Small solutions of quadratic congruences and small fractional parts of quadratic forms. *Acta Arith.* **37**, 241–8.

Schlickewei, H.-P. (1979). On indefinite diagonal forms in many variables. *J. Reine Angew. Math.* **307/8**, 279–94.

Schmidt, W. M. (1964). Metrical theorems on fractional parts of sequences. *Trans. Amer. Math. Soc.* **110**, 493–518.

—— (1976). *Equations over finite fields: an elementary approach.* Lecture Notes in Math. **536**. Springer, Berlin.

—— (1977a). On the distribution modulo one of the sequence $\alpha n^2 + \beta n$. *Canad. J. Math.* **29**, 819–26.

—— (1977b). *Small fractional parts of polynomials.* Regional Conference series no. 32, Amer. Math. Soc., Providence.

—— (1979a). Small zeros of additive forms in many variables. *Trans. Amer. Math. Soc.* **248**, 121–33.

—— (1979b). Small zeros of additive forms in many variables. II. *Acta Math.* **143**, 219–32.

—— (1980a). Diophantine inequalities for forms of odd degree. *Advances in Math.* **38**, 128–51.

—— (1980b). *Diophantine approximation.* Lecture Notes in Math. **785.** Springer, Berlin.

—— (1982a). On cubic polynomials I. Hua's estimate of exponential sums. *Monatsh. Math.* **93**, 63–74.

—— (1982b). On cubic polynomials II. Multiple exponential sums. *Monatsh. Math.* **93**, 141–68.

—— (1982c). On cubic polynomials III. Systems of p-adic equations. *Monatsh. Math.* **93**, 211–23.

—— (1982d). On cubic polynomials IV. Systems of rational equations. *Monatsh. Math.* **93**, 329–48.

—— (1982e). Simultaneous rational zeros of quadratic forms. *Séminaire Delange–Pisot–Poitou* 1980–81, 281–307.

—— (1984a). *Analytische Methoden für Diophantische Gleichungen.* DMV Seminar, **5**. Birkhauser, Basel 1984.

—— (1984b). Bounds for exponential sums. *Acta Arith.* **44**, 281–97.

—— (1985). The density of integer points on homogeneous varieties. *Acta Math.* **154**, 243–296.

—— (to appear). Small solutions of congruences in a large number of variables. *Canad. Math. Bull.*

Seidenberg, A. (1974). Constructions in algebra. *Trans. Amer. Math. Soc.* **197**, 273–313.

Serre, J.-P. (1977). Majorations de sommes exponentielles. *Astérisque* **41–2**, 111–26.

Smith, R. A. (1970). The distribution of rational points on hypersurfaces defined over a finite field. *Mathematika* **17**, 328–32.

Tartakovsky, W. (1935). Über asymptotische Gesetze der allgemeinen Diophantische Analyse mit vielen Unbekannten. *Bull. Acad. Sci. USSR,* 483–524.

Tenenbaum, G. (to appear). Sur la concentration moyenne des diviseurs.

Thanigasalam, K. (1982). New estimates for $G(k)$ in Waring's problem. *Acta Arith.* **42**, 73–8.

Vaughan, R. C. (1977). Homogeneous additive equations and Waring's problem. *Acta Arith.* **33**, 231–53.

—— (1981). *The Hardy–Littlewood method.* Cambridge University Press, Cambridge.

—— (1984). Some remarks on Weyl sums. *Colloq. Math. Soc. Janos Bolyai* **34**. *Topics in classical number theory*. Elsevier, North Holland, Amsterdam.

—— (1985). Sums of three cubes. *Bull. Lond. Math. Soc.* **17**, 17–20.

Vinogradov, I. M. (1927). Analytischer Beweis des Satzes uber die Verteilung der Bruchteile eines ganzen Polynoms. *Bull. Acad. Sci. USSR* (6) **21**, 567–78.

—— (1950). An upper bound of the modulus of a trigonometric sum. *Izv. Akad. Nauk SSSR Ser. Mat.* **14**, 199–214; Amer. Math. Soc. Translations no. 94 (1953).

—— (1951). General theorems on the upper bound of the modulus of a trigonometric sum. *Izvest. Akad. Nauk. SSSR Ser. Mat.* **15**, 109–130; Amer. Math. Soc. Translations no. 94 (1953).

—— (1954). *The method of trigonometric sums in the theory of numbers*. Translated, revised and annotated by Davenport, A. and Roth, K. F. Interscience, New York.

—— (1957). Trigonometric sums involving values of a polynomial. *Izvest. Akad. Nauk. SSSR Ser. Mat.* **21**, 145–70.

—— (1971). *The method of trigonometric sums in number theory*, Moscow.

Waerden, B. L. van der (1949). *Modern algebra*, Volumes I and II. 2nd edn. Frederick Ungar, New York.

Weyl, H. (1916). Über die Gleichverteilung von Zahlen mod Eins. *Math. Ann.* **77**, 313–52.

Zygmund, A. (1968). *Trigonometric series*, vols. I and II combined. Cambridge University Press, Cambridge.

Index